About Island Press

Since 1984, the nonprofit Island Press has been stimulating, shaping, and communicating the ideas that are essential for solving environmental problems worldwide. With more than 800 titles in print and some 40 new releases each year, we are the nation's leading publisher on environmental issues. We identify innovative thinkers and emerging trends in the environmental field. We work with world-renowned experts and authors to develop cross-disciplinary solutions to environmental challenges.

Island Press designs and implements coordinated book publication campaigns in order to communicate our critical messages in print, in person, and online using the latest technologies, programs, and the media. Our goal: to reach targeted audiences—scientists, policymakers, environmental advocates, the media, and concerned citizens—who can and will take action to protect the plants and animals that enrich our world, the ecosystems we need to survive, the water we drink, and the air we breathe.

Island Press gratefully acknowledges the support of its work by the Agua Fund, Inc., The Margaret A. Cargill Foundation, Betsy and Jesse Fink Foundation, The William and Flora Hewlett Foundation, The Kresge Foundation, The Forrest and Frances Lattner Foundation, The Andrew W. Mellon Foundation, The Curtis and Edith Munson Foundation, The Overbrook Foundation, The David and Lucile Packard Foundation, The Summit Foundation, Trust for Architectural Easements, The Winslow Foundation, and other generous donors.

The opinions expressed in this book are those of the author(s) and do not necessarily reflect the views of our donors.

Introduction to

Restoration
Ecology

Introduction to
Restoration Ecology

Evelyn A. Howell

John A. Harrington

Stephen B. Glass

ISLANDPRESS

Washington | Covelo | London

Library of Congress Cataloging-in-Publication Data

Howell, Evelyn A.
  Introduction to restoration ecology / Evelyn A. Howell, John A. Harrington, and Stephen B. Glass.
    p. cm.
  ISBN-13: 978-1-59726-189-0 (cloth : alk. paper)
  ISBN-10: 1-59726-189-0 (pbk. : alk. paper) 1. Restoration ecology. I. Harrington, John A. II. Glass, Stephen B., 1946- III. Title.
  QH541.15.R45H69 2011
  639.9--dc23
                    2011026551

Printed on recycled, acid-free paper

Manufactured in the United States of America
10 9 8 7 6 5 4 3 2 1

Keywords: Island Press, ecological restoration, restoration ecology, restoration practitioner, interdisciplinary, community ecology, landscape ecology, climate change, adaptive restoration, site inventory, restoration management, pest species, user impacts, volunteer-based restoration

Contents

CASE STUDIES

Chapter 10

Chapter 11

Chapter 12

Chapter 13

Chapter 14

Acknowledgments

This book would not have been possible without the help and encouragement of many people. First, we would like to thank our colleagues at the University of Wisconsin–Madison Landscape Architecture Department and the University of Wisconsin–Madison Arboretum for their support and advice. Special recognition goes to the many students who have participated in our in our restoration ecology courses over the years. It has been a pleasure and a privilege to work with such an outstanding and enthusiastic group of both undergraduate and graduate students.

We are most grateful to Barbara Dean of Island Press for nurturing the idea for this book, sticking with us through thick and thin, and encouraging us to keep going. We would also like to thank Erin Johnson of Island Press for assisting us with the many stages and details of production. We also thank Jude Berman and are grateful for her exceptional developmental editing, which helped transform raw editorial material into an actual textbook. Also deserving of our thanks are Betsy Dilernia for excellent and exacting copyediting, Sharis Simonian for production expertise, and the rest of the staff at Island Press for doing a great job of handling the artwork and marketing of the book. Autumn Sabo also helped edit early versions of the text.

Thanks also go to the Society for Ecological Restoration for sponsoring, with Island Press, the Science and Practice of Ecological Restoration Book Series, of which this textbook is a part. We are proud to be associated with SER and this book series.

We would also like to thank Cate Harrington of The Nature Conservancy for reviewing portions of the manuscript and securing images, Doug Hadley for assistance with graphic production, and Math Heinzel for aide with information technology. We also wish to thank the numerous individuals and agencies that helped secure images and case studies: Sue Ireland and the Kootenai Tribe of Idaho; Keith Bowers and Jean Wisenbaugh, Biohabitats, Inc.; Brad Woodson and John Peters, McHenry County Conservation District; Mark Godfrey, John Wagner, Jon Fisher, Susan Miller, and Jennifer Molnar, The Nature Conservancy; Brian Ickes, USGS; Jeb Barzen; Robert Hansis, Wisconsin Department of Natural Resources; Megan Foss; Nicole Van Helden; and Katie Beilfuss. Thanks to Dr. Matthias Drösler, Technische Universität München, for hosting John on a visit to the Kendelmühlfilzen area bogs of Bavaria in 2005. We also thank our colleagues in Australia, Noel Grundon, Tien McDonald, Tanya Simmons, and Nigel Tucker, for their hospitality and for sharing their restoration expertise and experiences with us.

Steve thanks Sharon Dunwoody for her encouragement, guidance, and support, without which a portion this book would not have been possible. John similarly thanks Cate Harrington. Evelyn would like to acknowledge two of her mentors, her major professor, the late UW–Madison Botany Professor Grant Cottam, and her former colleague in the UW–Madison Department of Landscape Architecture Professor Darrel Morrison, for pointing the way toward restoration ecology, and her family for their loving support.

Introduction

Restoration ecology is a relatively young discipline and one of the most exciting and promising approaches to the conservation of native species, communities, and ecosystems practiced today. It is a global enterprise, with participants coming from all walks of life—ranging from academicians and conservation professionals to dedicated volunteers of all ages. Not only does the restoration of native plant and animal communities help us to reverse our legacy of the destruction of the world's biodiversity, the act of participating in restoration helps us reengage with the natural world, thereby greatly enriching our lives. It is therefore not surprising that interest in restoration is strong.

One of the signs of the emergence of restoration ecology is the number of courses in schools and colleges that are dedicated in whole or in part to this topic. Although many books and articles have recently been written about the theory and practice of restoration, there are few textbooks, and very few meant as an introduction to the field. We have written this book based on our experience in teaching restoration ecology at the college level.

Our target audience is students of restoration ecology. However, we also believe that students in the fields of landscape architecture, botany, land planning, and conservation biology, as well as related natural resource programs, will find this text useful in their studies. Our purpose in writing this textbook is to introduce students to the theory and practice of restoration, and, in so doing, to provide a guide to designing the landscapes of the future.

▶ MEET THE AUTHORS

Evelyn Howell, plant ecologist, and John Harrington, landscape architect, are professors in the Department of Landscape Architecture at the University of Wisconsin–Madison. Together, we have taught a four-course restoration ecology sequence since 1984. The courses include lectures, labs, seminars, and workshops, and feature hands-on experience in designing, planning, and implementing restorations. Students come from across the campus and work together in interdisciplinary teams. Evelyn has received two college and university awards for excellence in teaching. Steve Glass, restoration ecologist, is the restoration planner and fire manager at the University of Wisconsin–Madison Arboretum and has been active in all phases of restoration there since 1989.

Collectively, we have worked on forest, prairie, savanna, and wetland plant community restoration plans for the U.S. National Park Service, The Nature Conservancy, the International Crane Foundation, and numerous additional state, local, and private conservation organizations. Each of us has been involved in both research and practice.

▶ FOCUS AND PERSPECTIVE

In writing this book, our focus is on providing a framework that can be used to guide restoration decisions anywhere on the globe—both now and in the future. Restoration ecology is transdisciplinary, drawing from the humanities and the biological, physical, and social sciences, as well as from engineering, agriculture, and landscape planning and design. Our goal is to prepare students to be able to work with the uniqueness, uncertainty, complexity, messiness, and constraints inherent in any restoration. Natural systems are dynamic and diverse across space and through time. The human contexts within which we conduct restorations also reflect diverse cultures and histories. As a result, no two restorations follow exactly the same course; it is not possible to use the same solutions for every restoration problem. In addition, restoration theory and knowledge continue to increase. Students need to be prepared to adopt new assumptions and practices as ideas and circumstances change.

In order to account for the wide range of global restoration situations, we have chosen to emphasize the general principles and logic that are common to restoration ecology wherever it is practiced, rather than trying to cover the wide variety of restoration circumstances in detail. In other words, our emphasis is on a restoration problem-solving process, just as many science and design texts concentrate on the scientific method or the design process, respectively. We focus on the reasoning that leads to restoration decisions, not the decisions themselves.

In order to illustrate the concepts we present, we use examples from around the world. However, because each of us has more than twenty years of experience with plant community restoration projects within North America, many of the case examples and discussion draw upon restoration examples in this part of the world. For more examples and information about your own regions, we urge you to supplement and customize this text by exploring the wealth of restoration resources that are now available.

▶ CHAPTER ORGANIZATION

We have organized the text around a restoration process that we have tested and revised in our restoration ecology courses over the past three decades. Chapter 1 sets the stage for what is to follow by introducing students to the origins of restoration ecology and to the scope of modern practice, as well as to the kinds of issues that the field addresses. We then introduce the restoration process around which the text is organized, and continue with a discussion of the theoretical and practical challenges restorationists face, as well as the many conservation opportunities restoration provides.

Chapters 2–5 describe the information-gathering stage of the restoration process. Chapter 2 introduces the concept of the "community/ecosystem model," a description of the restoration target community, and presents an overview of the ecological theories that underlie such models. In Chapter 3 we introduce the adaptive restoration approach, and describe the important roles played by research in restoration, ranging from the investigation of reference communities in creating community/ecosystem models to exploring different implementation and management tools. Chapters 4 and 5 introduce site inventory and analysis, the step whereby you investigate the current biological, physical, cultural, and social conditions of the restoration site and its context, in order to match the restoration to the site.

The next five chapters describe the master (Chapter 6), site (Chapter 7), implementation (Chapter 8), monitoring (Chapter 9), and management (Chapter 10) plans that form the heart of a restoration project. The master and site plans describe what the designers of a project hope to achieve—first in broad terms with the master plan, and then in more detail with the site plan. The plans include information about the desired conditions of biological communities, and physical site features, as well as how people will use and enjoy the site. These two plans set the stage for what follows in a restoration project, as they describe what the restorationists hope to achieve, the assumptions about native communities and ecosystems, and the people–environment interrrelationships that underlie their vision. In the implementation plan, as the name implies, we describe how to move a site from the conditions you find it in at the start of a restoration project to the conditions described in the site plan. In Chapter 9 we discuss how to check in on a restoration to evaluate how well it is meeting the site plan vision. Chapter 10 explains the actions that may be needed, once the site plan vision has been achieved, to manage the restoration site so it continues to meet that vision through time.

In Chapters 11 and 12, we focus on two impacts, introduced in Chapter 10, that may move an established restoration away from the site plan vision—pest species and continuing human impacts. Chapter 11 explores one of the most pervasive management issues encountered in conservation: dealing with pest species. We describe how to anticipate the arrival of species that may interfere with the continued success of a restoration, what measures to take to prevent their arrival on site, and how to minimize their impact, should they appear. Chapter 12 discusses how to minimize and control the impact of humans coming from site visitors or offsite land uses.

Chapter 13 expands on the "working with people" theme we emphasized in exploring the restoration process in Chapters 6–10. Chapter 14 highlights a series of restoration projects to show how the planning process has (or has not) worked in practice. We conclude the book (Chapter 15) with some final thoughts about the exciting, stimulating, and promising future of restoration.

▶ HOW TO USE THE BOOK

We have organized each chapter using a set of features that we hope will help you learn and understand the concepts we introduce. Each chapter begins with a set of Learning Objectives, so you can anticipate what you will be able to do after learning the material. Throughout the chapters you will find Case Studies and Sidebars. The Sidebars supplement the text by providing additional information about the ideas, methods, and planning processes that underlie restoration. There are tables, diagrams, outlines, and photographs to illustrate the general principles presented in the text. You can use all these features to investigate and visualize restorations in more detail. Several of these examples come from projects completed by students while taking our restoration ecology classes.

At the end of each chapter, you can use the Key Concepts to check that you have picked up our main points and have achieved the chapter objectives. Food for Thought questions will engage you through hypothetical and real-life situations, and test your understanding of the concepts. These questions are at the end of all chapters, and they also appear in some of the Sidebars, where you can put the concepts we discuss to work in solving practical restoration problems. There is a list of Resources for Further Study for every chapter, if you would like more information. We have selected these because we have found them to be useful in our own work. You will find the references cited in the text at the end of the book, organized by chapter.

▶ A NOTE TO READERS

We hope you will find restoration ecology to be as fascinating, challenging, and rewarding as we do. Be sure to take the time to think about and apply the concepts that we introduce, and to explore the resources at the end of the chapters. We also encourage you to explore the many online resources that are available, and to do so with a critical eye. However, no text or website can really convey the excitement of restoration ecology. This is a profession and a discipline in which much of the joy and satisfaction come from being in the field, working in nature with people. We hope you will be able to find and work with a conservation group in your area. The best way to learn about restoration ecology is to do it! And by so doing, you are helping make the world a better place.

Restoration Ecology
Composing the Landscape of the Future

LEARNING OBJECTIVES

After reading this chapter, you will be able to:

- Discuss the historical roots and current scope of restoration ecology.

- Describe the similarities and differences between complete, ecological services, and experiential restorations.

- Explain the design and planning process used by restoration practitioners.

- Appreciate the theoretical and practical challenges restorationists face, as well as the many ways in which restoration ecology can contribute to composing the landscape of the future

Many students of ecology and the environment, such as yourself, find the world in which we live to be full of a seemingly infinite number of environmental conflicts, not to mention what promises to be a time of rapid climate change, the consequences of which are as yet unknown. In light of the numerous environmental and social dilemmas we face, such feelings and concerns are neither unwarranted nor surprising. Indeed, Aldo Leopold, in his essay "Round River," sized up the psychological state many of you find yourselves in when he wrote: "One of the penalties of an ecological education is that one lives alone in a world of wounds.... An ecologist must either harden his shell and make believe the consequences of science are none of his business, or he must be the doctor who sees the marks of death in a community that believes itself well and does not want to be told otherwise" (1949, 197).

Many of these conflicts have been set in motion by the activities of people. Humans are part of the natural world and are not alone in modifying the environment. In order to perform the essential functions of life (obtaining energy and nutrients, for example), all organisms change, and are changed by, the physical and biological environment. People have been altering the landscape for millennia—by building dwellings and transportation routes, harvesting wild plants and animals, and developing agricultural practices that range from adding nutrients to soils, building terraces and canals for irrigation, and domesticating plants and animals, to both setting and controlling wildfires.

What makes the actions of modern humans so problematic? For one thing, compared with most other species, modern humans are more cosmopolitan; there is nowhere on Earth, except perhaps the depths of the oceans, that humans or their by-products have not touched. People are now capable of moving themselves and other organisms across the globe and in all directions in a matter of a few hours. Especially since the 1940s, humans have created synthetic chemicals, and now, with the advent of genetic engineering, novel organisms with which the biosphere has had no previous evolutionary experience.

We do live in a "world of wounds"—a world where such practices as logging, dam building, and agricultural and urban/suburban development contribute to the loss of biodiversity and ecosystems, species extinctions, and habitat fragmentation and the destabilization of ecosystem processes. The conterminous area of the United States has lost 95–98% of its old-growth forests, 53% of its wetlands, more than 70% of its riparian forests, and nearly all its prairies and savannas. In addition, 98% of its rivers and streams are degraded to the point where they no longer qualify as wild or scenic rivers. This is to say nothing of losses along coastal zones and the degradation of ocean waters and fish populations, or what the future consequences of global warming and climate change may bring. There is obviously plenty of "land doctoring" needed (Figure 1.1).

However, as a student with an ecological education, you do not, as Leopold suggested, live alone in a world of wounds. In fact, many of us are interested and involved in doctor-

FIGURE 1.1. **The Impact of Increased Stormwater Runoff.** Urban stormwater runoff has eroded a ditch through the soils of the Curtis Prairie restoration in Wisconsin. (Photo reprinted by permission from Steve Glass.)

ing and caring for the land and water. Environmental conservation has a long and storied history in North America. With its overarching goal of conserving and restoring species, ecosystems, and landscapes, it is an antidote to those economic and social activities that "wound" the environment. Conservation practices are implemented across a broad spectrum of activities that include social activism, preservation of high-quality natural areas, restoration of degraded and damaged ecosystems, management of both protected and restored sites, and efforts to educate and enlist the general public in conservation activities. In the context of an increasingly environmentally aware population, we can put technology to work to advance the success of conservation.

Dr. Richard Hobbs, who teaches restoration at the University of Western Australia and is editor-in-chief of the journal *Restoration Ecology*, has been quoted as saying he is involved in ecological restoration "because it's not all negative." He likes the fact that his students typically approach him a few weeks into the semester and say, "Most of the news in the world is so bad, it's great to be doing something where things can actually get better." Hobbs says: "Restoration is a magnet for getting many people involved at the local level. Ecologists too often think of people as if they were part of the problem. That's not the whole truth. People are part of the solution" ("Restoring Our Wetlands" 2005, 9).

Many people would seem to agree with Hobbs and have been captured by restoration's magnetic appeal. For example, as of 2010, the Society for Ecological Restoration (SER) had 2,300 members from 37 countries and 14 international chapters around the world. SER supports *Restoration Ecology*, which publishes peer-reviewed research articles aimed primarily at the academic audience. The profession's oldest journal, *Ecological Restoration*, has a more practical and applied orientation, with project profiles by and for restoration practitioners, plus peer-reviewed research reports that help bridge the gap between research and practice. The Australian journal *Ecological Management and Restoration*, published by the Ecological Society of Australia, has a similar mission of "linking science and practice."

▶ 1.1 WHAT IS RESTORATION ECOLOGY?

The purpose of this textbook is to explore one emerging sector on the conservation continuum—restoration ecology. Throughout the text you will find that the term "restoration ecology" or simply "restoration" refers to both the theory and the practice of restoration. We prefer this usage to separating theory ("restoration ecology") from practice ("ecological restoration"), as we believe that the discipline is a seamless blend of both endeavors—theory informs practice and practice informs theory.

The definition of restoration presented in the *SER Primer on Ecological Restoration* has been adopted by a number of groups, including The Nature Conservancy: "The process of assisting the recovery of an ecosystem that has been degraded, damaged, or destroyed" (2004, 4). This definition seems simple; however, as you will discover, restoration ecology is a multifaceted discipline and, as a result, is more complicated than it might at first appear.

Although the term "restoration" implies that you are restoring, or putting something back, modern restoration ecology looks to the near-distant past primarily because our understanding of natural communities comes in part from historic records and past experience. In doing restorations, you are not striving for perfect duplication of some former state; instead, you are reestablishing dynamic, ever-changing entities and processes. As Donald Falk, former SER director, wrote: "Restoration uses the past not as a goal but as a reference point for the future. If we seek to recreate the temperate forests, tall grass savannas, or desert communities of centuries past, it is not to turn back the evolutionary clock, but to set it ticking again" (1990, 71). Indeed, even if human activities had not converted land to agricultural, urban, or industrial uses, or had not fragmented the landscape or influenced global nutrient cycles or energy flows, the ecosystems of the present would not resemble those of the past.

Natural communities are variable in space and through time. You cannot return an ecosystem to what it was like a decade ago, let alone hundreds of years ago, nor should it be your goal to do so. You can, however, attempt to provide conditions under which the products of history—plants, animals, communities, and ecosystems—can continue to thrive. An appreciation of the past can help you understand what these conditions might be.

▶ 1.2 EARLY RESTORATION EFFORTS IN NORTH AMERICA

As a profession, restoration ecology dates only from the 1980s with the establishment of SER; however, it has its roots in many fields, ranging from traditional ecological knowledge to sustainable resource management. Restorations have been attempted in one form or another for many years. From early on, the act of restoring the land has been an interdisciplinary effort.

Naturalists such as Henry David Thoreau and John Muir wrote about the need to conserve and protect the environment (see Sidebar). But it was perhaps George Perkins Marsh, in his classic *Man and Nature*, who made one of the first explicit calls for the restoration of lands damaged by human activities when he urged that forests be replanted and restored. In his view, "The objects of the restoration of the forest are as multifarious as the motives that have led to its destruction and as the evils which that destruction has occasioned" (1965, 250). Marsh envisioned the benefits of restoration to include flood prevention, erosion control, moderation of climatic extremes, recharging of the groundwater, and the supplying of natural resources for human use.

Many of today's restorations had their origins in projects conceived of and designed by garden writers, landscape architects, and horticulturists in Europe and America who were active during the period from 1870 to 1940. These were men and women— Frederick Law Olmsted, Ossian Simonds, Jens Jensen, Wilhelm Miller, Edith Roberts, Elsa Rehmann, Harriet Keeler, George Aiken, and George Perkins Marsh, among others—who began by advocating for native plants in their writings, and who filled their landscape designs with them. Later, new ideas from the emerging field of plant ecology began to infuse their thinking and their designs. Projects emerged that closely resembled many modern restorations.

One of the earliest attempts in the United States to restore some of the character and functions of a native plant community may have been the work of landscape architect Frederick Law Olmsted. In 1878, the Boston Park Commission asked Olmsted to design a plan for the Boston Back Bay Fens. Prior to development, the Back Bay Fens had been a large salt marsh. By Olmsted's time the area had become badly polluted by raw sewage. Instead of seeing a problem, Olmsted saw an opportunity to restore the salt marsh. He used an engineering approach to redirect the sewage flow, installed gates to regulate the saltwater level, and carefully selected native plants to re-create a salt marsh in an urban environment.

Working in the midwestern United States during the late 1800s and early to mid-1900s, Wilhelm Miller, Ossian Simonds, and Jens Jensen, in particular, promoted an understanding

Sidebar | John Muir

John Muir (1838–1914) remains, long after his death, one of the most influential advocates for wilderness preservation. Born in Scotland, he came to the United States with his family as a boy, spent his childhood in Wisconsin, and then moved west to California. Muir not only wrote of his deep personal and spiritual connection to nature in popular books such as *My First Summer in the Sierras* (1911) and *The Story of My Boyhood and Youth* (1913); he also took action. In 1892, he cofounded the Sierra Club, the oldest conservation organization in the United States. He helped save the Yosemite Valley in California and Sequoia National Park. Following is an excerpt from a speech he gave to the Sierra Club, describing the establishment of Yosemite National Park:

> The Yosemite National Park was made October 1, 1890. For many years I had been crying in the wilderness, "Save the forests!" but so far as I know, nothing effective was done in the matter until shortly before the park was organized. In the summer of 1889, I took one of the editors of the *Century Magazine* out for a walk in Yosemite and in the woods and bowlder-choked [sic] canons around it; and when we were camped one day at the Big Tuolumne Meadows, my friend said, "Where are those wonderful flower gardens you write so much about?" And I had to confess—woe's me!—that uncountable sheep had eaten and trampled them out of existence. Then he said, "Can't something be done to restore and preserve so wonderful a region as this? Surely the people of California are not going to allow these magnificent forests, on which the welfare of the State depends, to be destroyed?" Then a National Park was proposed, and I was requested to write some articles about the region to call attention to it, while the *Century* was used for the same purpose, and every friend that could be found was called on to write or speak a good word for it. The California Academy of Sciences became interested, and began to work, and so did the State University. Even the soulless Southern Pacific R.R. Co., never counted on for anything good, helped nobly in pushing the bill for the park through Congress.
>
> National Park around Yosemite Valley went through. And in a little over a year from the time of our first talk beside that Tuolumne camp-fire the bill organizing the park passed Congress, and a troop of Cavalry was guarding it. (1896, 275–276)

of the value of the native landscape—its plants, wildlife, and physiography—in the context of both preservation and renewal through landscaping. Jensen especially promoted and applied natural landscape treatments to the Chicago parks and large estates bordering Lake Michigan, with the intent of mimicking the landscape that was present when Europeans began to settle there. Over time his understanding of the landscape increased, through friendships with plant ecologists such as Henry Cowles of the University of Chicago, and by studying natural remnants on excursions to the country. The Lincoln Memorial Garden in Springfield, Illinois, was perhaps the culmination of his efforts. Jensen enlisted the help of the Boy Scouts and Girl Scouts, the Friends of Our Native Landscape, and other conservation groups in planting prairie and forest seeds and seedlings. Today this landscape continues to grow and be managed as a representation of natural landscaping and early Midwest restoration.

In the early 1930s, University of Wisconsin faculty members William G. Longenecker (horticulturist and landscape architect), Norman Fassett (taxonomist and botanist), and Aldo Leopold (wildlife ecologist) were looking for natural areas where they could teach students about plant and animal communities (or "associations," in the parlance of that period) and the interactions of individual species. Longenecker remembered that the trio wanted their students to study plants and animals "not only as individuals, but as units in an organism—the pine woods or the bog is an organism dependent for its very existence upon the various parts" (1941, 5). Since remnants of Wisconsin's native plant and animal communities were becoming scarce and were not easily accessible to university faculty and

FIGURE 1.2. Curtis Prairie, in the University of Wisconsin–Madison Arboretum. The world's oldest restored plant community; this is as it appeared in 2010. (Photo reprinted by permission from Steve Glass.)

students, their solution was to create an outdoor laboratory where students could study the ecological processes and functions of plant and animal communities. Their opportunity to create such a teaching experience came in 1934, with the dedication of the university's new Arboretum, Wildlife Refuge, and Forestry Preserve, a few miles south of the main campus.

Some of the plant associations the Wisconsin professors desired already existed on site as intact or degraded remnants, including a native wet prairie and a grazed oak-hickory woodland that Longenecker thought needed to be replanted with the appropriate trees, shrubs, and wildflowers. In the center of the property was an 80-acre old-field and horse pasture they decided to return to tall grass prairie. Their decision laid down a fertile path that resulted in the emergence of the Arboretum's John T. Curtis Prairie, the world's oldest restored ecosystem, and the practice of ecosystem restoration (Figure 1.2).

Another step along the road to restoration ecology took place in 1935, when Ambrose Crawford undertook the conservation and rehabilitation of tropical rainforest in Lumley Park Alstonville in New South Wales, Australia. A year later in Broken Hill, New South Wales, Albert Morris undertook what was "probably the first ecological restoration project in Australia using a local ecosystem as a reference," according to Dr. Tein McDonald, *Ecological Management and Restoration* editor (2008, 165). Designed for the Zinc Corporation, the purpose was to both plant and facilitate the re-colonization of native species in order "… to counter desertification occurring in and around Broken Hill" (McDonald, 2008, 166).

▶ 1.3 RESTORATION ECOLOGY TODAY

Modern restoration projects range in size from a few hectares to thousands of hectares, and include wetland, aquatic, and terrestrial systems across the globe. Restorations are conducted by private individuals, conservation groups, and government agencies; they range from working with totally degraded sites, such as mine tailings, bare ground, and old-fields, to relatively undisturbed natural area remnants. Large projects are becoming increasingly more common, due to the recognition by state and national governments that restoration ecology is a significant conservation approach.

1.3.1 Scale and Scope

The scale and scope of restoration projects vary considerably. Some projects require high-end technology and fleets of earth-moving equipment, such as the plan to restore the hydrology of the Florida Everglades (estimated cost almost $8 billion), or the Chesapeake Bay restoration project (estimated cost over $28 billion), which is composed of thousands of small river and stream restorations covering mile-after-mile of riparian habitat in parts of Pennsylvania, Virginia, and Maryland.

However, restoration activities also include less expensive and less technical endeavors, often showcasing the work of private citizens such as those volunteering to restore habitat and recover biodiversity on a few acres in their local natural areas. On this local level, the work attracts individuals who wield shovels, pruning saws, and drip torches to clear their local landscapes of pest species, just as it calls to others who spend their summer and fall weekends collecting, cleaning, and sowing native seed. Perhaps it is this easy accessibility that attracts so many people to restoration. Maybe its popularity is due to the feeling that just one person really can make a difference. And finally, many people are probably drawn to restoration work because they recognize that humans are a part of nature and not an alien intrusion.

1.3.2 Transdisciplinary Focus

Perhaps because of its historic roots in conservation, urban planning, landscape design, and education efforts, modern restoration ecology draws on a variety of different disciplines and skills in order to set the evolutionary clock ticking again. It is transdisciplinary and most often uses a team approach in which individuals with different skills and knowledge work together.

From the biological and physical sciences (especially ecology), restoration draws upon information about geology, physical geography, soils, plants, animals, microbiology, hydrology, and climate. It integrates the intricate and synergistic interactions these various factors have with one another.

From the world of design (landscape architecture, civil engineering, architecture, and art), restoration derives methods for spatial and logistic problem solving and communication. The design arts also provide the form-giving methods and skills, as well as the engineering concepts, required for transforming degraded landscapes and damaged ecosystems to satisfy the aesthetic and emotional needs of people.

The humanities (history, human geography, anthropology, philosophy) and the social sciences (sociology, psychology, economics, political science) also play important roles in helping restorationists understand such important factors as the history of the land, the legal and regulatory frameworks within which projects are accomplished, the principles behind building and managing budgets, the sociopolitical nature of working with others, and the ethics of restoration-related activities. The code of ethics (moral principles and rules of conduct) followed by a profession such as restoration ecology derives from the set of philosophical assumptions held by a society about how the world works and the nature of the relationship of humans to that world. In restoration, ethics provide a framework for decision making that helps you make choices. Examples include choices about whether people have the "right" to interfere with nature; what is thought to be beneficial for the ecosystem of interest; which ecosystems are in need of, or are amenable to, restoration; and which parts of the ecosystem primarily require human assistance.

Ironically, the same human activities that degraded the environment (especially agriculture and forestry) now provide us with some of the tools and methods for restoring the land. For example, grassland restorationists often use tractors to prepare the soil and plant prairies, and combines to harvest seeds. Likewise, the fire-fighting equipment developed to suppress forest fires is now used intensively to conduct prescribed fires in grasslands, woodlands, and wetlands across the globe.

▶ 1.4 TYPES OF RESTORATION

Contemporary restorations are undertaken for many different reasons, each of which is tied to the social values and resources of the people involved. Most projects can be classified into three types: (1) complete restorations, which strive to include all the attributes of historic natural communities; (2) ecological services restorations, sometimes referred to as process-based, structural, or functional restorations, which often include a simplified array of native species and may result in novel communities; and (3) experiential restorations, which are created to provide pleasure for people.

While it is often convenient to separate these three types, they are not mutually exclusive. For example, over time, the emphasis in a long-term project may shift, as the needs of a community change, or as more information becomes available. Complete restorations are generally the most complex and difficult to achieve, especially on very degraded sites. On the other hand, ecological services and experiential restorations are simpler and usually require less effort to achieve an initial success. Perhaps what begins as a structural restoration may eventually become a complete restoration over time. Also, remember that all three restoration types are created for and by people, although they may vary in the degree to which people may visit a particular site.

1.4.1 Complete Restoration

Complete restoration aims to restore the complexity of all of the attributes (composition, structure, functions, processes, services, and aesthetics) of a plant community, ecosystem, or landscape to the best extent possible. It is the most idealized form of restoration and may be the most difficult goal to achieve. Early examples of attempts at complete restoration include the Curtis and Greene Prairies at the University of Wisconsin Arboretum in Madison. Contemporary examples include the restoration of former boreal hayfields in Norway, and the restoration of vernal pools in California. In Brazil, The Nature Conservancy (TNC), a conservation agency operating across the world, is in the midst of an ambitious project to restore the Atlantic Forest in South America. This moist tropical forest is found in Brazil, northern Argentina, and southeastern Paraguay. TNC is concentrating on earning the cooperation of landowners—many of them farmers and ranchers—to link remaining remnants by restoring forested corridors. In 2008, TNC launched an ambitious project to plant one billion native trees in the forest. The idea is that by restoring a canopy to degraded and/or denuded land, the species from the remaining forest patches will be able to colonize this new habitat. (See Chapter 14 for more information about this project.)

CASE STUDY The Nachusa Grassland Project

The goal of complete restorations is to achieve all the attributes of historic natural communities. One of the most common applications of this approach is to start with an existing remnant and restore the land around it to create a larger, and therefore more diverse and fully functioning, community. With larger size comes more wildlife habitat and the resources needed to support species that require large territories (see Chapter 2). The Nature Conservancy is increasingly turning to complete restorations to enhance the preservation of core natural areas. One thriving example is the Nachusa Grasslands (Figure 1.3).

Located in northern Illinois, the Nachusa Grasslands Project mixes the preservation of grassland remnants with the restoration of former agricultural lands, to restore a landscape that is one of the last examples of the prairie that once covered most of the state. Since 1986, through acquisitions and easements, TNC has created a more than 1,000-ha of grassland, with more than 600 species of native plants and 180 species of birds, including the grasshopper sparrow. The grassland is being assembled from a network of small sites, each of which currently has its own set of restoration plans. The goal is to eventually connect all of the units to create one grassland complex that can be managed as a unit, using natural processes such as grazing and fire (see Chapter 10).

The project relies on the cooperation and assistance of owners of private property, who, through easements, allow the restoration and preservation of portions of their land, as well as monetary donations from individuals and foundations. Perhaps the most important key to the success of the restoration, however, is the loyal group of volunteers who have donated thousands of hours to restoring the prairie. Volunteers participate in all aspects of restoration, including collecting and planting seed, removing brush and pest species, monitoring plants, conducting prescription burns, leading public tours, and serving as land stewards. They are responsible for the hands-on implementation, monitoring, and management of specific portions of the grasslands. The Nachusa Grasslands stewards include many volunteers who have worked on the restoration for more than a decade. Some stewards have to travel for two hours to reach the site from their homes, and often do so several times a week. ■

FIGURE 1.3 The Grasshopper Sparrow (*Ammodramus savannarum*). Nachusa Grasslands, a project of The Nature Conservancy, mixes the preservation of grassland remnants with the restoration of former agricultural lands. Increasing the acreage of remnants through restoration enhances its value for wildlife habitat, as well as the ability to reintroduce natural disturbances such as grazing and fire. The grasshopper sparrow is only one of a number of grassland birds that have rapidly declining populations thought to be due to the loss of habitat. (Photo reprinted by permission from John Harrington.)

1.4.2 Ecological Services Restoration

Ecological services or process-based restoration seeks to establish those structural and functional attributes of communities/ecosystems that support or satisfy human needs (food, for example), or that support species we value, such as endangered or symbolic plants and animals. Ecological services restoration goals are often more attainable than those envisioned in a complete restoration project, or at least they tend to take a shorter length of time to reach. They may also serve as an early phase in a long-term complete restoration project. Marsh's call for forest restoration and Olmsted's Back Bay Fen salt marsh restoration in Boston are nineteeth-century examples of this type of restoration. Modern-day examples include the creation of artificial coral reefs in Indonesia to restore habitat for the reef animals, and the planting of a few species of dominant prairie grasses, without the wildflowers (forbs) and less common grasses that would be present in a true prairie, to provide habitat for grassland birds in the American Midwest. Projects are also under way to combat global climate change.

Sometimes ecological services restorations result in novel communities—groups of species that have no counterpart in "nature." Examples of novel communities with which most of you will be familiar are the "weed" communities that develop on abandoned urban lots, and the so-called old-fields that result after formerly cultivated agricultural land is abandoned. Such communities include both native and non-native (exotic) species. Novel community restorations are usually designed for settings where human disturbance has altered the site conditions to such a degree that there is no natural analog, and/or efforts to restore a "natives only" community would be impractical, at least in the short run. Examples include efforts to restore plant cover to sites such as (1) industrial brownfields—sites that contain hazardous waste or other pollutants left behind by industrial or waste disposal operations; (2) mine-tailings deposits—the crushed rock left over after such metals as iron, lead, zinc, copper, gold, and silver have been extracted; and (3) abandoned properties in urban areas containing soil contaminated by lead from the paint used on the buildings or decades of automobile exhaust in the days when gasoline contained significant amounts of lead. Some authors use the terms "rehabilitation" and "reclamation" to refer to these kinds of projects. Because these words often refer to a broad array of projects in addition to restorations (such as turning a former landfill into a turf-covered dog exercise park), we prefer "process-based" or "functional" restorations to avoid confusion.

As you will see in Chapter 2, communities are constantly changing, so every restoration is novel in a strict sense. What is distinctive about the restoration of novel communities is that the restoration goals are established from the beginning of the project to focus on structure or function, or on a composition that includes exotics rather than on a native species composition per se.

CASE STUDY | Restoring Wetlands for Carbon Storage in the Bavarian Alps

In 2001, the United Nations launched the Millennium Eco-system Assessment Project to assess the impacts of humans on natural systems, and to investigate strategies for conservation and sustainable use. It describes four major types of services:

- Provisioning (Food, Fresh Water, Wood, Fiber)
- Regulating (Climate, Floods, Disease)
- Supporting (Nutrient Cycling, Soil Formation)
- Cultural (Aesthetic, Spiritual, Educational, Recreational)

The first three categories are provided by ecological services restorations. (The fourth category relates to experiential restorations.)

One of the "supporting" functions that is attracting a lot of attention right now is the carbon cycle (see Chapter 2). Human impacts, especially the burning of fossil fuels, are contributing to ever-increasing levels of carbon dioxide in the atmosphere, which in turn contributes to climate change. One idea for reducing the atmospheric concentration is to store carbon in the living biomass, or soil organic matter, of natural plant communities, particularly old-growth forests or peatlands. In the Bavarian Alps, a project is under way to restore wetlands for this very purpose (Figure 1.4).

Peat is a type of wetland soil composed of partially decomposed mosses and plants. The lack of oxygen and often high acidity in peat-forming wetlands means that the release of the carbon contained within the plant debris is very slow. This results in more litter being deposited every year than is decomposed, thus leading to thick layers of peat. The thicker the peat, the more carbon is stored. If the wetland area is drained, decomposition increases, with a consequent release of carbon to the atmosphere.

Over hundreds of years, much of the historic peatlands in northern Europe have disappeared. For example, estimates are that at least 90% of the wetlands in the Bavarian Alps have been degraded, usually by draining the soils to support various activities, especially peat mining. Efforts are now under way to restore the carbon-storing peat.

Three main steps are involved. (1) In most cases, the soils must be rewet in order to create the low oxygen conditions necessary for peat formation. (2) Native plants and *Sphagnum* and other mosses must be reintroduced. (3) Pest species must be controlled. Some of the strategies being employed are blocking drainage ditches and creating terraces to retain rainwater; removing invading shrubs and trees; introducing plants by sowing seed, planting seedlings, or using mulch containing native seeds; and reintroducing *Sphagnum* moss.

Peatland restorations to date have met with mixed success. One factor that influences the ease of mined peatland restorations is whether the peat was cut or milled, with the restoration of milled lands being the most difficult. Rewetting strategies are complicated by the topographic variability of the disturbed sites. In some cases, rewetting can cause portions of a former wetland to flood. *Sphagnum* moss seems to be difficult to restore, and the best success comes from spreading vegetative parts in a thin layer on bare peat and covering the layer with a straw mulch. (See Chapter 8 for more information.) ∎

FIGURE 1.4 Ongoing Bog Restoration in the Chiemsee Region of Germany. These areas were historically mined for peat, releasing much of the stored carbon into the atmosphere. Restoration efforts are being studied for their ability to use restored peat bogs as carbon sinks, as well as for a better understanding of gas exchange between bog and atmosphere. (Photo reprinted by permission from John Harrington.)

FIGURE 1.5 **Amsterdamse Bos, Near Amstelveen, Netherlands.** Planted from the 1930s through the 1960s, this woodland park was modeled on native vegetation communities of northern Europe and German "people parks." The land on which the park is constructed is reclaimed; the forest was not native. (Photo reprinted by permission from John Harrington.)

1.4.3 Experiential Restoration

The goal of experiential restoration is the revitalization of landscape attributes that bring pleasure to humans. These plantings typically simplify the number of species and exaggerate the more charismatic qualities of an ecosystem or plant community. The works of landscape architect and park designer Jens Jensen in the early twentieth century are good examples of such designs that are inspired by nature and use native plants (see Sidebar). Another example is Amsterdamse Bos, near Amstelveen, Netherlands. This is a woodland park planned and planted from the 1930s through the 1960s. The landscape of marsh and woodlands was modeled after native vegetation communities of northern Europe and German "people parks." However, in this case, the land upon which the park is constructed is reclaimed (Figure 1.5).

Today, plantings of native plants along highway corridors, and efforts to develop park systems with a plant community theme, are typical exercises in experiential restoration and echo the earlier initiatives. Experiential restoration often serves as a transition from more human-dominated landscapes into increasingly more wild ecosystems.

▶ 1.5 OUTLINE OF THE RESTORATION PROCESS

Restoration ecology is an applied profession, similar to medicine, conservation, engineering, planning, architecture, or landscape architecture. Professionals in these fields both generate and use knowledge, as well as other information, to solve problems.

Moreover, restoration ecology is a design profession. Restorationists influence the three-dimensional form and function of a landscape by shaping the size, shape, composition, structure, processes, and interactions of plant and animal communities. As a restorationist, the problems you face are complex, and require an understanding of physical and natural systems, human–environment relationships, political and legal systems, and the use of several kinds of tools and technologies.

CASE STUDY   The Alfred Caldwell Lily Pool

Many restorations are created for the enjoyment of people. These include natural landscapes surrounding homes or office buildings, as well as public parks designed for hiking, bird watching, and the like. The 0.6-ha Lily Pool (Pond) in Lincoln Park, Chicago, Illinois, designed by Alfred Caldwell, is a wonderful example of an experiential restoration (Figure 1.6).

In the late 1930s, Alfred Caldwell, a disciple of landscape architect Jens Jensen, was asked to redesign a pool that had been built in 1889 to raise water lilies. Caldwell's goals were to create a sanctuary for nature in the middle of urban Chicago and to express the structure of the native midwestern landscape through design. The heart of his garden oasis was a lagoon designed to look like a meandering glacial river cutting through native limestone rock (created by stonework), surrounded by native trees, shrubs, and wildflowers. The design included a waterfall as the source of the "river," and walkways paved with stone that moved visitors through the site.

The Lily Pool indeed proved to be an oasis for people, and especially for birds. In fact, the spot became so popular with birdwatchers that it became known as "The Rookery." However, by the late 1990s, as can happen to most restorations unless they are actively managed (see Chapter 10), the impacts of birdwatchers, other visitors, and of the birds themselves had caused many of the features of the restoration to deteriorate. Exotic pest species had overrun the native plantings, the soils had eroded, the lagoon had silted in, and the stonework of the landscape and pathways was crumbling.

In 2001, a partnership of the Chicago Parks District and a Friends Organization undertook a restoration of the restoration. Using historic photographs and Alfred Caldwell's original site and implementation plans, their efforts included pest species removals, the planting of thousands of native plants, and the repair and redesign of the path system to improve accessibility, using universal design principles. By 2006, their work had proved so successful that the Alfred Caldwell Lily Pool was designated a National Historic Landmark. ■

FIGURE 1.6. The Alfred Caldwell Lily Pool. Designed by Alfred Caldwell in the 1930s, the Lily Pool emulates the prairie landscape of the U.S. Midwest. (Photo reprinted by permission from John Harrington.)

The ability to solve problems is an important survival skill for all kinds of organisms, and certainly for people. Many different problem-solving approaches exist, ranging from trial and error, in which many things are attempted (often at random) until one seems to work, to more systematic methods using a progression of steps based on a conceptual framework.

All restoration projects follow a plan, whether it is written down for others to follow and learn from, or whether it remains in the memory of a single restoration ecologist. Simply stated, a restoration plan sets a goal, and recommends a set of strategies that will take a site from the condition it is in at the start to a condition that matches the goal. The plan also includes checks and balances, which are a built-in series of opportunities to evaluate whether the chosen strategies are working, or whether a midcourse correction needs to be made. Having an explicit, systematic process is particularly important for the many restoration projects that are conducted in public settings. Transparent decisions, justified with explanations and processes that allow for meaningful public involvement, go a long way in gaining support for projects and in resolving conflicting interests.

Each restoration site is unique and requires a similarly unique process for its restoration. However, any restoration process can be simplified into several basic steps, which are briefly discussed here and outlined in Figure 1.7. (You will learn more about each of the steps in subsequent chapters.) The process is systematic in that it outlines a number of topics you should consider before taking action. By following all the steps, you will produce five plans: (1) the master plan, (2) the site design plan, (3) the implementation plan, (4) the monitoring plan, and (5) the management plan. Each of these includes text and graphics. Together the set of plans constitutes the full project documentation.

Although we are presenting the steps in a linear fashion, it is important to understand that the process is more complex than this linearity implies. Usually, you move back and forth among the steps as a project proceeds, generally with a different focus at each iteration (Figure 1.8).

The process and resulting plans remain flexible and subject to revision as new information is discovered, and/or site or other conditions change. For example, the outline implies that the goals of a restoration are set before the site conditions are known ("propose initial project goals" precedes "site inventory and analysis" in the process outline). In practice, when you work as a restorationist, you will usually have some idea about the features of a site before or at the same time as you are articulating its purpose. In fact, the preliminary goals help direct what to inventory. Then the results of the inventory help to further refine the goals. Thus, you may know at the outset that a site is currently wooded, so planning a forest restoration would be appropriate. But not until you go on a site visit and identify the kinds of trees that are present will you be able to identify the appropriate type of forest. It is even possible that, if the existing trees are not native, your project will shift to restoring grassland to the site instead.

Also, two themes to keep in mind as you follow the restoration process are (1) the value of involving people (site owners, residents, and neighbors; other interested members of the community) in the establishment, planning, implementation, and care of restoration projects and (2) the importance of learning through restoration by documenting projects, and establishing field trials and more formal research projects.

1.5.1 Preliminary Research: Steps A and B

The first two steps in the restoration process have several purposes: (1) to establish the general principles for guiding the project and articulating what you hope to accomplish;

A. Develop the Project Purpose and Use-Policy: Propose Initial Project Goals

B. Gather Background Information
    1. Develop community/ecosystem models to guide the restoration.
    2. Review previous restoration projects and current practice.
    3. Establish documentation, communication, and research procedures.
    4. Establish procedures for adoption and revision of plans.
    5. Review legislative requirements and necessary permits.
    6. Conduct site inventory and analysis.

C. Create the Master Plan: Determine the Desired Outcome and Practical Constraints of the Restoration
    1. Review the results of the site inventory and analysis.
    2. Approve the project purpose and the site use-policy.
    3. Determine the general physical layout of the restoration.
    4. Determine the community/ecosystem restoration goals.
    5. Determine the goals for people (site users, restoration volunteers, project supporters).

D. Create the Site Plan: Describe the Details of the Desired Outcome
    1. Collect and analyze additional site information, if needed.
    2. Determine restoration objectives.
    3. Refine the layout of the communities/ecosystems.
    4. Determine the composition, structure, and function of each community/ecosystem unit and the onsite locations of species.
    5. Locate any infrastructure needed to achieve use-policy objectives.

E. Create the Implementation Plan: Describe What Actions to Take (If Any) and Where and When to Achieve the Restoration Goals and Objectives
    1. Determine site preparation needs and approaches.
    2. Resolve logistics, including permitting and resource acquisition.
    3. Select implementation procedures, including acquiring plants and determining planting sequence, rates, and methods.

F. Create the Monitoring Plan: Determine Whether the Goals and Objectives Continue to Be Met
    1. Specify what is to be checked and how, based on restoration objectives.
    2. Specify where and when monitoring is to occur and who will do the checking.
    3. Determine how monitoring data will be analyzed and how the results will inform the implementation and management plans.

G. Create the Management Plan: Ensure the Goals Continue to Be Met After the Restoration Objectives Have Been Initially Achieved
    1. Anticipate/identify potential impacts if the management is or is not implemented.
    2. For each process/problem, evaluate several potential strategies (solutions), including the hands-off approach.
    3. Choose and develop a prescription for the strategies of choice.

H. Implement, Monitor, and Manage the Restoration

I. Periodically Review the Restoration Plans and Procedures and Modify According to Previously Adopted Guidelines, as Needed

FIGURE 1.7 Outline of the Restoration Process

FIGURE 1.8 The Flexible Flow of Steps in
the Restoration Process.

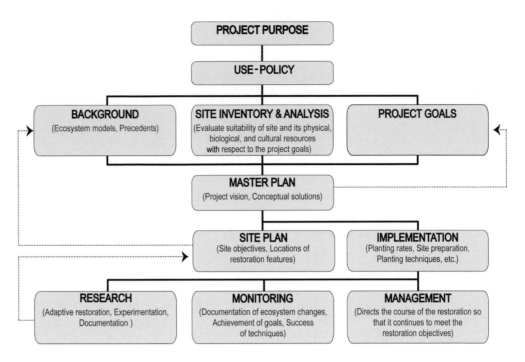

FIGURE 1.8 The Flexible Flow of Steps in the Restoration Process.

(2) to propose a use-policy describing the ways people will be able to interact with the site; (3) to gather and synthesize as much existing information as possible about past restoration efforts, as well as the composition, structure, and functions of the kinds of natural systems you believe may be appropriate to the site; and (4) to determine the site's current natural and cultural conditions.

Restoration ecology is a relatively new profession. Although we have learned enough over the past decades to be able to share information with you in a textbook such as this one, there is much we do not know. Many questions remain both about restoration practice and about the nature of the communities/ecosystems that you hope to restore. The "adaptive" approach to restoration is designed to enable you to work with uncertainty and with the dynamics of natural systems. It is important for you to take time in the early stages of a project to establish flexible procedures.

Although the focus of many restorationists is on the conservation of native plants and animals, people are always an important component of any project. It is important that you communicate at the outset with the members of the human community who potentially have an interest in the project, and establish ways to keep people involved while at the same time protecting the restoration.

The development of community/ecosystem models is one of the important features of this stage of the process. These models are central to creating the restoration vision and determining project strategies. The models are based on ecological theory and information about both contemporary and historic remnants.

"Site inventory and analysis" refers to collecting and evaluating information about the natural and cultural features of a restoration site. You start these activities early in a restoration project by gathering general site information, and repeat them in more detail throughout all the remaining steps of the process. Matching species to the environmental conditions of a site is crucial to the success of a project.

1.5.2 The Master Plan: Step C

In this step, you develop the initial vision of what the restoration is hoping to achieve, and then translate it into a set of goals that describe the site as it will be after the completion of the restoration. In addition, you describe the ways in which people will be able to interact with the site in a use-policy. Often, especially if the site is large, you, together with neighbors and any other interested parties, will consider at least three different visions and weigh the opportunities and constraints of each before settling on one. Having to think of alternatives ensures that you really think about a particular site, and not just apply a "standard" solution. The chosen alternative in the master plan guides the site plan.

1.5.3 The Restoration Site Plan: Step D

The site plan gives actual three-dimensional form to the project vision. It refines the restoration goals by creating measurable objectives you can use to assess the progress of the restoration through time. The plan explains exactly which species should be present, where they should be located, and in what abundance. It also describes desired ecosystem services and the nature of necessary disturbance cycles (fire, floods, etc.).

1.5.4 The Implementation Plan: Step E

The implementation plan provides detailed instructions for achieving the objectives of the site plan, given the conditions of the site at the start of the project. For example, if species need to be added, it describes *how* this should be done (e.g., by hand planting, using a mechanical planter, using seeds, transplants), *when* it should be done (e.g., in the dry season or in winter), and *where* on site the species should be added. And, if the site needs to be modified before planting can proceed, the implementation plan explains how. If species are to be removed, the plan discusses different procedures, and so on. Once again, as in Step C, the plan describes alternative strategies and the reasons for choosing some over others. In many cases, before settling on any one strategy, the plan may suggest establishing experiments or demonstration field trials to test several techniques on a portion of the site and evaluating the results, before applying the one that best fits the situation.

1.5.5 The Monitoring Plan: Step F

In order to determine the progress of a restoration, you establish a monitoring plan for evaluating the extent to which the goals and objectives are being met as implementation proceeds. The plan describes measurement and analysis procedures and the rationales behind using them. It addresses two different restoration stages: (1) the time period from implementation to the achievement of the restoration goals and objectives, and (2) the time period after restoration has ended and management begins.

1.5.6 The Management Plan: Step G

Management begins when the restoration goals and objectives have been achieved. Ideally, you outline the management plan at the same time you are creating the site plan, implementation plan, and monitoring plan. Using the community/ecosystem model, with an understanding of how the site is to be used, plus information about the nature of the land uses that surround the site, you anticipate the kinds of changes (impacts) that could occur over time. The plan then suggests alternative strategies to deal with each potential impact.

1.5.7 Installation and Continuous Review: Steps H and I

The last two steps of the outline involve actually carrying out the restoration—that is, implementing the site plan—and then continuing to evaluate its success, making changes in procedures or even in the goals and objectives, if necessary.

▶ 1.6 ASSUMPTIONS, CHALLENGES, AND OPPORTUNITIES FOR RESTORATION

Underlying each restoration effort, regardless of its size or scope, is a series of assumptions, challenges, and opportunities. It is important to recognize these ideas and facts, whether you are a beginning student or long-time practitioner, because they will affect how you think about and relate to the environment, your work, and other people. These ideas also provide you with opportunities for new research and exploration of the interactions and interdependencies of humans and the world environment.

1.6.1 Basic Assumptions

In every enterprise in life, you make assumptions about the way the world operates. Without making some assumptions, you would be paralyzed into inaction. On the other hand, having assumptions and either not recognizing them, or not challenging them, can be unhealthy for any individual, social institution, or practice. In restoration projects, it is important to be clear about the assumptions you have made about the systems you hope to restore.

Restorationists operate with many assumptions in carrying out projects. An important step in improving the discipline is to recognize and discuss these critical underlying suppositions.

One assumption that drives restoration is that many parts and processes of the Earth are damaged, destroyed, or missing. As a result, the Earth's natural capital (the total accumulation of the goods and services provided by global ecosystems) is diminished, to the detriment of the well-being of the planet and its human and non-human inhabitants. Restorationists assume that solutions exist to repair the damage to ecosystems and their value to the world. We also assume that people have some capacity for caring for the planet and repairing damaged parts of the Earth's systems, thereby increasing or at least stabilizing the natural capital.

By repairing the damage to the Earth's systems, we also enter into a relationship with the planet that helps repair our connection to nature and our own communities, thereby increasing our own personal and social capital. This set of assumptions constitutes what Hull and Robertson might call "a restoration ethic." In their view: "Restoration blurs the distinction between culture and nature. It makes an open continuum out of the more simplistic and polarizing human–nature dichotomy. Where the preservation ethic polarizes humans and nature as distinct and separate entities, a restoration ethic opens up a broad middle ground where it is acceptable for humans and nature to interact. It encourages us to look forward to the potential of nature rather than backward to what nature was in the past" (2000, 301).

1.6.2 Challenges

Despite the many restoration projects that have been completed, are in process, or are being planned, restorationists face many challenges, some of which have an especially complex nature. It is not surprising that complex problems are inherent in restoration ecology, where uncertainties exist about weather, sociopolitical conditions, global warming, and the very historical uniqueness of the communities/ecosystems being restored.

Here are some of the more important challenges that interfere with the success of restorations: (1) the fact that natural systems are constantly changing; (2) the fact that we have an imperfect understanding of natural systems; (3) the lack of available information about previous successes and failures; (4) the fear that natural remnants will be destroyed, because people assume restoration projects can replace them; (5) the fact that restoration is situational—there is no single restoration formula, and each project is very time consuming; (6) the fact that project stakeholders often have conflicting desires; and (7) the lack of sufficient resources to support long-term projects.

Constantly Changing Natural Systems

Natural systems are variable in space and through time. No two forests are identical in composition or structure, and the same forest will be different in 2035 than it was in 2010. Therefore, as a restorationist, you face the challenge of having a fuzzy, moving target. This sometimes makes it difficult to know if a project is working in the short run, and often makes it hard to predict the extent to which it will continue to work in the long run. In effect, it may be the case that a restoration is never fully realized.

Long-term changes, such as climate change, are still not completely understood. Ecologists have suggested more than one scenario about how climate change may affect the composition and structure of plant and animal communities and the functioning of ecosystems. As a result, ecologists recognize ecosystems as operating in nonlinear, complex ways that are influenced by past and future conditions, their current surroundings, and stochastic events.

Restoration cannot be viewed as returning a site to a narrowly defined community or ecosystem structure. Instead, restoration plans must be flexible enough to allow for changes in ecological understanding, as well as in the circumstances of a specific site, in order to achieve long-term sustainability (with minimal human input). A fundamental challenge for the discipline is that you, as a restorationist (similar to a medical practitioner), need to act with the best available contemporary information, and be ready to change course if conditions warrant.

The Limits and Evolution of Our Understanding of Communities and Ecosystems

From both research and applied perspectives, we have a poor understanding of many aspects of natural communities. For example, we know little about the relationships between the physical soil matrix and soil biota that underlie most restoration projects. Similarly, our knowledge about plant pollinators and wildlife as they relate to plant community restoration remains sketchy. We have limited knowledge of competitive interactions and the sequence of colonization and the natural processes that lead to community change. Ecosystem services, such as the cycling of nutrients, the maintenance of water quality, and the production of biomass, are often important restoration goals. Yet we do not fully understand how many of these services operate, while many others are likely to be unknown to us. Although we have general models to describe such ecosystem functions, we have much to learn about the details of particular sites. Although this makes determining the success of restorations especially problematic, it represents an immense opportunity to learn about ecosystem functioning through restoration.

Much of our understanding of natural communities and ecosystems comes from the study of remnants (especially in the Americas)—fragments of natural communities that have been relatively undisturbed by modern development (within the past 150–200 years)—and by historic collections and memoirs. Many community remnants exist because they are in situations or settings that had little interest for humans—land too steep to plow or that readily floods, for example—and therefore may not be representative of the community as a whole. Others, even those that seem natural, have been influenced by fragmentation (leading, for instance, to the loss of species), changes in fire regimes and hydroperiods, changes in climate, and invasions of pest species. It is therefore possible that the conclusions we draw based upon studying such reference communities are incomplete. Add to this the uncertainties of future climate changes and other, as yet undiscovered ecosystem interrelationships, and it becomes clear that we will be challenged to keep up with new developments in the discipline throughout our careers.

Lack of Documentation and Experimentation

Despite the publication of numerous books and journals, and websites devoted to restoration ecology and its related disciplines and practices, information remains hard to come by. Unfortunately, most restoration efforts have been poorly documented. Records have either not been kept at all or have been kept only through the planting period. And monitoring data are often not analyzed or shared publicly. The absence of shared information means that valuable data and advice about successes and failures cannot be passed on to others.

Few restorations have incorporated experiments to test our understanding of either restoration theory or practice. Predictive models and testable hypotheses are relatively rare, making it difficult to generalize from one restoration project to another, or to adapt learning and experiences from one ecosystem to another. This hinders the sharing of information among practitioners and researchers, and thus the development of new ideas that such learning might generate. As a result, our understanding of restoration has been slow to develop.

Moreover, we particularly lack information about the social aspects of restoration. Researchers are making efforts to learn more about the relationship between people and restoration, but there is much more work to do. Similarly, there is a significant gap in our knowledge about indigenous land management practices—key pieces of knowledge when trying to understand what ecosystems were like and how they operated prior to becoming degraded. Here, too, work is just beginning.

Limitations and Dangers of Mitigation

As restorationists, we are faced with the perennial danger of people using the possibility of mitigation: restoring an area as a means of leveraging the destruction of an existing natural area. Common in the United States, especially in wetlands-related work, this practice is written into state and federal law. The dilemma is, of course, that restorations cannot replace the diversity and complexity of a naturally evolved community. In such cases, restorationists often have to choose between accepting a job and the income it provides (often accompanied by the knowledge that if you don't do the work, someone else will), and the loss of an acknowledged area of natural quality. The concern is that by engaging in the practice of restoration, society will come to believe that a restored site is just as valuable as a natural remnant.

Mitigation provides justification for destroying quality remnants as long as another location is available for its "restoration." Some politicians showcase ecosystem restoration projects, while simultaneously passing legislation and establishing programs that destroy or impair existing high-quality environments. One example is the wetlands banking programs, which are used throughout the United States. Highway departments will restore wetlands in one area and bank or reserve these as replacements when they build a road through an intact wetland area. The problem with this scenario is that studies show these restored wetlands do not have the species diversity, structure, functions, or services that the destroyed wetland had and will most likely never come close to achieving them.

Some corporations use a similar tactic when they promote the restoration of one area, but pollute or destroy other areas. This activity, known as "green washing," makes it difficult for other environmentalists to support restoration activities, and puts restorationists in a moral and financial dilemma when given the opportunity to work with government agencies or corporations who operate in this manner. The bottom line is that restorations should be viewed as a complement to, not a substitute for, preservation, and we should always be clear with the public about the limits to restoration.

The Situational Nature of Restorations

Each site has a unique history in terms of how it started (founder events or initial conditions); how it was affected by weather and other chance events, including catastrophic disturbances (hurricanes, floods, wind throws, fires); how it will react to global climate change; and how it was used and managed over time by both animals and humans. Although we have some understanding of how communities respond to such natural disturbances as fires, windstorms, or floods, we cannot yet predict the occurrence of these events for more than a few days in advance. Therefore, ecologists consider ecosystems to be operating in nonlinear, complex ways that are influenced by past conditions, their current surroundings, and stochastic events.

Restoration projects are no different. The implication is that there is no single formula for success. An establishment technique that works on one site might not work on another, or on a seemingly similar portion of the same site in a different year. Because of the situational nature of restorations, you need to create project goals that allow for a range of variation in community composition and structure, and to build flexibility into the plan for responding to unanticipated situations. It also means you must develop an understanding of the principles behind the design of each project in order to adapt to local conditions, rather than relying on a one-size-fits-all approach. As a result, restorations take much time and effort. On the other hand, it is important to remember that the individual challenges of each project represent opportunities for problem-solving innovations.

Conflicts Among Users

Sometimes restoration projects run into difficulty because the people involved have different visions about how to proceed. A now classic example of this kind of conflict occurred in northern Illinois in the 1990s (Mendelson et al. 1992). The managers of the Chicago Forest Preserve District set out to restore Chicago's forest preserves to oak savanna (a plant community composed of scattered "open-grown" trees growing in a matrix of grasses and wildflowers). Their rationale was that the historic vegetation of the preserve had been oak savanna, and the current forest cover was the result of human disturbance—namely, the cessation of wildfires brought about by modern human settlement. Therefore, they reasoned, to properly protect the conservation values of the preserve, the best thing to do would be to restore its historic conditions. The implementation of the plan involved cutting down many trees, and almost immediately, opposition arose from other groups of environmentalists who believed the forest should be left alone, to let nature take its course. The opposition induced the government to halt the savanna restoration until the conflict of visions could be resolved.

Shortage of Resources

No list of challenges would be complete without mentioning the problem of securing sufficient resources with which to carry out restoration projects. Time, money, and personnel are often in short supply, and the situation is magnified by the fact that restoration projects often take many years to implement and require lifetime management. As restorationists we often spend as much time writing grants to public and private agencies and contacting individual donors, community groups, and volunteer organizations as we do being engaged in restoration activities on the ground.

1.6.3 Opportunities

In spite of the challenges restoration ecologists face, the field holds great promise. Restoration provides numerous opportunities for repairing and enhancing degraded landscapes. These opportunities include (1) protecting existing natural remnants by restoring land on their boundaries and creating buffer areas, (2) providing missing habitat links, (3) restoring degraded remnants, (4) restoring ecosystem processes, (5) generating knowledge and education, (6) repairing and encouraging human capacity, and (7) creating resilient social-ecological systems.

Enlarging Preserves and Creating Buffer Areas

Recent developments in the study of spatial and temporal landscape patterns have directed conservation organizations toward protecting large, diverse tracts of land. This is not to say that small, high-quality remnants are not still considered worthy of protection, but rather that ecologists have come to recognize that large areas are more likely to retain high levels of diversity and maintain natural processes over the long haul. Among other things, large preserves make it possible to maintain patches of different ages within communities, thus providing diverse composition and community structure. They also create "interior" environments isolated from the surrounding landscape and provide habitat for animals requiring large territories. We have come to recognize that large preserves are less vulnerable to invasion by pest species, and more resistant to catastrophic disturbances (wind throw, fire, etc.).

Because of the highly fragmented nature of most landscapes, large tracts of relatively undisturbed communities are rare. Small high-quality sites are usually surrounded by land that is in various stages of development (logging, farming, etc.). In such situations, restoring the disturbed land surrounding a natural remnant both increases our ability to conserve the original community and presents a situation in which the remnant can serve as a source for local materials, thereby enhancing the chance of success for the restoration.

Providing Missing Habitat Links

One of the reasons fragmentation is problematic for conservation efforts is that it greatly impedes the processes of emigration and immigration. Depending on the nature of the surrounding land and/or the distance between suitable sites, it may be impossible for pollen, seeds, or animals to successfully move from one habitat to another. This lack of mobility can lead to loss of species through inbreeding depression, a situation in which detrimental mutations are expressed within a population due to mating between close relatives, and to slow recovery following a major disturbance due to a lack of native colonists. By acquiring and restoring strips of land that bridge the gap between community fragments, we can create corridors that provide opportunities for safe passage, thereby greatly enhancing the biodiversity within the linked fragments. The existence of such corridors will prove to be especially important in the face of global climate change, allowing organisms to immigrate to more favorable habitats.

Of course, it is also possible that the restored corridors serve as an entry point for disease and/or pest species. Isolated communities may offer some protection from such invasions. Therefore, it may be important to establish both linked and unlinked groupings.

Restoring Remnants

Some of the most successful restorations to date in terms of both structure and composition have involved moderately degraded remnants—sites that still contain many native elements. Sites retaining the skeletons of natural communities, especially if they are relatively free from undesirable species, often require less drastic interventions than do projects started from scratch. In general, the more native species and fewer pest species present, the better, and the more closely the physical features and disturbance cycles (fires, flood events) resemble those of natural areas, the more successful the restoration will be. Many such opportunities exist, from grazed woodlots to rivers (dam removal and reverse channelization) to previously unmanaged public land. Under the right circumstances, by restoring remnants, you can gain large benefits for relatively little effort.

Restoring Processes

Ecological services restorations, such as flood abatement, accelerated nutrient cycling, and the provision of habitat for an endangered species, often do not require the same degree of complexity as do complete restorations, and they hold promise for solving land use and conservation problems. Flood control, in particular, has been approached in this way at a number of scales, ranging from the restoration of river channels, such as the Kissimmee River in Florida, to the establishment of native species rain gardens by individual homeowners.

Many land management agencies are reintroducing the process of fire and trying to simulate historic fire regimes in areas where humans have eliminated fire for the past 50–60 years (Figure 1.9). The goal is to restore the structure of savanna, grasslands, wetlands, and

FIGURE 1.9. Prescribed Fire in a Long-leaf Pine–Wiregrass Community in South Carolina. This savanna-like community was once widespread across the southeastern United States. As a result of fire cessation and logging, this community now occupies a fraction of the area it once did. (Photo reprinted by permission from John Harrington.)

fire-adapted forest types. This change can result in increased benefit to native plant and animal species, and may restore nutrient cycling and hydrologic regimes that are beneficial to human societies.

Knowledge Generation and Education

The process of restoring sites, as well as the restorations themselves, provides invaluable contributions to education and research. British ecologist Anthony Bradshaw once called restoration "an acid test for ecology" (1987)—a means to demonstrate what we know and don't know about ecosystems. As outlined above, restorationists use conceptual models of communities to establish project goals, and to guide the implementation of the project. The models express either explicitly or implicitly assumptions about the ecology of populations, communities, and landscapes. The extent to which a project succeeds or fails can add to our understanding of natural systems—as long as records are kept. Documented failures can give as much information as successes, and can generate more hypotheses to be tested.

Restorations can prove especially useful in providing sites for manipulative research. In most instances, intact remnant communities are governed by policies that prohibit destructive sampling or deliberate additions of nonlocal species and, in some cases, even translocated endangered species. Such policies are meant to conserve the irreplaceable historic biodiversity and functions. Restorations, especially those that have been started substantially from scratch, have less history to lose, such that the promise of information gained through manipulative research may outweigh the losses caused by experimental disruptions. Similarly, restorations can provide settings for people to interact with natural systems without fear of destroying fragile remnants, and without having to travel far from home.

Repairing and Encouraging Community Capacity

Since restoration is a human activity that involves manipulation of the landscape, some scientists suggest that restoration ecology may be more of a test of our understanding of mutually beneficial interactions of humans and ecosystems than an "acid test" of how well we understand the functioning of ecosystems. If this is true, then restorations make a great contribution in that they provide people with the opportunity not only to reconnect with nature but also to build relationships within the civic community. Many believe that the very act of restoring a site to native vegetation is a way to build connections between neighbors, and to reestablish a link between people and nature.

The nonprofit United Indian Health Services is forging such links in northern California. By restoring and maintaining traditional native culture—and the landscape it depends on—it seeks to ensure the wellness of patients. One of its shining examples of such work is the Potawot Health Village in Arcata, California, a facility that serves the Yurok and several other tribes on the state's north coast. The clinic occupies a 40-acre former pasture on the outskirts of Arcata. A portion of the site is now a restored coastal prairie, where staff members have planted and are tending native plant communities, wildlife habitat wetlands, and a community garden.

In Far North Queensland, indigenous peoples help restore forest communities by sharing their knowledge of the cultural and propagation requirements of rainforest tree species with nonindigenous Australians. These examples, and many others from around the world, illustrate that restoration projects enable people to teach, to learn from, and to share their ideas with others.

▶ 1.7 CREATING RESILIENT SYSTEMS

In a rapidly changing world, restoration may also prove to be a model of how to manage uncertainty, find the connections between social and ecological issues, and turn challenges into opportunities. For example, restoration is already a model of how to manage across watersheds and landscapes that are shifting from rural to urban. Restorationists can build on this experience and expertise by developing detailed protocols for maintaining and managing resilient social-ecological systems in fragmented and densely populated landscapes. The restoration of urban and suburban settings requires not only the ability to manage things like plants and animals, storm water, the safe use of prescribed fire, and small, fragmented parcels; it also requires the skills to cultivate ecological literacy, encourage appropriate social and recreational uses of the restoration, and build community support (both financial and sweat equity) for the project.

KEY CONCEPTS

- Restoration ecology is a biological diversity and ecosystem conservation discipline, involving theory and practice. Although the term "restoration" implies that we are "putting something back," restoration ecology looks to the near-distant past primarily because our understanding of natural communities comes in part from historic records and past experience. In doing restorations, practitioners are not striving for a perfect duplication of some former state; they are reestablishing dynamic, ever-changing entities and processes.

- There are many kinds of restoration projects, including those that strive to include all the attributes of historic natural communities (complete restorations), those that are process-based, structural, or functional and often involve a simplified array of species (ecological services restorations), and those that are created to provide pleasure for people (experiential restorations).

- The practice of restoration uses an iterative problem-solving approach that results in a unique solution for each problem.

- Restoration challenges include a lack of documentation and experimentation, a scarcity of information about natural systems, the potential of restoration being used as an excuse for destroying remnants, and the fact that there is no single formula for success, so that each situation requires a unique solution.

- The opportunity of restoration is that the process can be used with good success to enlarge existing preserves, create buffer areas, provide habitat links, revitalize damaged remnants, provide ecosystem services, create sites for research and education, and reconnect people to nature as well as to one another.

FOOD FOR THOUGHT _____

1. How do you define restoration? What does it mean to you? Describe your experience with restoration, as a student, volunteer, land manager, or restoration ecologist.

2. Should we restore ecosystems? Why not let nature take its course? How "natural" is a restored community?

3. What assumptions beyond those mentioned in this chapter do restoration ecologists make about their work?

4. What are the differences between complete restoration, ecological services restoration, and experiential restoration? How are they similar? How would you classify the types of restorations found in your community?

5. What additional examples of challenges and opportunities might restoration provide in your community?

6. Restoration has been compared to improvisational theater—someone gathers characters (species) on a stage (site), gives them a sketchy idea of their roles in the cast (community), and lets the play unfold. What are the merits of this analogy? Where might it fall apart (if you think it does)?

7. Think of a community with which you are familiar. Using the improvisational theater analogy, assign the characters and roles for the restoration play of that community.

Resources for Further Study

1. For a wealth of information about the discipline and practice of restoration, check out the website of the Society for Ecological Restoration. http://www.ser.org Here are two of the publications mentioned in this chapter, as well as many examples of current projects from around the world:

    Society for Ecological Restoration International Science & Policy Working Group. 2004. *SER International Primer on Ecological Restoration*. Version 2. http://www.ser.org/pdf/primer3.pdf

    Clewell, A., J. Rieger, and J. Munro. 2005. *Society for Ecological Restoration International: Guidelines for Developing and Managing Ecological Restoration Projects*, 2nd ed. Tucson, AZ: Society for Ecological Restoration. http://www.ser.org/content/guidelines_ecological_restoration.asp

2. The Nature Conservancy is an example of a conservation organization that uses both preservation and restoration techniques. Its website is a good resource for information about many current projects, including the Nachusa Grasslands and the South American Atlantic Forest:

    http://www.nature.org

3. For examples of the historical roots of restoration, check out the following:

    Marsh, G. P. 1965. *Man and Nature*. Edited by D. Lowenthal. Cambridge, MA: Harvard University Press. First published 1864.

    Roberts, E. A., and E. Rehmann. 1996. *American Plants for American Gardens*. Athens, GA: University of Georgia Press. First published 1929.

4. If you would like to explore different examples of contemporary restoration practice in more depth, you'll find the following publications of interest. The first is an excellent overview of wildlife restoration, and the second provides an overview of contemporary practice, illustrated with examples from around the world.

    Morrison, M. L. 2009. *Restoring Wildlife*. Washington, DC: Island Press.

    Clewell, A. F., and J. Aronson. 2008. *Ecological Restoration: Principles, Values, and Structure of an Emerging Profession*. Washington, DC: Island Press.

The Community Model
Ecological Theory

It has often been said that one of the most important decisions to be made at the beginning of any project is to figure out where to begin, and the best way to start is to determine what you know, what you don't know, and what you hope to accomplish. Some of the first steps in the restoration process—developing the project purpose and use-policy, proposing project goals, developing community/ecosystem models—are designed to address these questions by helping to articulate (1) your understanding of biological communities and ecosystems and the varying roles humans play within them, and (2) what you hope to achieve with the project. Because restoration projects take many years, even decades, to establish, it is important to be clear about the assumptions you have about the systems you hope to restore, so that in the future, as understanding deepens or even changes in light of new information, managers can modify the restoration plan accordingly.

FIGURE 2.1. The Restoration Process: The Community Model.

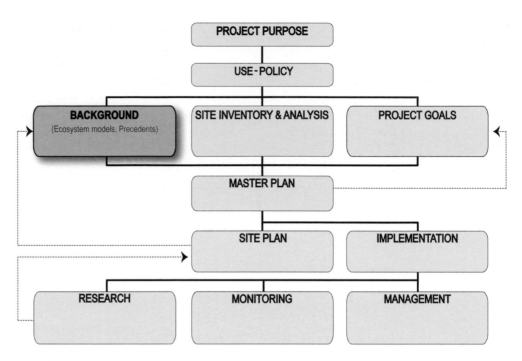

In this chapter and in Chapter 3, you will find information about two key components of restoration plans and the types of background material to consider before creating restoration plans (Figure 2.1). We begin by introducing a tool—the conceptual community/ecosystem model—to help you organize your understanding of natural communities/ecosystems in order to inform your restoration decisions. As discussed in Chapter 1, the science of ecology is one of the roots of the field of restoration ecology. Because community/ecosystem models (and many implementation and management practices) are based on scientific theories, we continue with a brief overview of some of the major ecological theories that are important to restoration. Chapter 3 presents a flexible, adaptive approach to restoration designed to accommodate information updates, and describes several kinds of research that will help you create a community/ecosystem model, as well as design a successful restoration project.

▶ 2.1 THE CONCEPTUAL COMMUNITY/ ECOSYSTEM MODEL

A conceptual community/ecosystem model is a description of the composition, structure, functions, interactions, and dynamics of the restoration target community. It presents your understanding of the nature of biological communities, as well as of the adaptations and interactions of the characteristic species within them. As a conceptual model, it is a representation of ideas, not a particular place, although, as you will see in Chapter 3, you can use actual sites as "reference stands" to help create your conceptual models.

We use the term "community/ecosystem model" because, as you will see, we consider "community" to be a concept that emphasizes organisms and the effects of their interactions with the biological and physical components of their environments. The focus is on the organisms. On the other hand, we consider "ecosystem" to be a concept that focuses on processes and interactions, not on the organisms per se. Depending on the purpose of a restoration (species, functions), the appropriate model could emphasize the community, the

ecosystem, or both. Throughout the text, we will use the term "conceptual community/
ecosystem model" and "community/ecosystem model" interchangeably.

Community/ecosystem models are essential in order to clarify potential areas of confusion about a project's purpose that result from different visions of what exactly is being restored. For example, in many parts of the North Central region of the United States, savanna was the most widespread community in the decades prior to the arrival of settlers from Europe or from the eastern U.S. coast and now has come to be one of the most rare. Not surprisingly, many conservation organizations have tried over the last several decades to restore savanna. However, a closer look at the variety of savanna projects reveals that there are differences in the visions of what savannas are, based on different interpretations of canopy densities. Depending on the authority, a site with a 40% canopy cover could be considered a savanna or a woodlands (Figure 2.2). Unless the restoration plan designer is

FIGURE 2.2. Canopy Gradient from Prairie to Forest. Which one of these scenes is a savanna community? The top image has very few trees present; most experts would classify this as prairie (<10% canopy cover). The second image shows a site with a few scattered trees; most would agree that this is savanna (10–30% canopy cover) and that the bottom image depicts a forest (80–100% cover). However, experts disagree about how to classify the site shown in the third image from the top (30–80% cover). Sites such as this one have been called savanna, woodland, or forest. And some would argue that savanna is not a community but a transitional zone or ecotone. (From *The Tallgrass Restoration Handbook* [Figure 4.1, p. 50], edited by Stephen Packard and Cornelia F. Mutel. Copyright © 1997 Society for Ecological Restoration. Reproduced by permission of Island Press, Washington, D.C.)

clear about what a savanna is from the outset, problems can arise—if, for example, the public believes there will be more trees on site than the plan calls for. Similarly, many debates have occurred over whether savanna is a "real" community, and therefore worthy of management interventions to preserve, or rather a temporary transition between a prairie and a woods, and therefore a situation to be left alone to change in due course.

The purpose of a conceptual community/ecosystem model is not to resolve such debates, but rather to ensure that the assumptions and visions that underlie your project are explicit. In addition to helping avoid conflicts, the model helps the next generation of site managers to decide, as ecological understanding changes, whether or how to follow or modify the original restoration plan to follow (or not) the original vision.

In addition to setting forth assumptions about ecological communities, a conceptual model also helps you (1) organize your thinking, (2) determine what information about a site and its context you will need to gather in order to create the restoration plan, (3) predict and evaluate the effectiveness of different restoration approaches, and most importantly (4) guide the restoration site, implementation, monitoring, and management plans. (See Sidebar.)

Conceptual models can take many forms, including diagrams, flow charts, and even equations, describing such things as the rates of carbon sequestration or litter decomposition. One of the more common formats used for community/ecosystem models is a narrative description, consisting of text, photographs, figures, and tables. Depending on the purpose of a restoration, a narrative model may provide a general description, or a detailed discussion of community composition, structure, and dynamics, or ecosystem processes. For example, the following case study is an excerpt from the beginning of a conceptual model prepared for a forest restoration in southern Wisconsin.

Sidebar | **Questions Addressed by Community/Ecosystem Models**

Fundamentally, restoration involves choosing strategies and implementing techniques that will take a site from what it is like today to what you would like it to be in the future. The community/ecosystem model includes the information required for addressing the kinds of questions you must think about to accomplish this goal. For example:

- Do communities have a necessary minimum/maximum number of species, organisms, or ecological roles/interactions (food chains, nutrient cycles, etc.) in order to be viable or sustainable?

- Are some species, life forms, or roles pivotal or essential for the survival of a particular community, and if so, what are they?

- How variable are communities in composition, structure, or function across space and through time? What leads to variation?

- Is there a minimum size necessary for a site to sustain a particular community?

- How does temporal and/or spatial scale influence restoration?

- How much, and in what ways, does the site context matter?

- How do communities respond to different kinds of disturbances?

For most ecological communities, the answers to these questions are incomplete at best, so the model represents a summary of our current understanding.

| CASE STUDY | Excerpt from a Conceptual Model |

Forests are communities that are dominated by mature trees forming closed stands (more than 50% canopy cover). In southern Wisconsin, the trees are deciduous hardwoods. In many ways, the canopy trees shape the forest environment, influencing the amount of light that reaches the forest floor, the wind and humidity levels of the interior, and the fertility of the soil. The forests often contain three distinct vertical layers: (1) the canopy, consisting of the mature trees; (2) the midstory, formed of shrubs, vines, and tree saplings; and (3) the understory, with herbs, ferns, sedges, grasses, vines, and tree and shrub seedlings.

Seasonal cycles are evident. Due to the combination of changing sun angles and the seasonal loss of leaves, most light reaches the forest floor in March and April, and then again in October. In spring, the forest greens-up from the bottom, starting with the herbs on the forest floor, and ending with the canopy. Many of the wildflowers on the forest floor bloom before the canopy leafs out. Similarly, the trees and shrubs often bloom and set fruit before their leaves emerge. This allows wind-pollinated species to expose their flowers directly to the wind and presents the flowers of insect- and bird-pollinated species more visibly to their pollinators. The spring expansion of leaves and flowers often requires a certain number of warm days. The spring can be interrupted by cold weather and the count resumed with the next warm front. By midsummer, in many tree and shrub species, the buds for next year have been formed. In early fall, the plants prepare for dormancy. Growth ceases, and starches and fats are stored. Nutrients are

FIGURE 2.3. **Mesic forest.** The major trees of the mesic forests of southern Wisconsin form a dense canopy that casts a deep shade over the interior in midsummer, such that midstory shrubs are very sparse. The majority of the understory herbs bloom in the spring; some are visible above ground for only a short time in spring when leaves are not yet on the trees; others are adapted to growing in deep shade. The dominant tree species, sugar maple (*Acer saccharum*) and basswood (*Tilia americanum*), reproduce in the shade, thus maintaining their presence in the community for many generations, barring disturbance. (Photo reprinted by permission from John Harrington.)

FIGURE 2.4. **Xeric Forest.** The southern Wisconsin xeric forest is dominated by oaks—bur (*Quercus macrocarpa*), red (*Q. rubra*), and white (*Q. alba*). The canopy is quite open, often resulting in an abundance of shrubs and vines that create a dense midstory (especially in the absence of fire). The understory herbs include many species that bloom in late spring and summer, as well as many species that produce nuts and berries. Ground fires influence the character of the understory. Frequent fires resulted in an abundance of sedges and grasses; less frequent burns led to more shrubs and herbs. (Photo reprinted by permission from John Harrington.)

FIGURE 2.5. Wet Forest. The major type of wet forest in southern Wisconsin is the floodplain forest, found along the banks of the major rivers. The dominant trees, including silver maple (*Acer saccharinum*) and American elm (*Ulmus americanum*), form a loose canopy that lets in enough light to support a tall, lush herbaceous understory that is well-supplied with nutrients carried in by the spring floods. Midstory shrubs are sparse—perhaps because floodwaters are damaging to the stems—but vines such as poison ivy (*Toxicodendron radicans*) and carrion flower (*Smilax herbacea*) are abundant. (Photo reprinted by permission from John Harrington.)

transferred from the leaves to storage places in the roots and other tissues. The chlorophyll in the leaves breaks down, making photosynthetic accessory pigments visible, thus producing fall color. Finally, the leaves fall to ground, forming a layer of litter. Decomposition of the litter will take place, beginning the following spring, releasing any nutrients that remain in the fallen leaves. Depending on species and environment, decomposition may take one to several years.

There are three main forest types: the mesic (Figure 2.3), xeric (Figure 2.4), and wet forests (Figure 2.5). (Depending on the project, this general description would be followed by a more detailed discussion of one or more forest types.) ■

▶ 2.2 ECOLOGICAL THEORY WITHIN RESTORATION ECOLOGY

"Theory" is one of those words that can cause confusion and misunderstandings because it carries several, sometimes very different, meanings. For example, a theory can be defined as a supposition or conjecture—in other words, a guess, as in, "I have a theory about why Jane likes peanut butter." Another definition of a theory is an explanation of phenomena, derived from an explicit, self-correcting process (e.g., the scientific method) that is generally accepted by scholars, as in "DNA is the basis of heredity." The two most important differences between these definitions concern how the theory is created and/or changed, and the nature and strength of its support.

When used in the first sense, a theory can be informal and personal, based on a hunch or on intuition coming from observations by one person accumulated over a lifetime. Such theories are often very fluid; they change quickly as new information becomes available. Many restoration practitioners develop these kinds of informal theories and use them to create and implement restoration plans. These ideas are often hard to communicate to others and can be highly situational. However, they may become hypotheses for establishing the formal theories described by the second definition.

As understood in the second sense, a theory can also change over time, but only after close examination following a set of agreed-upon rules. The idea is that as long as new information is consistent with the existing theory, the theory remains unchanged. But if enough new information does not really fit, the theory is modified or even discarded in favor of a new explanation. Theories are based on many observations made by many people over time. There is a formal process by which an explanation becomes a theory, and it generally requires a consensus of many of the experts in the field. Scientific theories, such as those ecological theories and concepts that underlie the practice of restoration ecology, represent the best explanations currently available. Yet, as we will see, many uncertainties and questions remain.

As we saw in Chapter 1, and as is apparent from the name of the discipline, the science of ecology is one of the roots of the field of restoration ecology. Now we will take a look at some of the major ecological theories that underlie restoration ecology. Because most contemporary restoration practice concerns communities and ecosystems (either as an aim, or for the habitat or other functions they provide), our focus is on these perspectives, rather than on organisms or populations. Our aim is to set forth some general principles that help to address the issues restorationists need to think about in designing any project:

1. The Nature of Communities: Concepts and Explanations from Community Ecology
2. The Ecosystem: Focus on Functions and Processes
3. Landscape Ecology
4. Climate Change

Our coverage of ecological theory is of necessity very brief and far from complete. At the end of the chapter, you will find a list of suggested resources for more detailed information.

▶ 2.3 THE NATURE OF COMMUNITIES: CONCEPTS AND EXPLANATIONS FROM COMMUNITY ECOLOGY

Most restorations involve communities, so understanding what characterizes a community is an important step in designing a restoration. The simplest, most often used definition is that a community is a group of interacting species living together in the same place at the same time. Following this definition, we usually describe communities in terms of the species that are present, how abundant they are, both individually and collectively, and how they are arranged, vertically and horizontally, in space.

This definition and these descriptions, however, do not really capture the essence of the concept. For example: What brings species together to form communities? Do communities change through time? If so, how and why? If not, why not? The answers to many of the questions posed in the Sidebar depend on understanding what a community is beyond its composition and structure. In this section, we will explore the idea of community as it has developed in the field of community ecology. The topics include (1) community organization (variation in space), (2) community dynamics (variation through time), (3) community stability (maintenance of the status quo), and (4) the historical nature of communities.

2.3.1 Community Organization: Variation in Space

Ecologists have long observed that communities are variable in space; that is, the identity, number, and abundance of the species that are present are different at every site. No two woods, grasslands, marshes, or deserts are exactly the same. In order to understand the nature of this spatial variation, it is helpful to begin by investigating how species form a community. Three factors seem to be involved. Communities are composed of species that are able to survive in the environmental conditions of the site, disperse to (reach) the site, and interact successfully with one another.

Environmental Tolerance

A community consists of those organisms capable of surviving in the environmental conditions present. This idea seems obvious—if the temperature, amount of oxygen, or, in the case of plants, the amount of light and soil nutrients is not sufficient to sustain an organism, it will either leave or die. As environmental factors vary from place to place, so does community composition.

One way to group organisms according to this organism–environment relationship is by describing their range of tolerance, the extent of the level of a particular environmental factor within which an organism can live. The relationships are often shown using graphs that describe not only survival per se, but relative success over the range (Figure 2.6). Some generalist species can survive over a broad range of conditions, while others, the specialists, have more narrow ranges.

To make things even more interesting, the same species can have a broad range of tolerance to one factor, such as moisture, but a narrow range of tolerance for another, such as light. For most species, the interactions of environmental variables are important. Certain plant species that are usually found in the shade of a woodland can do well in full sun if soil moisture is high, but not under dry conditions. In the shade, they can tolerate dry soils. In other words, their tolerance of high light conditions increases as soil moisture increases. Other species tolerate one set of conditions as juveniles, and another as adults—think of tadpoles and frogs, for example. Once one knows the range of a particular important environmental factor found on a site, it is possible to determine which species might survive there, and thus be potential community members (Figure 2.7).

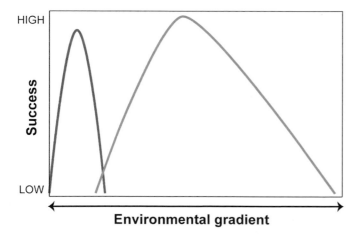

FIGURE 2.6 **Range of Tolerance.** The curves represent how well two different species do in response to varying levels of a particular environmental factor, such as temperature, a critical nutrient, or water. The levels of the factor over which a species can survive are referred to as the "range of tolerance" of that species. One species, a habitat specialist (narrow curve), can survive over only a limited range of the factor, and therefore is generally found in fewer places. The other species, a habitat generalist (wider curve), has a wider range of conditions under which it can be found.

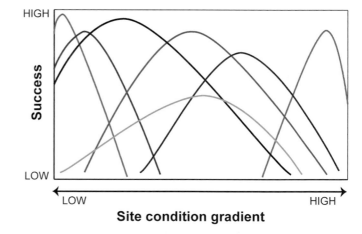

FIGURE 2.7 **Potential Community Members.** The curves represent the tolerance ranges of a number of species with respect to a particular environmental factor. Those species whose ranges match the range of conditions of the factor found within a particular site are potential community members. Important environmental factors include moisture levels (wet to dry), light (full shade to full sun), and soil nutrient levels.

Ability to Reach a Site

Which of the set of possible species (those adapted to the environmental conditions of the site) are actually present depends on which are, or were, able to reach the site. Which species arrive on site depends, in turn, on the dispersal mechanisms of the species, coupled with the distance the species must travel and the nature of the habitats the species must cross to get there. For example, windblown seeds, or animals capable of flight, can cross more potentially dangerous territory, and they often travel farther than animals that travel on foot. Therefore, the composition of a community at any one location is a product of the features of the site's particular physical environment and of its location, both past and present, with respect to sources of potential colonists.

Species Interactions

The third factor in forming a community involves the interactions that occur among individuals of the different species involved. These can be relationships that benefit all parties to the interaction—for example, pollination of plants by insects, or seed dispersal by fruit-eating mammals. Or they can benefit some at the expense of others; for instance, herbivory in grasslands by bison benefits the bison but decreases the vigor of some of the dominant grasses. This decreased grass vigor, in turn, increases the overall plant species diversity by decreasing the competitive ability of the grasses.

Table 2.1 lists some of the major categories of species interactions that can occur within communities. Ecologists have studied these interactions—predator-prey, competition, mutualism—in many communities and groups of organisms throughout the world. It is important for restorationists to be familiar with the current understanding of these interactions within the communities with which they are working.

Most of the relationships are essential, an obvious example being that in order to survive, animals must obtain energy and nutrition by consuming other organisms, in whole or in part. But others are not as obvious. For instance, at first glance, it might appear that some plant species, especially those that are wind-pollinated and dispersed, can do well without interactions with other members of the community. However, plants must obtain essential mineral nutrients, such as nitrogen, from the soil. Nitrogen in forms available to plants is released into the soil by the actions of organisms that decompose organic litter and other waste, or convert atmospheric nitrogen gas to ammonia or nitrates through the process of nitrogen fixation. Also, the uptake of soil nutrients and water is usually facilitated by relationships between plant roots and fungi called mycorrhizae. And, of course, competition for light, water, and minerals often occurs among plants in a community.

In some cases, relationships are species-specific. Some plants can be pollinated by only one kind of organism, in the absence of which they will fail to reproduce. In many other cases, the situation is more complex, with several species playing similar roles. In such cases, it is possible that species with similar relationships to others within a community are interchangeable. The composition and structure of a restoration will reflect all three of these organizational factors, and it often takes some detective work to uncover the complexities involved.

Table 2.1 • Species Interactions That Help Shape Communities

Predator-prey (trophic relationships)	One organism benefits at the expense of another
Tertiary consumer (carnivore)	Snake
Secondary consumer (carnivore)	Toad
Primary consumer (herbivore)	Grasshopper
Producer (plants)	Big bluestem grass
Detritivore	Dung beetle
Competition	Interactions that occur when two organisms or populations use the same limited resources; the more similar the resource use, the stronger the interactions
Guild (a set of species that are potential competitors because they exploit the same resources)	
Guild members can coexist if they exploit different parts or sizes of a shared resource, if they are active at different times of day, or if resources are more abundant than demand	Tropical fruit-eaters guild Hole-nesting forest birds guild Prairie herbivore guild
Competitive exclusion (one member of a guild eliminates another from a site)	House sparrows prevent songbirds from nesting
Guild members coexist, but populations of one or both are reduced in size	
Mutualism	Both organisms benefit from an interaction
	Animal-plant pollination and seed dispersal
	Mycorrhizal associations between plant roots and fungi provide water and inorganic nutrients to plants and food to fungi
	Ants and swollen thorn acacias: acacias provide shelter and food for ants; ants defend plants
	Leaf-cutter ants and fungi: ants provide leaves that they feed to fungi; fungi produce special structures the ants use for food

CASE STUDY   The Restoration of Clapper Rail Habitat in California

Professor Joy Zedler and her colleagues wanted to find out why the cordgrass (*Spartina foliosa*) in a California salt marsh restoration was not growing tall enough to provide needed habitat for a rare bird, the light-footed clapper rail (*Rallus longirostris levipes*). They discovered that the structure of the grasses was influenced by both the environment (soil) and species–species interactions (herbivory and predator-prey). It turned out that the soil, composed primarily of coarse sand dredge material, caused nutrients, especially nitrogen, to drain from the site, thus stunting the growth of the grasses (Boyer et al. 2000). To compound the problem, scale insects, which feed on cordgrass and thus reduce the vigor, were unusually abundant on the restoration. The restorationists determined that this outbreak was likely due to the fact that the beetle that preys on the scale insects and therefore helps check the growth of its population, like the clapper rail, requires the presence of tall grasses. ■

Community Variation in Space: Implications for Restoration
In order to plan a restoration for a particular site, it is important to understand:

1. How the current and predicted future range of environmental factors on site matches the range over which a specific community type is found. This range will set the limits on potential species composition.

2. Which competitive, trophic, and mutualistic relationships most strongly influence community structure, especially species richness and diversity, but also vertical and horizontal spatial patterns. These relationships influence how interchangeable the organisms participating in these major interactions are.

3. Whether potential colonists (species with appropriate environmental tolerances and dispersal abilities) are present on or off site, and how these species fit the restoration model.

Many ecologists have attempted to identify "assembly rules," general principles that predict the composition and structure of various kinds of communities, which are based on these principles of species–species and species–environment interactions. Under the right circumstances (willing clients, adequate resources, the ability to embed experiments; see Chapter 3), restoration projects could contribute greatly to this quest to understand underlying processes of community organization.

2.3.2 Community Dynamics: Variation Through Time

Communities are different not only in composition and structure from place to place, but also through time. For example, Noe Oak Woods in the Arboretum at the University of Wisconsin–Madison is different in 2011 than it was in 1820 or 1950. According to data collected by the Public Land Survey (see Chapter 4), the site now known as Noe Woods was likely an oak savanna in the 1830s. By 1950, it was an oak woodland, dominated by black (*Quercus velutina*) and white (*Quercus alba*) oaks. In 2011, the site is still a woodland, but the tree layer is losing oaks and gaining Norway maple (*Acer platanoides*), a naturalized exotic tree species introduced to North America from Europe. In this case, both the composition and structure of the vegetation on the site changed over the years. Sometimes, as in the maple woods of the North American deciduous forest, the species composition of a community remains the same over many years, but the individuals within the populations come and go over many generations of births and deaths.

One idea that captures the importance of the dynamics of the natural world is the concept that organisms and communities are systems—that is, that they are made up of interacting components (organs in the case of organisms, species populations in the case of communities). More specifically, living systems are open systems, exchanging both matter (gases, nutrients, waste products, etc.) and energy (light, energy within organic chemical bonds, and heat) with their environment, and adaptive, meaning that the makeup of the components or the nature of the interactions can change according to changing circumstances.

The systems concept is a way of looking at the world that emphasizes understanding the processes and interactions of the natural world (see Sidebar).

Sidebar | The Systems Concept

A system is a set of regularly interacting and often interdependent components that together form a unified whole, sometimes with properties that are more than the sum of its parts. The interactions of the components of a system are always stronger than with anything outside of the system. Systems are often depicted graphically as a series of boxes and arrows, with the boxes representing the components and the arrows denoting the magnitude and direction of the interactions. Such diagrams can very helpful in thinking through the relationships among different organisms within a community.

We can test our understanding of the systems concept with systems models that use set equations to model relationships. For example, we can compare the amount of actual stormwater runoff from a site with the amount that is predicted by a systems model. If the prediction is close to the actual results, we can conclude that our understanding of what influences runoff is a good one. Recently, a new kind of model is emerging, the complex adaptive system, in which each component (agent) of a system can change in response to the environment and to what the other agents are doing; the interactions are decentralized, often unique, yet seemingly organized. Examples include flocks of birds, ant colonies, and (maybe) communities.

The systems perspective has generated a large body of theory and concepts that are widely used in ecology, not always only in systems models. For example:

Open system: A system with a penetrable boundary; it exchanges matter and energy with its environment.

Inputs: Additions to a system that set the interactions of the components in action.

Output: The results of the interactions within a system following the addition of the inputs.

Feedback: The outputs of a round of system interactions that are communicated to the inputs, thus changing the result of the next set of interactions.

Steady state: A system that remains stable in terms of the makeup of its components and/or the nature of the interactions.

Resilience: The ability of a system to recover its initial structure and functions after a disturbance.

Nested hierarchies: The idea that the components of any system are themselves systems. For example, a species population is made up of individual organisms, a community is made up of a collection of populations, and a landscape is composed of a number of communities.

The systems concept is fundamental to the concept of ecosystem, which we will discuss later in the chapter. It is of interest here because not only does the physical environment influence the survival of organisms (as discussed above), but organisms can change the environment. Here are four examples involving plants, lichens, animals, and microbes:

- Perhaps you have walked from an open field into a forest on a sunny day and experienced a drop in temperature and light intensity. This shift occurs because the tree canopy blocks and absorbs the sun's radiation, causing the microclimate to be different for the organisms in the understory than in the canopy.

- Exposed rock on a relatively dry, sunny slope can be colonized by lichens and some kinds of moss, which over time trap litter and particles eroded from the rock, thus creating soil. The presence of soil, in turn, helps increase moisture levels. If seeds fall on the rock after the soil has begun to develop, some species capable of surviving in shallow soils may germinate, and over time might continue to alter the microclimate in favor of reduced winds and more moisture.

- On May 18, 1980, the volcano Mount St. Helens in Washington erupted, and forests, along with most of the wildlife, were swept away by the heat, gases, debris, and ash flow generated by the blast. In the aftermath, vegetation was stripped from large areas and/or covered by as much as 25 cm (9.8 in) of ash and other debris. Among the survivors were northern pocket gophers (*Thomomys talpoides*), small burrowing mammals. The gophers create tunnels to feed on underground plant parts and, in the process, deposit soil from below ground onto the surface, creating small mounds. By virtue of being underground at the time, the gopher population was largely protected from the immediate effects of the eruption. After the volcano quieted down, the gophers resumed their activities, which resulted in the mixing of relatively fertile soil from below with the volcanic ash, creating a new soil medium for plant growth (Anderson and MacMahon 1985).

- Looking at the bigger picture, most scientists believe that the oxygen atmosphere is largely the result of photosynthesis by microorganisms that occurred more than a billion years ago. During this period of geological history, the Proterozoic era, photosynthetic microorganisms, the cyanobacteria, increased in numbers and released large amounts of oxygen into the atmosphere. This increase is thought to have led to the extinction of many forms of life, as oxygen can damage certain organic compounds. However, the presence of an oxygen atmosphere eventually led to the development of the complex organisms with which we are familiar today.

The dynamic interchanges between the physical environment and organisms occur at several spatial and temporal scales. Whether or not a change matters to the composition and/or function of a community or ecosystem depends on the spatial or temporal scale at which it occurs, and the physical size or life span of the organisms or communities involved. We are all familiar with temporal environmental patterns that occur daily or yearly—the sun rises and sets, a rainy spring season is followed by a dry summer season, and so forth. Scientists can also identify longer recurrences, such as multiyear bursts of hurricane activity followed by 20–30 years of relative calm, droughts in the northern U.S. Great Plains that have appeared every 160 years, or the four major glaciations that have affected the planet over many millions of years. For organisms with life spans that are longer than the length of the natural environmental cycles, these changes become "predictable" in an evolutionary sense. They act as selection factors that favor adaptations enabling a species to respond to, for example, day and night, or periodic droughts, wildfires, or winds.

Organism–environment interactions and intersecting environment and life history cycles are important drivers of the dynamic fabric of natural communities. At any one time, a community at a specific location is composed of species that are present as:

- Adults only (e.g., mature trees or migratory birds)
- Juveniles only (seeds, seedlings)
- Both adults and juveniles (adult birds and nestlings)
- Organic remains (downed log)
- Dormant or inactive states (buried seeds, hibernating animals)

Descriptions of community composition and structure are often snapshots representing an instant in time. It is important to remember that communities are dynamic. Over time, the composition of each of the compartments will change, often (but not always) following predictable cycles. In order to restore communities, it is important to understand how the biological and environmental cycles operate.

Succession

Ecological succession refers to how communities change through time. Specifically, it is a shift in the presence or relative abundance of species populations over time in a given location under a relatively stable climate. As a result of ecological succession, communities replace one another.

The history of Noe Woods changing from savanna to forest (described earlier) is an example of succession at work. The process begins when new habitat is revealed (perhaps through a geological event, such as the eruption of a volcano or the recession of a glacier), or when a catastrophe (referred to as a disturbance), such as a severe fire or tornado, or flood, strikes an existing community, changing its composition, structure, or function to such an extent that the existing community is largely destroyed. Ecologists recognize two different types of succession:

- *Primary succession*: Community change in a location that has never supported a biotic community.

- *Secondary succession*: Community change in a location where an existing biological community has been disrupted by a disturbance.

Some species, "the pioneers," are specialized to colonize and survive in the environmental conditions found on a site immediately following a disturbance. For plants, this generally means conditions such as full sun, thin or no soil, few existing established plants, and/or sparse available moisture. Other species, the "climax species," are adapted to survive in the environmental conditions found in long-established communities (see Sidebar).

Sidebar | Pioneer and Climax

Pioneer species are usually considered to be short-lived, capable of producing many offspring, poor competitors, and capable of dispersing over large distances (in plants, seeds blown by wind or carried by birds). Pioneer plants often so modify the environment that their offspring cannot survive there. In contrast, climax species are usually considered to be long-lived and good competitors. Their offspring can survive in the environmental conditions created by the adult plants.

You will also see the terms "pioneer" and "climax" used to describe communities—those found on a site immediately and many years after succession begins, respectively. In the absence of any externally generated disturbance, pioneer communities generally change rapidly in composition and structure. Climax communities, in contrast, change slowly and are thought of as being relatively stable.

A BIT OF HISTORY

Succession has been a topic of interest to plant ecologists (especially in North America) since the early part of the twentieth century. Ever since Henry Cowles (1901), working in the sand dunes of Lake Michigan, described what he viewed as a temporal sequence, ecologists have been proposing mechanisms to explain succession. Cowles described changes in the dunes as involving an initial colonization of bare sand deposited by waves on the shore by grasses that stabilize the sand with their rhizomes and roots; when the sand is sufficiently stable, shrubs and then trees move in, if propagules are available and weather conditions allow.

F. E. Clements (1936) introduced the terms "pioneer" and "climax," and described succession as a process of community maturation proceeding according to a predictable series of well-defined sequential stages similar to the life cycle stages of a developing organism. According to this theory, the mature stage—the climax—is the same for every site within a climatically defined region. Clements's idea was widely adopted. Over the years, however, his explanations have been challenged, because data that do not fit his predictable step-by-step sequence have accumulated.

The definition we are using describes succession as a change in species composition. Several other features of communities have also been studied to see how they change through time. Diversity, and processes such as rates of primary production and nutrient retention, have also been shown to increase in many cases as succession proceeds.

The most direct way to document succession is to conduct a series of longitudinal studies (chronosequences) in which data are collected from specific sites through time. Because changes in many kinds of communities occur over hundreds of years, a time frame obviously much longer than the life spans of scientists, such studies are difficult to achieve. Nevertheless, by using the longitudinal studies that are available, by doing comparative analyses from observing sites of different ages since the start of succession, and by looking at changes through time in various environments, ecologists have identified some general themes.

Here is an example. The results of several investigations of successional changes in the deciduous forest biome on bare ground (often abandoned agricultural fields) or rock in upland situations have identified a pattern of shifts in the dominance of different life forms. Often the first plants to dominate a site are annuals, followed by short-lived and then long-lived herbaceous perennials, then by shrubs, by trees with light-tolerant seedlings, and finally by trees with shade-tolerant seedlings. However, this sequence does not always hold true—sometimes a plowed field will be colonized by tree seedlings right along with herbaceous annuals—and even when the general pattern of life form changes holds, there is variation in the details from site to site, regarding, for example, which species are involved.

One way to view restoration is as an attempt to direct the course of succession in order to achieve and then maintain a desired goal, whether that goal is expressed in terms of species presence, a particular community structure (diversity, abundances), or the rates of processes, such as carbon storage or the uptake of pollutants. In considering restoration in this fashion, it is helpful to understand what the course of succession is likely to be at a certain location. Based on observational studies as well as field-based experiments, several ideas have emerged (note the relationship to the community concepts discussed earlier):

1. Change happens differently on every site, but the eventual outcome is limited by the regional climate to those assemblages of species and community types adapted to the environment.

2. Patterns of change depend on what colonists are available, from either onsite or offsite within dispersal distance. The availability of colonists is influenced by distance, and also by the timing of the disturbance with respect to environmental and biological cycles. In many situations, the dispersal ability of potential colonists is seasonal, involving such things as seed set, molting patterns, and the maturation of young.

3. Because timing is influential, chance events are often key. For example, a strong wind from the east may carry the seeds of a species from a distant site with it, allowing that species to colonize a site instead of a species growing next to the site on the western edge.

4. The mechanisms driving change involve species interactions (especially competition), colonization, and environmental changes brought on by the organisms themselves.

Let's look at this last set of interactions a little more closely.

Most successional sequences seem to involve one of two interactions. In the first, often referred to as facilitation, early colonists (generally pioneers) modify the environment so that it becomes less suitable for themselves and more suitable for other organisms. Species composition changes with time because the organisms change the environment, moving the site outside the range of tolerance of some species and into that of others. The change tips the survival balance in favor of the later-arriving species. The composition of the community becomes more or less stable when the species-mediated environmental changes slow down, and the site is composed of species that can survive as both juveniles and adults. Here are some examples of the kinds of environmental shifts that can occur during succession:

- Ground-level light levels change from high to low, such as when a site shifts from grassland to forest.
- Soil moisture goes from wet to moist, as when a site shifts from an open marsh to a wooded swamp.
- Soil changes from unstratified rock particles, or soils homogenized and disrupted by agriculture, to stratified layers with mineral and organic materials as well as microbial populations.
- Wind velocities diminish, especially at ground level, with increasing height and density of vegetation.

The practice of using nurse crops in restoration plantings is based on facilitation (see Chapter 8). Restorationists also refer to the idea of facilitation in determining the sequence with which species should be introduced in a planting—for example, establishing sufficient shade before establishing forest ground layer species.

With the second interaction, referred to as inhibition, the first organisms to reach the site (often, again, the pioneers) inhibit the establishment of others, rather than enabling them to survive, as is the case with facilitation. Community change can occur eventually when the early colonists die and a different species gets to the habitat opening first. If the species occupying the site are able to live long, survive disease, or avoid disturbance, they can occupy the site for a long time.

The concepts of facilitation and inhibition were proposed by Connell and Slatyer (1977), and you will find them in widespread use today. You may also encounter the "initial floristics" concept (Egler 1954). Initial floristics is similar to inhibition (and to the tolerance model, also proposed by Connell and Slatyer), in that the first colonists on a site establish and prevent further colonization until they are removed by a disturbance.

One example of a mechanism by which inhibition operates is allelopathy, a process whereby plants produce chemicals that are harmful to other plants. If the first colonists produce such chemicals, they prevent others from establishing. In effect, rather than making the environment more favorable for new species, as is true during facilitation, this form of inhibition makes the environment less suitable for others.

One situation where allelopathic inhibition may be operating is in some bracken-grassland communities of northern Wisconsin. The bracken fern (*Pteridium aquilinum*), after which the community is named, is known to produce allelopathic chemicals. Many of the bracken grasslands contain pine stumps—evidence of logging activities that took place more than 100 years ago. In most logged sites, forest has grown back. In the logged areas with bracken grasslands, few, if any, trees survive. One hypothesis is that the herbaceous species of the bracken grassland are inhibiting the expected change to woodland or forest.

Implications of Community Dynamics for Restoration

It is likely that most successional sequences are the result of a complex of species interactions involving patterns of both facilitation and inhibition. This means that an understanding of the nature of individual species, and of their roles and interactions within communities, greatly enhances the success of a restoration.

To begin to predict the outcome of successional changes at a particular site, it is important to know:

- What is and was on the site.
- What is nearby (potential colonists).
- What community types are possible, given the current and future regional climate.
- What are the likely impacts of any initial species to be introduced as part of the restoration on the colonization and establishment of later-successional species.
- What the weather and disturbance patterns are likely to be over the time frame of interest.

In Chapters 4 and 5, we will discuss how to address these issues for a particular site. In Chapters 7 and 8, we will discuss how the information can inform restoration implementation practices, such as by guiding the selection of species to be planted, in the light of what species can be expected to establish on their own.

2.3.3 Community Stability

Ecologists have no doubt that natural communities are constantly changing through time, so why do the composition and structure of the vegetation on many sites seem to be relatively constant? Part of the answer lies in the scale of observation and the attribute being measured. For example, taken as a whole, the size of a bird population may be very similar from year to year, but during that period, individuals are born and die. At the scale of an individual, a lot of change has occurred, but at the scale of the population, if the rate of births plus immigration equals that of deaths and emigration, the size of the population will not change. If the individuals within a population are long-lived, as is the case with many forest trees, and the community is not affected by a disturbance, it may take hundreds of years to notice a turnover, even at the scale of an individual. But part of the answer may also lie in the three concepts of disturbance cycles, system resilience, and keystone species.

Disturbance Cycles

As we have discussed above, events such as fires, floods, and windstorms, to the extent that they severely damage or destroy an existing community, are considered to be the initiators of successional change. Each disturbance is a discrete event—with a well-defined beginning and end—that can be measured in terms of the size of the area affected, the time interval over which it operates (generally, though not always, short), and the change(s) it brought about.

But can some types of disturbance events serve instead to slow or even stop change? In many cases, the answer is yes. With regard to many communities and ecosystems, natural "disturbances" might best be thought of as environmental factors, similar to temperature, light, and precipitation, that shape species adaptations. Such processes as floods caused by rapid snowmelt, wildfires ignited by lightning, and windstorms are natural to many communities. These environmental change agents occur with enough frequency that species populations adapt, for example, to be able to quickly disperse to and survive in disturbed environments (pioneer species), or to survive for extended periods rooted in waterlogged soil (trees in a floodplain forest).Arguably, in such cases it is the cessation of fire or the installation of flood control that becomes the agent of change. The disturbance is the absence of the fire or flood.

One example of disturbance-dependent communities is the grasslands of midwestern North America. For hundreds of years, the North American Midwest has had environmental conditions that are easily within the tolerance ranges of many tree species. Based on the general successional pattern, one would expect that woods would eventually replace grasslands under these conditions, except perhaps in places where other factors, such as the soils, were not favorable to trees. Yet grasslands were widespread prior to European settlement, largely due to frequent wildfires (many of which were set by Native Americans), to which the grasses are adapted and many of the trees, at least in their seedling stages, are not. Because the fires occurred regularly (i.e., they were cyclical with a periodicity well within the length of the life spans of these perennial plants), the resistant grasslands remained relatively stable. With the advent of European settlement and the cessation of the wildfires, many grasslands not converted to cropland or otherwise developed became colonized by trees and converted to woods. In other words, successional change resumed.

Similarly, spring floods are natural occurrences along major river systems in the American Midwest and help maintain the composition and structure of floodplain communities; the relatively predictable drought cycles of the western plains of North America contribute to the maintenance of grasslands in that area. Moreover, a given factor may be a regular part of the environment for one system, but an unusual event in another. Most western forests in the United States are adapted to fire; most tropical rainforests are not. Also, although plants may be adapted to spring floods, recently summer floods, the consequence of development, have become common. Few native species are adapted to prolonged flooding in summer. The consequences of these "unnatural disturbances" are often negative with regard to community/ecosystem survival.

Sometimes an agent such as fire can be either a disturbance or a stabilizing factor on the same site. For example, as mentioned earlier, the fires that help maintain dominant vegetation of grasslands in the North American Midwest can have at least short-term detrimental effects on the populations of ground-nesting birds of the community if they occur during the nesting season. If a cycle of frequent fires is interrupted for a time and then reinstated, the results can be problematic. In certain woodland situations, for instance, frequent ground fires serve to reduce the fuel load available while maintaining community stability. If such fires are stopped for long enough to allow fuels to build up, the next fire to come through can destroy the whole stand, at least in the short term.

The role of disturbance events must be included in any community/ecosystem model. A goal of many restorations is the reintroduction of natural "disturbance" processes or, in situations where this is not possible (such as areas where fire is banned), to simulate the effects (by exposing the seeds of chaparral plants to smoke in a laboratory). It is important to understand the natural disturbance regime to successfully restore these communities.

System Resilience

The concept of resilience comes to ecology from systems theory. A community is considered to be resilient if it returns to normal after a disturbance. Since the introduction of this idea, ecologists have wanted to investigate whether some types of communities/ecosystems are more resilient than others, and what circumstances contribute to or detract from resiliency—the characteristics of communities and/or the kind, intensity, frequency, and duration of the disturbance.

One hypothesis is that the greater the species diversity and the more complex the web of species interactions, the more resilient a community will be. The idea is that such diverse and complicated systems have a built-in redundancy, so that if even if a disturbance wipes out a set of species, others remain to take over the functions that would otherwise have been lost.

There have not been many studies to date that directly examine the relationship between diversity, complexity, and resilience. One field trial, conducted over 12 years at the Cedar Creek Long-Term Ecological Research Site in Minnesota, found a link between diversity and resilience, as measured by stable rates of production in the face of climate extremes. The more diverse the plots in the study, the more stable were the rates of production (Tilman et al. 2006). However, other studies, using different sites and study approaches, have not demonstrated a consistent relationship between diversity and resilience.

Given that the nature of community resilience remains an open question, research questions focused on this topic would be good choices to incorporate into restoration projects.

Keystone Species

Do some species more than others influence a community's nature and/or rates of change? The answer appears to be yes. Most communities are composed of a few species, which, because of their relatively high abundance or large physical stature, play a leading role in influencing either the microclimate of a site or species interactions, or both. Think, for example, of how the height and density of the canopy trees in a forest influence the vegetation in the midstory and understory layers below, as well as the nature of the habitat available for animals. As long as the trees persist, the rest of the community remains relatively stable.

The idea that dominant species are influential makes intuitive sense. Community stability can also be affected by the presence or absence of less prominent members of a community, much as the presence of the keystone at the apex of a simple arch keeps the arch standing.

One way a single species may maintain community stability is by acting as a keystone predator (Figure 2.8). By keeping densities of potential competing prey populations below the level at which they would directly compete, the predator prevents competitive

FIGURE 2.8 **Keystone Species.** By grazing on dominant grasses within a prairie, bison reduce the ability of the grasses, through competition, to suppress other plant species within the community, thus helping to maintain a more diverse plant community. Bison have a role in nutrient distribution and cycling and expose new areas for colonization. Bison also influence the behaviors of predators and carrion feeders. (Photo reprinted by permission from John Harrington.)

exclusion, thus maintaining community diversity. This situation has been demonstrated in a variety of communities, including prairie (bison) and the oceanic intertidal zone (sea stars); it was even reported in 1859 by Charles Darwin in his lawn (mower): "If turf which has long been mown, and the case would be the same with turf closely browsed by quadrupeds, be let to grow, the more vigorous plants gradually kill the less vigorous, though fully grown plants; thus out of twenty species grown on a little plot of mown turf (three feet by four) nine species perished, from the other species being allowed to grow up freely" (1979, 120). Remove the keystone, and the arch (or community) becomes unstable or changes to a different state.

Implications of Community Stability for Restoration

In order to accomplish a restoration, you need to make sure that the goals and objectives described in the master and site plans are both realistic and flexible. Once you have initially achieved the goals and objectives of a restoration, your aim is for the restoration to continue to meet these standards over the long haul. The management plan explains how this can be done (see Chapter 10). Just as an understanding of how and why restoration communities might change through time is necessary for you to create realistic restoration goals and anticipate management needs, so, too, is an understanding of the factors that can contribute to stability. For example:

1. If the community is adapted to a recurring pattern of disturbance, your management plan should address either how to restore the cycle or, if this is not possible, how to simulate the effects. There are many examples of prescription fires used to maintain the structure of fire-dependent communities. We will discuss this tool in more detail in Chapter 10.

2. If the community is shaped by the presence of keystone species, your restoration site plan must ensure that they are present, and the management plan needs to be sure that they remain. If the site cannot support a keystone species, then your plans should discuss how to mitigate its absence. For example, if a prairie restoration is not large enough to support a bison population, you can use periodic mowing of the dominant grasses to simulate grazing.

2.3.4 The Importance of the Historical Nature of Communities

An important theme that underlies this discussion of variation in time and space is the idea that communities are historical. The particular set of species and the nature of the ecosystem processes that occur at present on any site are unique products of the past. The nature of the changes that occur through time depends on the environmental and biological conditions in place at the time the change was set in motion. Each event continues to influence the site for many years.

Consider, for example, the long-term influence of an individual hemlock tree (*Tsuga canadensis*) growing in the deciduous forests of parts of North America. A tree influences a site for:

- Potentially up to 1000 years while it is actively growing.
- Many years as an upright snag, or downed tree (coarse woody debris).
- 500 years or more while the area is influenced by whatever grows on the log and tip-up mound of soil created when the snag falls, which is often another tree—hemlock or yellow birch (*Betula alleghaniensis*), for example.

Let's consider a population. A fire that burns through a prairie during the breeding season of the ground-nesting prairie chicken may have a very different effect on the future population structure of these birds than a fire that occurs in late fall, after the young have fledged. A spring fire may not only wipe out most of the cohort of young for that year, it can also kill brooding adult females, thus further reducing the reproductive potential of the population for several years to come. Most birds (including the young of the year) can escape a late fall fire through flight; however, such an event may also spell trouble, if it is on such a scale that it eliminates all the available winter habitat within the range of movement of the population.

Implications of the Historical Nature of Communities for Restoration

There are two major implications of the historical nature of communities to the practice of restoration. (1) Understanding the history of a site helps you interpret the current condition and the nature of potential future changes. (2) The initial conditions of both the site and the implementation will have long-lasting effects on the restoration. How to uncover the history of the natural communities of a site will be discussed in Chapter 3.

▶ 2.4 THE ECOSYSTEM: FOCUS ON FUNCTIONS AND PROCESSES

With the adoption of the systems model in ecology has come the concept of the ecosystem. An ecosystem is defined as all the living and dead organisms in an area, together with the atmosphere, the hydrosphere (water systems), the lithosphere (rocks and minerals), and their interconnections. The focus here is not on the organisms per se, but rather on the roles they perform. In other words, the spotlight is on functions and processes.

You can create ecosystem models for a small site or at the scale of the entire biosphere. Ecological services restorations are particularly informed by ecosystem models (e.g., restorations that focus on protecting surface and groundwater from urban stormwater runoff, or on removing a portion of the carbon dioxide [CO_2] emitted by gas- and oil-powered vehicles and power plants). The idea is that organisms interacting with one another and with the physical environment provide services that help sustain human societies. Ecological services restorations are designed to conserve the functions of ecosystems, not the actual organisms.

Two of the major kinds of system interactions are (1) energy flows, or one-way transfers through a system; and (2) biogeochemical cycles, or reciprocal transfers within a closed system. In a cycle, materials can change form, but they are not lost or gained. The interactions that are of particular importance to ecological restorations are energy flows and biogeochemical cycles

2.4.1 Energy Flows

Energy enters an ecosystem primarily through the sun's radiation, and leaves in the form of heat. Living systems, through the trophic structure (predator-prey interactions of the food chain) of a community, are intermediaries in the flow. By incorporating and storing energy in their biomass, they delay the ultimate conversion to heat. The producers of the system (usually plants) convert light energy into chemical energy stored within their tissues. As energy is transferred from organism to organism through the herbivores and predators of the food chain, some energy goes to waste materials, some goes to maintenance and is lost to the system in the form of heat, and the rest is incorporated into the bodies of the organisms, resulting in growth.

Here are two ways in which the concepts that underlie energy flow become important for restoration:

1. Some communities consist of organisms that retain energy in their biomass for long periods, such as in the trunks and stems of long-lived trees, and can be energy storage sites, a function that may be a valuable service.

2. Because energy is lost to the system at each step in the food chain, there are energetic limits to the number of links in the chain and therefore to species richness.

2.4.2 Biogeochemical Cycles

Biogeochemical cycles describe the reciprocal transfers of chemicals between ecosystem compartments, including the biota, atmosphere (Earth's gaseous surroundings), lithosphere (Earth's crust), and hydrosphere (Earth's watery surface). Nutrients are essential for the survival of all organisms, and it is important to understand how different nutrients become available, as well as how particular site conditions may influence availability. Systems models of biogeochemical cycles often consist of compartments (boxes), representing storage units (reservoirs or pools) of a given nutrient, and fluxes (arrows), representing the processes that move the nutrients between the pools, and in so doing often change their form, such as from a gaseous to a solid state.

Important characteristics to consider about a biogeochemical cycle include:

1. The identity and size of the pools that serve as sources. They export the nutrients to other pools and thus diminish in size.

2. The identity of the sinks, the pools that gain nutrients.

3. The length of time a nutrient spends in a particular pool—the residence time—and the time it takes for a specific molecule to move through all the pools in the cycle.

4. Whether a cycle is local—that is, all the pools and fluxes occur on site or in the region—or global.

The chemicals that are of most interest in restoration are water, and three important mineral nutrients: nitrogen, carbon, and phosphorus.

The Hydrologic Cycle

The hydrologic cycle moves water from the atmosphere to the Earth by precipitation, and back again, by evaporation from the oceans or other bodies of surface water, or through transpiration by plants. Because the hydrologic cycle involves the atmosphere, it is a global process, and it is at that scale that it is a closed system. At the scale of a particular site, however, it may be better represented as a flow.

Water moves onto a site through precipitation, through overland flow from surrounding uplands, or through groundwater discharge (via springs, for example), and leaves the site by evaporation, transpiration, running over the surface and away, or groundwater recharge. An individual site may accumulate water, move surface water to or from the groundwater, or act as a throughway, depending on climate, topographic position, soil composition and structure, and the nature of its biota. As we will describe in more detail in Chapters 4 and 5, documenting and understanding the role of a site in the local hydrologic cycle is essential to the success of any restoration.

The Nitrogen Cycle

Nitrogen is an essential component of biomass and genetic material. The largest pool of nitrogen is in the form of nitrogen gas (N_2), which makes up about 78% of the atmosphere. Most plants and animals cannot use nitrogen in this form. The first step in the movement of nitrogen from the atmosphere to the biota is the conversion of molecular nitrogen (N_2) to ammonia—a form of nitrogen that is usable by plants. In terrestrial systems, this process, referred to as nitrogen fixation, involves free-living soil bacteria and *Rhizobium* bacteria found in plant root nodules (particularly in legumes). In aquatic ecosystems, notably in the oceans, cyanobacteria (blue-green algae) fix nitrogen. A small amount of nitrogen is also fixed as a by-product of lightning strikes. Animals obtain nitrogen by eating plants or other animals. When an organism dies, microorganisms break down the remains as decay proceeds, and release nitrogen back into the soil through ammonification (conversion to ammonia) or nitrification (conversion to nitrates). Together these decay processes are often referred to as mineralization—the conversion of organic matter to inorganic compounds.

Nitrogen can continue to cycle from organisms to soil and back to organisms for many years. It can also cycle back to the atmosphere through the process of denitrification, which involves microorganisms and low oxygen (anaerobic) environments, such as waterlogged soils.

A good portion of the nitrogen cycle in terrestrial communities occurs locally, as nitrogen moves through the food chain from plants to animals to decomposers and back to plants. In other words, the largest flux is between plants and animals and the microorganisms of the soil. Nitrogen changes form, but does not change in amount.

Nitrogen can also be added to a site through processes other than fixation. These include:

1. On developed lands, applications of fertilizer or other forms of organic matter, such as mulch.

2. Deposition from storm water that runs on to the site from adjacent fertilized land or from the fallout of industrial pollutants in the air.

3. Organisms moving to or through the site and leaving behind waste materials.

Similarly, nitrogen can leave a site via denitrification or by being carried off in storm water or filtering through the soil into groundwater, or (in small amounts) through the emigration of organisms. Because of human activities, many sites accumulate more biologically active nitrogen than would occur otherwise. Therefore, in some restoration projects, it becomes necessary to take steps to remove nitrogen from the soil (see Chapter 8).

The Carbon Cycle

Another nutrient cycle that has a large biological component is the carbon cycle. Carbon is a component of all organic molecules, and thus of the biomass of all organisms. Although the cycle includes the lithosphere and hydrosphere as well, much of the activity within the cycle involves exchanges between organisms and the atmosphere using the familiar processes of photosynthesis and respiration, simplified as follows:

$$\text{Photosynthesis} \longrightarrow$$
$$6CO_2 + 6H_2O \longleftrightarrow 6O_2 + C_6H_{12}O_6$$
$$\text{Respiration} \longleftarrow$$

Plants get carbon from the atmosphere in the form of CO_2, and convert it to biomass (represented above as a simple sugar). Animals get carbon from the biomass of the organisms they eat. Both plants and animals release CO_2 back into the atmosphere via cellular respiration, the means of releasing the energy stored in biomass in order to fuel the chemical processes of life.

Decay organisms get carbon from organic remains and also release carbon to the atmosphere via respiration. In aquatic systems, the carbon cycle involves similar transitions, but CO_2 must first dissolve in water to be accessible to the photosynthetic organisms.

Carbon can move between the biota and the atmosphere in a matter of hours. In many natural communities, the amount of carbon returned to the atmosphere each year by cellular respiration is about equal to the annual amount fixed by plants in the process of photosynthesis. Carbon can also be stored for many years in the tissues of organisms, especially the trunks of long-lived trees, in soils (in the bodies of microorganisms and humus), as dissolved organic compounds in the ocean, or for millions of years in sedimentary rocks such as limestone, in fossil fuels, or in ocean sediments. Eventually carbon is released from long-term storage in the deep ocean sediments during the movement of the Earth's tectonic plates formed of solid crusts floating on a more liquid mantle. As the plates separate or come together chemical transformations occur. For example, volcanoes found where one plate slides under another release large quantities of CO_2 into the atmosphere as the carbonate sediments are transformed by heat and pressure.

Human activities now constitute a significant part of the carbon cycle and have increased the amount of CO_2 in the atmosphere through the burning of fossil fuels. This atmospheric increase is one of the main factors thought to be contributing global climate change.

The Phosphorous Cycle

Phosphorus is an essential component of DNA and many other constituents of living cells and tissues. It is one of the limiting factors for the growth and survival of organisms, particularly those found in freshwater communities/ecosystems. Unlike the situation with the nitrogen and carbon cycles, the atmosphere plays only a very small role in the phosphorous cycle, and as a consequence most natural cycles are local. Large amounts of phosphorus are stored in rocks, in ocean waters and sediments, in soil humus, and in living plants and animals. The weathering of rocks releases phosphates to soil, where it is taken up by plants and is eventually circulated throughout the food web, returning to the soil through the actions of decomposers.

Implications of Biogeochemical Cycles for Restoration

The general outline of ecosystem nutrient cycles is a useful first step in understanding the processes involved. The different types of ecosystems that have been studied indicate that the details of the cycles vary; the relative sizes of the pools, the residence times, and the rates of the fluxes differ. It is important to understand how nutrients move through different kinds of ecosystems in order to determine whether the key components are present on a restored site and, if not, to ensure that a restoration plan addresses their absence.

In addition, the details of how individual organisms in a community influence the pools and fluxes are important. For example, the amount of nitrogen in the soil may favor some species over others, thereby influencing the success of restorations. One of the reasons prairie restorations on former fertilized agricultural land in the midwestern United States often face stiff competition from weeds is that high levels of soil nitrogen favor the weeds over the native prairie species. Common buckthorn, *Rhamnus cathartica*, an invasive exotic shrub in North American woodlands and savannas, has been found to increase the decomposition and ammonification (mineralization) rates of the litter on the floors of the forests it invades, thus reducing the litter layer and therefore changing the habitat. Ecologists speculate that this change has the potential of causing the local extinctions of litter-dwelling invertebrates. This effect is thought to be due to the high nitrogen content of the buckthorn leaves relative to that in the leaves of the native woodland shrubs.

More and more restorations are planned for providing nutrient cycle services. For example, the restoration of native communities, particularly forests, has been mentioned as one approach toward reducing the carbon load in the atmosphere. The idea is that such communities can act as a carbon sink by storing carbon in the wood of long-lived trees or the humus-rich soils of the forest floor. Various wetland communities contribute to hydrologic cycle functions by retaining storm water on site or serving as groundwater recharge areas. They contribute to the nitrogen cycle by trapping and holding nitrogen-rich sediments, and in the waterlogged soils by providing a habitat for the anaerobic denitrifying microorganisms to operate.

▶ 2.5 LANDSCAPE ECOLOGY

Ecological theory is often organized into sets of ideas appropriate to different levels of living systems. Thus, there are theories of organism–environment relationships that study, for example, the physiological responses of individuals to different levels of light, water, nutrients, and temperature. Other theories apply to groups of organisms belonging to a species (population ecology) and to groups of populations (community ecology). Landscape ecology focuses on landscapes, a collection of interconnected communities distributed across a region.

2.5.1 Landscape Ecology: Focus on Spatial Relationships Among Communities

The structure of landscapes can be thought of as consisting of four basic spatial components:

1. *Landscape mosaic:* A region composed of different communities, many with gradual transitions from one to the other (forming a continuum); some with relatively distinct transitions at the borders, often due to abrupt topographic changes or human land use. Mosaics may contain patches, a matrix, and corridors.

2. *Patch:* A small area, distinct in form and composition from the surroundings.

3. *Matrix:* The larger area surrounding a patch.

4. *Corridor:* A passageway linking patches in a matrix. For different species, it can serve as a habitat link, filter, or barrier.

Most natural landscapes are mosaics of different communities; many communities, in turn, are mosaics of units of different sizes and ages since disturbance (different successional stages). A large tract of hardwood forest, for instance, although it looks at a glance to be a homogeneous expanse of large trees, will, upon closer examination, be found to contain units (patches) with trees originating at different points in time. This arrangement is due to isolated windstorms, the burning patterns of wildfires, or diseases. In other cases, environmental variation across a tract of forest may lead to the site's having several distinct areas, each of which is dominated by different species.

Patches of the same community type are sometimes separated and embedded in a matrix of a different community type. This pattern is frequently due to environmental variation.

In the grassland regions of the North American Midwest, for example, early European explorers described the endless sea of grass as occasionally being broken by groves or lines of trees, often responding to the presence of stream valleys. In parts of Africa, clumps of trees are interspersed with grassland. In California, serpentine rock, which creates soils rich in magnesium and iron silicates, occurs in patches interspersed with other parent materials. This leads to mosaics of communities composed of species able to tolerate the serpentine soils (serpentine grasslands) together with oak woodlands and California annual grasslands.

Because restorations are often patches within a large mosaic of human-developed land dominated by non-native species and human-mediated disturbance regimes, thinking about the ecology of patches within a landscape can be instructive. Patch geometry—the size and spatial distribution of community patches, and patch edge-to-interior relationships—and patch isolation influence community structure and function.

Patch Geometry and Community Composition

As discussed earlier, ecologists have long understood that the physical characteristics of a site influence the species that can survive there. Sandy soils support different species than clay soils; north-facing slopes have different species than those facing south. Similarly, there is generally a positive correlation between the size of an area and the number of species it can hold. This relationship, often described by a species-area curve, is likely because, all other things being equal, the larger the size of an area, the greater the number of habitats available (Figure 2.9).

The edge is a zone at the boundary of a patch that separates the patch from the matrix. In many circumstances, a patch edge has different microclimate features than the interior of the patch or the matrix on either side of it. For example, the edge of a patch of forest in the midst of cultivated cropland is generally hotter, drier, and more susceptible to wind damage than the forest interior. These microclimate differences mean edges have different species than the interior; this, in turn, changes species interactions.

Ecologists have long recognized that edge habitats are different from interiors, and have even designated some species as being "edge species" because of their ability to do well in these environments. What the landscape ecology perspective has added to this discussion is the idea that the size and shape of patches influence the proportion of the patch that is edge rather than interior habitat. This feature is referred to as the edge-to-interior ratio.

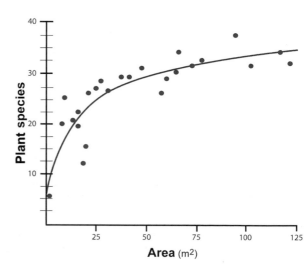

FIGURE 2.9 A Species-Area Curve. In general, the larger the area, the more species it will contain.

In general, small linear patches have more edge habitat in relationship to the amount of interior than do more circular, larger patches. So, depending on the kinds of habitats and species favored by the goals of a project, you can change the geometry of the project to favor edge or interior (Figure 2.10).

We have seen that among the variables that contribute to the makeup of the species in a community at a particular site are the availability and success of colonists. The ability of a species to actually get to a site depends on a number of factors, including the distance of the site from the source of colonists, the presence of barriers that limit access, and the manner by which dispersal occurs. For instance, animals that travel on foot (and plant seeds for which these animals act as dispersal vectors) may not be able to cross large bodies of water, animals relying on heavy plant cover for protection from predation may not make it across a sparsely vegetated expanse, and windborne seeds may not land on a site if the prevailing winds carry them in the opposite direction.

In the 1960s, the ecologists Robert MacArthur and E. O. Wilson proposed a theory that uses the species-area relationship and the influence of distance on colonization to predict the pattern of species richness seen in oceanic islands. The main idea of this theory, known as the equilibrium model of island biogeography, is that the number of species—the species richness—on any island represents a dynamic balance between the processes of colonization and extinction. Over time, new species will continue to arrive, but the number that an island can support will be relatively stable. If the number of species on an island is at equilibrium, when a new species comes, either it or an existing species is likely to become locally extinct. Species richness remains in a dynamic equilibrium. The result of this dynamic balance is the prediction that large/near islands have more species than far/small islands—a relationship that seems to hold in many (but not all) places.

If we consider the patches within a matrix to be analogous to the islands in an ocean, we can use the model to predict the relative species richness of community patches within many different kinds of landscapes. However, as is the case with many models, the

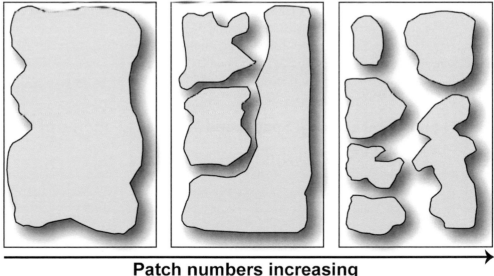

Patch numbers increasing

Patch size increasing

FIGURE 2.10 Habitat Patch Size and Edge Effect. The size and shape of landscape patches influence the amount of edge versus interior habitat is available. The large patch on the left with relatively smooth borders has a lower edge-to-interior ratio than the small patches with more indented borders on the right.

assumptions about the factors used in the model—immigration and extinction rates, and patch size and distance—are simplifications. For example, the model assumes that the islands are equivalent except for size and distance, and that the matrix is inhospitable. In most restoration settings, many additional variables contribute to species richness, including habitat differences and variation, disturbance and successional patterns, and the establishment of interrelationships among species. In addition, the focus on species richness means that the fates of individual species are not considered. Moreover, the model assumes that the islands eventually achieve equilibrium—that is, the number of species becomes stable, although the identity of species may change. In reality, restorations are rarely at equilibrium.

Nevertheless, the island biogeography model has been helpful in thinking about the early stages of the development of communities, and therefore is of interest to restoration. The emphasis on the influence of patch size and location reinforces the importance of considering the size and landscape context of restoration projects.

Patch Isolation: Invasions, Genetic Diversity, Catastrophes
The location of a community patch relative to its distance from other communities is important not just from the standpoint of contributing to overall species richness. Additional considerations include these:

1. The more isolated the community is, the less likely that a disease or aggressive invading species will reach it.

2. Isolation can lead to genetic problems if populations become inbred due to the lack of immigrants.

3. In a situation where several community patches are isolated from one another, it is less likely that a single catastrophe will wipe out all patches at once.

4. Many species populations are divided into subgroups, each of which inhabits one of a series of landscape patches. The patches often differ in the quality of habitat they provide, such that in some (sources), organisms can produce an abundance of offspring, whereas in others (sinks), reproduction is less successful and may even fail. The idea is that there is a one-way transfer of organisms from a source to a sink. However, the quality of habitat often varies from year to year, such that in some years, even the sink populations produce offspring. In addition, the sinks can provide refuge for nonreproductive individuals.

2.5.2 Landscape Ecology: Spatial and Temporal Patterns of Disturbance

In our discussion of disturbance cycles, we have seen the influence on community structure and dynamics of temporal patterns operating on a single site. Now we will take a look at the influence of temporal and spatial patterns operating on a landscape scale.

Consider, for example, species diversity. Some species are adapted to postdisturbance conditions—the pioneers—whereas other species are adapted to more steady-state conditions—the climax species. Without disturbance, the pioneers would be out of habitats, and many would be lost. Without sites that are free from disturbance for long periods, the climax species would be lost. For maximum biodiversity, it is important to have communities at various postdisturbance stages in a landscape, and links so that organisms can move from place to place as succession occurs.

Large-scale restorations can be designed as landscape mosaics with different kinds of communities and different successional stages interacting to greater or lesser degrees with one another and with different disturbance regimes. Many projects are instead patches within a matrix, the elements of which are under greater or lesser degrees of control by managers. In either case, it is important to consider the spatial and temporal dynamics that can influence communities.

2.5.3 Implications of Landscape Ecology for Restoration

The size, shape, and distribution of patches are important considerations for many, if not all, restorations. It is essential for you to know, for instance, what the minimum size of a community has to be to create sufficient "interior" habitat. If a site is not large enough, it may be better for you to change the restoration goals, rather than struggle to establish or maintain species on sites where the species-microclimate match is wrong. We will discuss strategies for evaluating the spatial context of a site in undertaking site inventory and analysis in Chapter 4.

Habitat quality is also an important point of concern. For example, it is possible that low-quality restored sites could become sinks for some species and thereby have a negative effect on nearby species populations. Also, if a restoration site is located near a high-quality natural area, it is critical to carefully consider what interactions the species introduced into the restored site might have with those in the natural area. Is mating between restored and natural populations likely? If so, would such interbreeding have a potentially positive or negative effect on the natural population?

▶ 2.6 CLIMATE CHANGE

The consensus among ecologists and other scientists is that our planet is experiencing a rapid global climate change, fueled by human activities and resulting in the release of so-called greenhouse gases (CO_2 in particular) into the atmosphere. One change involves warmer temperatures. Evidence to support this view includes everything from the summer melting of greater extents of polar sea ice than seen before, decreasing numbers of days during which ice on lakes in the northern regions of Europe and America are frozen in winter, to earlier spring blooming and breeding times exhibited by several plant and animal species. But temperature change is not the only climate factor being affected.

According to the most recent report (2007) from the Intergovernmental Panel on Climate Change (IPCC), as well as information from the U.S. Environmental Protection Agency (USEPA), global climate change models predict the following:

1. The average global near-surface temperature is rising, and estimates predict that it will continue to increase, perhaps on the order of several degrees, by 2100.

2. The warming trends will differ geographically, with most of North America and Europe warming more than South Asia or Australia, and by season, with winters warming more than summers.

3. Increased temperatures will have the effect of causing sea levels to rise from the melting of polar ice, with a resulting loss of coastal land area.

4. Precipitation patterns will change, with some areas experiencing annual increases and others decreases.

5. The intensity of extreme weather events, such as hurricanes, will rise.

6. Depending on the magnitude of the average temperature change, up to 30% or more of the world's species are at risk of extinction.

The above predictions represent results that are considered at present to be highly probable. It is important to note that they are based on best estimates, and that the science surrounding climate change continues to advance. One of the factors that contributes to uncertainty is that some of the effects of climate change involve feedback loops. For example, it is difficult to predict the magnitude of sea level changes because not only does the melting of polar ice caps cause such a rise, but also as ocean waters warm, they expand.

The implications of these potential climate changes for the Earth's biota are many and varied. The effects of global climate change are expected to be acute in several parts of the world. The Arctic and sub-Arctic areas are sensitive to temperature and are expected to experience the most warming. Coral reefs are sensitive to warmer ocean temperatures and increased ocean acidity, and both of these are rising from higher atmospheric CO_2. And tropical rainforests are sensitive to slight changes in temperature and precipitation. A summary of a few of the major types of impacts follows.

2.6.1 Community Composition: Range Shifts

The tolerance ranges of the species making up a particular community must match the environmental conditions of the site. As the environment changes, ecologists would expect to see a change in composition. Some species will likely remain, such as species with broad tolerance ranges, or those previously on the edge of their ranges for which the new climate becomes even more suitable. However, other species will no longer be able to survive, and still others may move in, if they can disperse to the area. The result will be new kinds of communities. The change may be more evident in some community types than in others, as the effects of global climate change will likely differ across the globe.

A historic instance of continental climate change that might prove instructive in considering the implications of contemporary climate change is the retreat of the glaciers that at one time extended over a good portion of the northern United States. The retreat ended, depending on location, between 18,000 and 10,000 years ago. Scientists who have reconstructed the revegetation of the region, based primarily on pollen preserved in lake sediments, have concluded that species moved into the region at different rates, resulting in new community types. Indeed, some scientists think certain species have continued to gradually extend their range since this glacial retreat.

Even if ecologists had information about the tolerance ranges of all species native to a region (which they do not), it would be hard to predict what the new communities would be like, for these reasons: global climate change involves a complex of interacting environmental factors likely resulting in unique climates, the microclimates of individual sites may not reflect overall climatic trends, and the dispersal abilities of species vary.

2.6.2 Community Composition: Species Interactions

As we have discussed, communities are also shaped by numerous species interactions, many of which may be modified as the species involved do or do not react to a changing climate. If the species do not react in the same way to the climate change, their interactions may be uncoupled, and one or more species may suffer as a result.

The pied flycatcher bird migrates in the spring from Africa to the Netherlands to breed. The parents rely on caterpillars to feed their nestlings. In recent years, spring temperatures have been coming earlier than before. As a result, the caterpillars have emerged as much as two weeks earlier than in previous decades. However, the birds arrive at the nesting grounds at about the same time they always have, and consequently their food supply has often diminished before it is time to feed their young. In this case, the birds raise fewer young, and the species may begin to decline (Both et al. 2006). It is likely that the rapid climate change we are predicted to experience throughout this century will lead to communities that are constantly changing as species do or do not adapt.

2.6.3 Change in Disturbance Patterns: Implications for Community Dynamics

If the intensity of storm events increases, as current climate models predict they will, and if regions of the world experience substantially more or less precipitation, the resulting new disturbance cycles will also influence the composition, structure, and dynamics of the emerging communities.

The IPCC also predicts an increase in the frequency of wildfires. In the middle of North America, evidence indicates that the pre-European settlement landscape was a shifting mosaic of communities. The patterns of the communities across the landscape were formed by decades-long climate changes (from hot-dry to cool-moist, for example) interacting with natural and human-caused wildfires. Hot-dry conditions and frequent fires favored prairie; cool-moist conditions and the resulting fewer fires favored woods. Any one location through time might change from prairie to savanna to woods to savanna to prairie, and so on, over the span of a few hundred years. Evidence demonstrates that some plant species were able to survive in place whether the area was woods or prairie (sometimes in a dormant state), while others could survive in the habitat patches that remained (refugia) and spread as the habitats increased. Because of modern human development, many scientists worry that refugia will be scarce, and that even if some exist, species will not be able to move across the inhospitable matrix of human developments.

2.6.4 Implications of Global Climate Change for Restoration

Although the implications of global climate change for restoration are uncertain, what does seem certain is that changes are under way. Restorationists are therefore faced with a dilemma: The species that are a good match for a restoration site today may not be a good match 20 or more years from now. Given that the goal of most restoration projects is to create communities/ecosystems that will survive for decades, and we are as yet uncertain what future climate conditions will be, how do we proceed? Here are a few suggestions:

1. Adopt a flexible, adaptive approach to restoration (see Chapter 3). Be prepared to change course several times as patterns of climate change proceed. The implication is that restorations will need to be actively managed and cared for over many years.

2. In restoration situations where you plan to add species to a site, consider selecting species that (a) have broad environmental tolerances; (b) may have a range of tolerance that is a marginal fit now, but will be central in a few years if current predictions about the direction of the climate shift hold true; and (c) if possible, select ecotypes (genetic varieties) of the species that come from regions with climates

that match the predicted future rather than current site conditions. Of course, the success of strategies (b) and (c) depends on the accuracy of the predictions.

3. To the extent possible, design restorations that either have a diversity of habitats on site or are connected to other natural patches, such that species can emigrate to more favorable habitats if conditions change.

4. In collecting information about the current environmental conditions of a site (see Chapters 4 and 5), use information averaged over the past 4 or 5 years, rather than longer-term averages, and/or look at the recent trends of precipitation, temperature, growing season length, and so on. Manage and select for species that fit these recent trends.

Another potential use of restorations is to provide sites in which the effectiveness of the assisted migration conservation can be studied with less risk to local populations. This is a controversial strategy (see Sidebar).

▶ 2.7 THE LINK BETWEEN THEORY AND PRACTICE

As we stated at the beginning of this chapter, the practice of restoration is grounded in ecological theory. Our assumptions and understanding of how organisms, populations,

---

Sidebar   Restoration and Assisted Migration

Conservationists have been moving species from one location to another for many years. A practice that may be familiar to many of you is the reintroduction of species to areas that they formerly occupied, but from which they have been absent for some time. One such example is the ongoing effort to return gray wolves to portions of their former North American range. This effort has been very successful, at least from the perspective of establishing viable reproducing populations. In fact, the number of wolves has increased beyond the conservation target in the upper Midwest to the point that managers are considering strategies to reduce the population of this once locally rare species. However, not all reintroduction efforts have been successful (Fisher and Lindenmayer 2000), and most are very expensive, leading some to question the use of this technique.

In recent years, conservationists have been discussing assisted migration—the movement of populations to new, as opposed to previously occupied, sites—as a strategy for saving species from extinction in the face of climate change. The concern is that, because of landscape fragmentation, species may not be able to move on their own to new geographic locations as the climate shifts.

Assisted migration is even more controversial than reintroductions. The issues that are being debated include the following:

- Our ability to predict the future course of local and even regional climate change is improving, but still relatively inaccurate. Also, we do not fully understand how species respond to all climate and microclimate factors and their interactions (see Chapter 4). This makes choosing appropriate relocation sites difficult.

- There is a risk that the relocated species will drive existing species to extinction, thus reducing overall biodiversity. Although we have some understanding of pest species characteristics, we are not yet able to predict with certainty which new colonists will act as pests (see Chapter 11).

- Species interactions within communities are one of the important threads that hold the fabric of a site together (see Table 2.1). Moving a species to a new site might result in its extinction, if it is not able to establish appropriate interactions. This means that assisted migration may need to involve whole sets of species, thereby increasing the risk to the host site's resident species.

communities, and ecosystems operate guide the direction of all the steps involved in restoration, from the choice of target communities to the kinds of long-term management situations that might arise. As ecologists continue to learn more about the biosphere, our approaches to restoration can be modified in light of new information. However, this will work only if we continue to keep up with these new developments.

As is true of many professions, restoration ecology benefits from the continuing education of its practitioners. It is also the case that the practice of restoration can contribute to the development of theories. In the next chapter, we will begin to explore the link between theory and practice in more detail.

KEY CONCEPTS

- The community/ecosystem model summarizes the restoration team's current understanding of the composition, structure, dynamics, and processes of the natural systems that are the focus of a particular restoration. The model is the foundation for the site design, implementation, monitoring, and management plans.

- Theories derived from scholarship underlie restoration ecology, and are particularly important in developing conceptual community/ecosystem models.

- The species that make up a community are present because they are adapted to the environmental conditions of the site, have been able to arrive on the site, and interact favorably with the other species that are present. Communities are variable in space because of geographically variable environmental conditions, as well as factors that influence colonization and species interactions.

- The composition and structure of communities often vary through time. Change occurs due to effects caused by the species within the community, or due to catastrophic disturbances coming from the outside. The composition and structure of communities may be relatively unchanged for long periods, due to the influence of keystone species or cyclical disturbances.

- Ecosystem ecology focuses on the interactions between organisms and the atmosphere, the hydrosphere, and the lithosphere. Energy flow and water and nutrient cycles are of particular interest.

- Landscape ecology looks at the spatial arrangement and interactions of communities.

- Climate change will likely influence communities in the years to come in ways that are not yet clear. This is a situation that needs to be considered by restoration planners.

FOOD FOR THOUGHT

1. Community/ecosystem models can be expressed in a variety of ways, including text, diagrams, and sketches. What aspects of a model are best expressed in these different ways?

2. Choose a community that would be a candidate for restoration in your region, and create a community model that addresses the questions regarding composition, interactions, structure, and dynamics posed at the beginning of the chapter.

3. Chose an ecosystem process (nutrient cycle, energy flow, hydrologic cycle), and describe the composition of the system compartments and the nature of their interactions in a community typical of your region.

4. What kinds of human impacts occur in your region, and what are the potential effects on the natural communities? How do human-caused impacts differ from so-called natural disturbances?

5. What course is climate change predicted to take in your region over the next 100 years? Select a species or community, and based on what is known about its response to temperature, precipitation, and so on, discuss what changes, if any, there will be in its geographic distribution 100 years from now.

6. We stated in Chapter 1 that one of the opportunities provided by restoration is that of enlarging the size of preserves. Based on the principles of landscape ecology, describe the species, communities, and processes in your area that are most likely to benefit from this restoration application.

7. What functions besides energy flow and biogeochemical cycles do ecosystems in your region perform? How might restoration influence them?

Resources for Further Study

1. To learn more about contemporary ecological theory, check out an introductory textbook, such as:

    Molles, M. C. 2009. *Ecology Concepts and Applications*, 5th ed. Dubuque, IA: McGraw-Hill.

2. The following books provide several excellent examples of how restoration ecologists are using and investigating theory.

    Falk, D. A., M. A. Palmer, and J. B. Zedler, eds. 2006. *Foundations of Restoration Ecology: The Science and Practice of Ecological Restoration*. Washington, DC: Island Press, 2006.

    Perlman, D. L., and J. C. Milder. 2005. *Practical Ecology*. Washington, DC: Island Press.

3. The Ecological Society of America (ESA) publishes the journal *Ecology*, which is an important source of the most recent developments in ecological theory.

4. If you are interested in learning more about succession theory, here are some of the classic papers referenced in the Sidebar on pioneer and climax species.

    Clements, F. E. 1936. Nature and structure of the climax. *Journal of Ecology* 24:252–284.

    Connell, J. H., and R. O. Slatyer. 1977. Mechanisms of succession in natural communities and their role in community stability and organization. *American Naturalist* 111(982):1119–1144.

    Cowles, H. C. 1901. The physiographic ecology of Chicago and vicinity. *Botanical Gazette* 31(3):145–182.

    Egler, F. E. 1954. Vegetation science concepts. 1. Initial floristic composition, a factor in old-field vegetation development. *Vegetation* 4:412–417.

5. For information at a global scale, check out the assessment reports on the website of

    The Intergovernmental Panel on Climate Change (IPCC): http://www.ipcc.ch/

    To see how climate change is being studied at both global and regional levels, see the website of the Center for Climatic Research, in the Nelson Institute for Environmental Studies at the University of Wisconsin–Madison: http://ccr.aos.wisc.edu/

Adaptive Restoration

Documentation and Research

3

> **LEARNING OBJECTIVES**
>
> *After reading this chapter, you will be able to:*
>
> - Appreciate the importance and significance of the adaptive approach to restoration.
>
> - Discuss the kinds of documentation and communication networks that inform the practice of restoration.
>
> - Discuss the role of research in contributing to adaptive restoration practice and theory.

One of the first steps in the restoration process is to learn as much as possible about the current resources of the site, the species/communities/ecosystems that might survive there, the interests of the various project stakeholders, and the planning and implementation techniques that have (or have not) worked in the past under similar circumstances.

Achieving a good match between the resources of a site and the habitat requirements of each species is a key factor in the success of any restoration. Some plant species can survive over a wide range of light intensities, precipitation amounts, and soil types, whereas others have more narrow tolerance ranges (see Chapter 2). Similarly, some animals are able to use a wide variety of food sources, while others can eat only a few. The more information you and your restoration team have about a site and the structure and functions of natural communities, the better equipped you will be for choosing species that will survive there. In order to include a rich variety of species, as would be the case in undertaking a complete

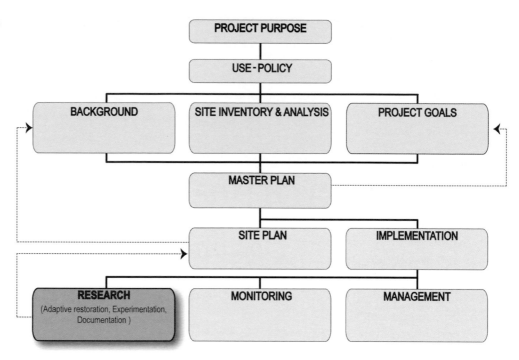

FIGURE 3.1 The Restoration Process: Documentation and Research.

restoration, it is important to understand the details of the microclimate variations on site, as well as the nature of the complex webs of species interactions that support biodiversity and sophisticated ecological services.

Becoming acquainted with what has and has not worked in previous projects helps narrow the choices you must make during a restoration, thereby saving time and energy. This kind of information is particularly useful for understanding why a restoration proceeded as it did. The more analysis about the success or failure of past restoration attempts, the easier it will be for you to tailor procedures to a specific situation.

Chapters 4 and 5 will discuss how to evaluate the resources of a restoration site. This chapter explores three practices that can contribute greatly to the essential background information for restoration planning (Figure 3.1). In many cases, the steps can be designed to yield a good deal of useful material for relatively little time and effort. The three practices are (1) project documentation, (2) establishing communication networks, and (3) incorporating research. First, however, we will introduce the concept of adaptive restoration—an approach that links research to the practice of restoration and highlights the importance of being flexible.

▶ 3.1 ADAPTIVE RESTORATION

Just as natural communities may continue to change and develop for decades after a disturbance, most restorations take a number of years to fully achieve the project objectives. During the course of a project, you may need to make midcourse corrections for a number of reasons, such as these:

1. Many factors that influence the success of a project are both unpredictable and out of your control (e.g., the weather, offsite land use changes, the introduction of new pest species). "Expecting the unexpected" is one thing you can count on in every

project. When an unanticipated event changes the site parameters (e.g., changes the course of a stream or removes 75% of the tree canopy), you must adjust the restoration plan accordingly.

2. Over time, the environmental conditions of a site may change such that species introduced to a site at the beginning of a project may no longer match the site, and others, not originally included in the plan, will now be more appropriate. This scenario is predicted in some global warming models, for example. It is important for you to be able to reset the expectations for the restoration's composition, structure, and function.

3. Efforts that have worked in the past may not work in the present situation.

4. General guidelines, based on theory and/or experience, are used to plan projects, but there might be several approaches to a situation that can work. The one that will be most successful may vary from site to site, or even within the site, and it is not always clear at the outset which it will be. You need to be free to switch techniques or even modify project goals in the face of early results.

5. Situations may arise that have few precedents and therefore little background information to guide your decisions. This is the case when a type of community is restored for the first time, or a new pest species arrives. Sometimes the first approach taken does not work as planned.

In certain situations—as when restorations are implemented as part of a mitigation requirement—practitioners may hesitate to adopt a flexible approach out of concern for exceeding the terms of the contract and/or being unable to meet specified criteria on schedule and within budget. This is a legitimate concern that needs to be addressed before a contract is written. Although the course of a restoration project is never likely to be completely predictable, the more information practitioners have, the more reliable the practice will become.

A flexible, adaptive approach to restoration prepares everyone involved to switch strategies if the situation warrants. In addition to recognizing the need for flexibility, this approach prepares you for the kinds of things you can anticipate by using an "if/then" logic—"If X occurs, then do Y; if A happens, then try B." The idea is to use the community/ecosystem model (see Chapter 2), as well as the kinds of research and documentation we will discuss below, to predict the behavior of the restoration under a series of different scenarios. In this way, should one of these situations occur, you will be ready to act.

▶ 3.2 PROJECT DOCUMENTATION

Strictly speaking, every restoration sets a precedent: A project that can serve as an example used to guide future efforts. It is instructive to visit existing restorations and observe in person the composition, structure, and functions of the communities/ecosystems involved. Such observations help you understand possibilities. Unless accompanied by some form of documentation, however, these observations are of limited value for fully appreciating and comprehending the restoration process and/or perfecting techniques. Seeing the results without knowing what the starting conditions were like, or what procedures were used, provides only a limited part of the story. So to be truly useful as a precedent, a restoration needs to be documented.

Documentation is the recording and storage of information so that it can be retrieved and used later. It can take a variety of formats, ranging from private informal notes and observations kept by individual restorationists, to publicly available formally adopted plans, to peer-reviewed scholarly publications. All forms of documentation are important and worth the effort involved. Ideally, the procedures for recording and storing information are set up at the beginning of a project, and cover ways to document all phases—the planning, design, implementation, and long-term management of the site.

3.2.1 Notes and Observations

One of the simplest forms of documentation, a form that can yield rich understanding, is the practice of taking notes on everything—from daily project activities, to observations on the progress of a restoration, to ideas about what to do next, given those observations. You can record notes in the field as work is going on, or later from memory back in the office. The advantage of notes taken in the field is their immediacy. They express unfiltered on-the-spot ideas and observations. The act of recording the notes does, however, take some time away from your activities at hand. If you take notes later in your office, your onsite activities will not be interrupted, and you will be able to benefit from perspectives gained throughout the course of the day. You could, though, forget some ideas or observations by the time you get around to documenting them.

There are several ways to record notes. Probably the oldest tool and one that is still preferred by many is a field notebook—a hardbound booklet with blank or ruled paper ready for handwritten comments and/or sketches. By numbering the pages and dating each set of notes, the notebook becomes a relatively permanent record of the course of a project. Portable digital devices (digital audio recorders or those with a keyboard) are also very useful for taking notes in the field. You can easily download the digital files onto office computers and store them as is, organize them into another file format, and/or use them as the basis for data summaries and analysis. If you use audio recorders, it is often useful to transcribe the files into text files. Software programs are available to perform this conversion, although at the time of this writing, they have room for improvement.

You can use another type of digital device, a global positioning system (GPS) unit, to record the locations where notes are recorded. A GPS unit basically triangulates locations using a series of satellite signals. Some GPS units are accurate enough to be able to pinpoint a location within a few meters (1 m horizontally and 1.5 m vertically) when satellite signals are strong.

3.2.2 Work Journals and Team Reports

Managers of projects that require the employment of hourly workers must keep track of the number of person-hours worked each day in order to create the payroll, and many project managers who rely on volunteer labor do the same. However, this information is generally not broken down in the work records by task performed. Experienced practitioners will have a good understanding of how long it takes to perform different procedures under varying site conditions. Teams who have worked together on projects over the years know how to most efficiently assign and stage tasks. But these rich understandings are often not transparent—they are often not explained to others.

One reason given by practitioners for not documenting daily restoration activities in detail is that doing so would take valuable time away from getting the job done. While it's true that recording information does take time, it can be argued that taking a few minutes to take notes each workday will save resources in the long run because of the information gained.

An approach that works well is to have field workers return to a central location about 20 minutes before the end of each day and prepare brief summaries of what was accomplished, including any relevant observations they have made, such as comments on the condition of the plant materials used, time needed to repair equipment, or modifications made to the implementation protocol because of unexpected conditions encountered. In situations in which there are several team members involved, and especially when people are working at different locations or on different tasks, it is very useful to have everyone come together at the end of the day to share information and observations. This practice gives everyone the chance to learn from different experiences, and also provides an opportunity to get a head start on organizing the next day's activities.

At the end of a project and/or during a time when restoration field activities slow down or are impaired by weather conditions, a review of the work journal entrees can identify issues to be addressed on site. This information may also prove useful in planning future projects.

3.2.3 Photographs

Photographs are becoming an almost indispensable tool in ecological restoration. Pictures can convey at a glance what it takes several paragraphs of text to describe. Digital photography has also made this tool relatively inexpensive to use—aside, of course, from the cost of the camera! Chapter 9 discusses the use of photographic imagery in tracking restorations through time. You can also use photography to record exactly how implementation procedures are carried out, what plant materials looked like at the time of planting, how much of the site has been affected by a prescription burn or a planned flood, or the presence of a particular bird or insect on site. Videos with audio tracks are particularly useful for capturing techniques; still images can provide sharp detail. Newer cameras include both video/audio and still image capabilities, which provide images of sufficient quality for the purposes of documentation.

When using photography, it is important to remember to record when and where the images are taken. Many digital devices have built-in GPS systems for recording this kind of information. It is always a good idea to record the date and location separately as well—in a field notebook, for example.

3.2.4 Records of Purchases

One form of documentation that may not be obvious initially is the practice of keeping purchase receipts and other records about the materials and equipment used during a restoration. People usually keep such records only for bookkeeping and tax purposes. Yet receipts and planting records can help support inferences about how an older restoration was implemented, especially in situations in which notes are scarce and the original planners have moved on.

In managing a site that was restored decades ago, for instance, you would need to know not only that plantings were called for and implemented—information that is included in most plans and work summaries—but also where the plant materials originated. Information learned since the time of the planting might indicate that points of origin are crucial to planting success. If the original materials came from a location outside of the currently recommended zone, that might be part of the explanation for the absence of the plants on site. Similarly, records of any physical site alterations that accompany restoration—such as topographic changes, the additions of soil amendments, or the repair of erosion gullies—are important.

CASE STUDY · Purchase Records

Restorationists working on the Lakeshore Nature Preserve, at the University of Wisconsin–Madison, are fortunate that many purchase records from the 1960s have been saved. Ironically, a different issue has arisen. The receipts and itemized lists for plants purchased from nurseries, as well as information on seed collection sites, exist, but there are no records on where in the Preserve (or whether) the species were actually planted. In addition, only limited information about what the Preserve was like at the start of restoration in the 1960s has survived.

Because of faulty documentation, the restoration team cannot be sure of the histories of plants included in the purchase or those now found on the site. Are the plants in a particular location the surviving remnants from hundreds of years ago, or were they planted? Were species planted in the locations where they are now found, or did they spread from an initial planting location? Due to the lack of answers to these questions, the restorationists have to take the time to investigate planting techniques and/or site preparation procedures their predecessors may have already determined but neglected to record. This situation reinforces the value of documentation and of keeping as many things connected to the project as possible. ∎

3.2.5 Information Management

Once information is collected, you need to create a system for organizing, accessing, and storing it. In other words, it is important to create a restoration archive: a system for the long-term storage and retrieval of various kinds of documents or artifacts. The issues to address include deciding how to provide physical protection, how to facilitate information retrieval, who should have access to the information, and whether or not to require the information to be stored in more than one format (e.g., digital as well as paper).

Storing records in an environment that prevents physical deterioration is one of the decisions you have to make in creating restoration archives. The types of documents generated by restoration projects include, but are not limited to:

- Original and revised restoration plans: text and plan graphics
- Photographs and/or videos
- Notes: field notes, meeting notes
- Work activity records
- Data archives: raw, summarized, analyzed
- Correspondence
- Outreach and publicity
- Budgetary materials: purchase histories and personnel records
- Monitoring and management information

The formats range from various kinds and sizes of paper to digital media.

Protecting paper-based information from moisture, light, dust, oily fingertips, coffee spills, and other kinds of damage is one challenge. Professional librarians and museum curators have, of course, worked with paper media for many years, and can provide excellent advice about how to manage the different kinds of information generated by a restoration project. There are many techniques and equipment available, ranging from the simple—files and file cabinets—to the complex—climate-controlled spaces, acid-free protective coverings, and the like.

Maintaining digital files in readable formats is another challenge, and one with a shorter history and changing standards. Within the past decade, standard storage formats have included floppy discs, zip discs, compact discs (CDs), flash drives, external hard drives, and offsite web-based services. As new formats replace the old, it becomes more difficult to retrieve the information. For example, disc drives are not included in many new computers. Also, the digital formats are (or can be) unstable and have a relatively short life span compared with paper documents. (Although some technicians claim that CDs can last 100 years or more, others have found the discs can be easily damaged and are vulnerable to decay.) Word-processing and data-handling software programs have also changed, and at present some documents are unreadable if they were produced with software that is now out of date. One way to avoid this kind of incompatibility is to save copies of documents using a more universal format, such as .pdf, .txt, or .rtf.

Digital photography has largely replaced film-based media, leading to similar concerns about format changes in the storage of images. Because of the rapid ongoing technical changes, it is a good idea to store critical information in both paper and digital formats, whenever possible.

Accessibility is another important issue with stored information. Notes, photographs, annotated plans, and purchase records are only useful to the extent that you can access them when needed. Because most restorations take several years to reach their goals and will need to be managed for many decades, it is essential to organize records so that they can be used in the short term, as well as be available through time.

One of the keys to access, of course, is knowing where to find the materials once they have been put away (in what file, on what shelf). A designated central location for information storage is optimal, supported by a cross-referenced filing system organized around major themes. It is not unusual for individual members of a restoration team to keep control of the information that he/she collects. Even though everyone may willingly share files when asked, this situation can lead, over time, to delays in finding information, and even to data being lost.

Restoration planners also need to consider who has access to project information. Should all information be available to everyone? Should preliminary data have restricted access? Should restoration techniques developed by for-profit restoration firms be trade secrets? For many public agencies and for projects funded with taxpayer dollars, the default position is to make as much information available as possible. Still, protecting fragile or rare elements is often grounds for limiting access.

▶ 3.3 ESTABLISHING COMMUNICATION NETWORKS: SHARING AND EVALUATING INFORMATION

The kinds of documentation discussed above provide essential information to the team in charge of a particular restoration. In order for the profession of restoration ecology to advance, however, it is important for its members to share information about the successes and failures of projects, the results of experiments and demonstration field trials, the development of new and innovative tools and procedures, and insights regarding new ways to use old tools. Sharing information expands everyone's knowledge base. Moreover, to the extent that dialogues develop, it creates opportunities for new ideas and better approaches to emerge from the discussions. Contributing in this way is one of the obligations of being a professional.

There are several avenues for sharing information—some casual, others more structured. Sometimes the most valuable exchanges occur during gatherings of colleagues. Several professional organizations concerned with restoration sponsor regular meetings, the Society for Ecological Restoration (SER) being a prime example. SER meetings feature presentations about recent discoveries and/or experiences of both practitioners and researchers, and provide display space for new products (e.g., field tools, data collection, and management technologies), as well as ample opportunities for conversations. Usually, summaries of the presentations are made available as published conference proceedings.

Such conferences are an excellent source of breaking news. The sessions often include discussions of work in progress, and opportunities to raise and exchange ideas about problems of immediate concern. Increasingly, organizations are beginning to use the Internet to communicate information and ideas. For example, SER launched the Global Restoration Network in 2007, a "web-based portal" to information about worldwide restoration efforts. However, not all of the material exchanged in these meetings, or on the web, has been checked for accuracy. The best source of reliable information about current theories and accepted practice is peer-reviewed publications, available in print, and increasingly online.

Peer review is the practice of asking experts in a field to review work in order to evaluate whether it meets the standards of the profession. For example, reviewers check to be sure that a descriptive report is sufficiently supported by evidence, or a discussion of the results of an experiment accurately represents the data, and that the experimental design used to generate the results followed accepted procedures. Formal evaluations conducted in this way are essential for ensuring that published books or articles contain reliable information.

Another very important means of sharing information is through visits to ongoing and long-established restoration projects. Such field visits are often part of professional conferences, and they are one reason conference locations change every year. Conference field trips usually feature presentations by the practitioners in charge of the sites and include time for question-and-answer sessions. There is no substitute for experiencing a site firsthand, and witnessing the results of different practice techniques. Especially in the case of projects sponsored by public agencies, you can even arrange a site visit outside the structure of a conference.

▶ 3.4 INCORPORATING RESEARCH INTO RESTORATION

Because every site has a unique history and landscape setting, and specific client needs, every restoration plan will be tailored to the situation at hand, using some variation of the restoration process presented in this text (see Chapter 1). Although an experienced restorationist will often have a good idea of how to proceed, a few questions almost always remain about which species will fit the site now and in the future, or which site preparation technique, planting procedure, and so forth will best suit each situation. This is especially true when the project involves communities that have had only a brief history of restoration attempts.

One way to proceed is, of course, by considering several possible restoration approaches and implementing one based on the experience gained from participating in previous restoration projects, hoping the choice yields a good result. Many projects have been and still are done in this way, and some of them are successful. However, all too often, the project does not go well. If the project has limited resources, for instance, making the wrong choice can mean settling for a partially restored site or trying again (after raising more funds), this time using a different technique. If something appears not to be working, you may be able to change course quickly and switch techniques before time and funding have run out, but the outcome is often not any more certain than it was with the original attempt. One way to lessen the risks of this kind of trial-and-error approach is to incorporate research into every restoration project. This idea is a hallmark of the adaptive restoration approach.

The word "research" is used in many different ways, from describing a search for relevant information in books and journals to performing tests in a laboratory. Simply stated, research in the broadest sense is the process of posing a question and systematically finding information for answering it. The product of research is a new, original synthesis of available information used to address the problem at hand.

We have already discussed two kinds of research used in restoration: the review of current literature on ecological theory for informing community/ecosystem models (see Chapter 2), and the study of restoration precedents. In Chapters 4 and 5, we will describe a third form of research: gathering and evaluating information about the current state of a site's resources, material that is fundamental to the ability to create functioning restorations. All of these activities, if done systematically, are forms of research.

In the sections below, we discuss four additional kinds of research that inform restoration projects: observing contemporary remnants, investigating historical communities, establishing scientific experiments, and establishing field trials. The first two activities are especially important in creating community/ecosystem models; experiments and field trials provide information that is critical in site design and in choosing restoration implementation and management approaches.

3.4.1 Observations of Contemporary Remnants

Much of our understanding of biological communities and ecosystem processes (described in Chapter 2) comes from visiting and/or studying places dominated by native species—plants and animals that can grow and reproduce without active care, and are present in a region due to forces other than direct or indirect human action. Although ecologists may do most of the work that goes into generating the ecological theory that underlies restoration, there is no better way than spending time in remnants for restoration practitioners to become familiar with the range of variation in composition, structure, and function found in the natural world. These "reference stands" can be contemporary, that is, living communities that are actively growing and functioning, or they can be historical, communities we come to understand by using historical investigation techniques in order to reconstruct scenes and events from the past. It is important to remember that the goal of restoration is not to use reference stands as blueprints to be duplicated in a particular restoration project. Rather, they are guides for determining what kinds of things are possible. It is by experiencing firsthand the sights, sounds, and feel of natural systems that restoration models come alive. These visits become the inspiration for creating dynamic and inspiring site plans (see Chapter 7).

Perhaps the best way to locate remnants to visit is through contacting private nature preservation organizations, such as The Nature Conservancy, the WWF (WorldWide Fund for Nature), and the National Audubon Society, or government agencies charged with protecting wildlands. Many parts of the world have preserves for the protection of native plants and animals. Many are open to the public, sometimes with restrictions, such as staying on trails. When visiting, be sure to check any use restrictions and ask for permission, if necessary, before wandering about.

Contemporary remnants are all that is left of much larger communities, which for a variety of reasons have remained while the surrounding land cover has been transformed by modern development practices. In North America, many native stands remain because they are on soils or slopes that have historically not been conducive to agriculture. Others were set aside as buffers for transportation corridors (especially railroads), or are blocks of land set aside but never developed. Some remnants have been spared development because of their spectacular beauty.

Most remnants are only a fraction of the size of their original extent and are missing species, some of which are extinct, while others require more habitat than the remnant can provide or cannot travel to the stands because of barriers. Because of this, remnants are not necessarily representative samples of the potential range of variation of a particular community type.

Nevertheless, by studying these living reference stands, you can learn a great deal about patterns and interactions. It is helpful to make observations about composition and structure, and also about the physical environmental features of the site. Here are the kinds of questions that frame the observations:

- What plant/animal species are present, and/or are most visible?
- Is there a pattern to the arrangement of the plant species?
- Are some in wet spots and others in dry areas?
- Are some on slopes and others in swales?
- Are some in sandy soils and others in loam?
- Are there sets of species that seem to be found together?
- What is the vertical structure of the community?
- Are species distributed in a clumped, regular, or random pattern?
- Is there a litter layer covering the soil?
- How tall/large are the most visible plant species?
- Is there a range of sizes/ages of individuals within a species?
- Is there evidence of reproduction—the presence of seedlings, nests, young animals?

A single visit to a site will be informative; however, because communities are constantly changing, it is better to revisit the site in all seasons and over several years. It is even better to visit different examples of the same community type.

Site Visit Approach: Gaining an Overall Impression of Vegetation

Chapters 4 and 5 discuss a variety of techniques for learning about the physical and biological resources of a site. Many of these approaches can be used with good results in studying remnants. But some require expensive equipment and take time to accomplish, and others require the collection of specimens and therefore may not be allowed in sensitive areas. In this section we present two observational site visit techniques that are particularly useful for getting an overall impression of natural plant communities: site immersion and quadrat sampling. Both techniques can also be used to obtain quick impressions of additional site resources, using the tools discussed in Chapter 5.

Site immersion. The purpose of the site immersion technique is to spend time exploring reference stands, looking for the essence of the community—the dominant species, group-ings, and patterns that stand out at the time of the visit. The idea is to use a variety of senses to get to know the community as a whole. The technique involves walking around the site and recording observations using notes, sketches, photographs, and checklists, all of which are designed to both guide your observations and leave room for interpretation. Examples of the kinds of observations that emerge from this technique include:

1. *Annotated species list*: A record of the species observed, plus observations about relative abundances, size, growth patterns, flowering status, the presence of adult or juvenile stages (tree seedlings, saplings), or any other feature that stands out.

2. *Descriptive overview*: An exercise that guides you to describe your overall impressions of the site's composition and structure, using prose, sketches, and photographs (see Sidebar; Figures 3.2 and 3.3).

3. *Guided checklist*: A series of questions designed to focus your observations on features of interest for a particular project, such as microclimate features of locations where certain species are found, the size/age class structure of trees or tree species, and evidence of fire scars, blow-downs, or other physical traces of past disturbances.

Sidebar | **Creating a Descriptive Overview of a Site**

One of the best ways to learn about natural com-munities is to make careful observations and record your findings in the form of notes, sketches, photo-graphs, or even checklists recording the presence (or absence) of specific features. Together, these impres-sions provide an excellent overview of the essence of a site. During the course of teaching a field class for students interested in restoration ecology, Dar-rel Morrison, Evelyn Howell, and John Harrington, at the University of Wisconsin–Madison, developed a worksheet that helps guide and sharpen field obser-vation skills. We use the worksheet to study remnants in order to create restoration community/ecosystem models, as well as in evaluating a proposed restora-tion site during the site inventory and analysis stages of a restoration project (see Chapter 5).

The worksheet uses a series of exercises, checklists, and questions to focus on understanding the structure and composition of a site (see Figure 3.2). The format can be readily adapted to fit the needs of restorationists in all parts of the world. Here are two sample responses to the first question of the worksheet, following student visits to a remnant prairie:

> *There was a magnificent display of shooting stars as well as blue-eyed grass. I found this community very pleasant because of the low plants and*

> *openness. It was breezy, which made it seem fresh. It was overall a nice place for relaxing.*

> *This wide-open community was pleasant to be in with absolutely no canopy of any type. The grasses and forbs were different shades of green giving an effect of a woven, textured carpet.*

Figure 3.3 shows two quick sketches, examples of responses to the second question.

The purpose of the first two questions is to record both initial and more detailed impressions of overall sense of place, using all the senses. The choice of things to report and sketch is left up to you. The worksheet continues with a set of more focused questions, asking about community composition and structure.

As you might imagine, different observers choose to highlight different things. Impressions change with the seasons and in different examples of the same kind of community, and often the observations of an individual will change with repeated visits. Taken as a whole, several themes relating to structure and composition begin to emerge. Understanding what features of an ecosystem contribute to creating a sense of coherence greatly aids our ability to develop experiential restorations.

Date: Time of Day:

Approximate Stand Size: Stand/Community Type:

Sky Condition:

(1) Find a location you feel is representative of the community. Sit down and briefly, in one or two paragraphs, describe your initial reaction to or perception of its character, based on your initial observations. Mention of any sounds and smells that you notice.

(2) Create one or more sketches that show the overall composition and structure of the site. Highlight significant features you want your sketches to emphasize, using labels and short notes.

(3) Physiographic Characters (landforms, rock outcrops, water bodies, etc.)

    Character

    Description

    Overall Impression

(4) Sounds

    Description (intermittent, constant, sounds like...)

    Overall Impression

(5) Smell

    Description (intensity, smells like....)

    Overall Impression

(6) Spatial Quality of the Stand

A–F: Circle the appropriate category and provide additional observations.

    A. Characteristic height of the canopy: (a) NA, (b) <10′, (c) 10′–20′, (d) 20′–40′, (e) >40′

    B. Estimated percent canopy cover: (a) <1%, (b) 1–20%, (c) 20–40%, (d) 40–80%, (e) >80%

    C. Average horizontal "hiding distance": (a) <20′, (b) 20′–50′, (c) 50′–100′, (d) 100′–500′, (e) 500′–1000′, (f) >1000′

    D. Degree of enclosure: (a) Extremely open, (b) Open, (c) Somewhat enclosed, (d) Extremely enclosed

    E. One or two words describing the spatial quality (small room, cereal bowl, cathedral-like, claustrophobic, etc.)

FIGURE 3.2 A Worksheet for Studying Remnants.

FIGURE 3.3 **Descriptive Overview Sketches.** These sketches are part of a descriptive overview of two Wisconsin plant communities, completed on late spring site visits by students and faculty in a University of Wisconsin–Madison field class. (Left sketch courtesy of Amy Jo Dusick, right sketch courtesy of John Harrington.)

The site immersion technique is similar to the walk-through used to inventory and evaluate the resources of a site prior to restoration, and to monitor the project during and after implementation (see Chapter 5). The purpose is somewhat different. These observations will help give three-dimensional form and complexity to the restoration model. If the remnants are in the same geographic region as the restoration, the technique also helps tailor the model to local conditions.

Quadrat sampling. A quadrat is an area of defined shape and size that is often used for collecting sampling data about the vegetation of a site. The idea is that a number of quadrats are distributed throughout a site according to a predetermined spatial layout, and information is collected about the plants that fall within each one. The kind of information collected varies, but it often includes data on number and abundance of species present, determined by counting stems and/or measuring ground cover.

During restoration projects, you can often use vegetation sampling techniques (with quadrats or several other approaches) to characterize the existing vegetation during site inventory and analysis, and to monitor the status of an ongoing project in meeting or sustaining the restoration goals. Chapters 5 and 9 discuss vegetation sampling in some detail.

You can also use the quadrat technique to become familiar with the plant communities in remnants in a more informal way than would be true when collecting restoration site data as part of a project. By distributing a set of quadrats in a portion of a remnant and noting the presence of species within each, you can obtain a quick quantitative understanding of a community to complement the broader and more qualitative understanding gained by the immersion technique. For example, you can explore:

- Which species are the most and least widespread at this scale.
- Which species, if any, seem to occur together.
- How many species are typically found in the area represented by the size of a quadrat.

Table 3.1 presents the data collected in an informal quadrat survey of a portion of Muralt Bluff Prairie, a dry, ridgetop grassland located in Green County, Wisconsin (see Sidebar). Conducted in early June, the study included fifteen 0.25-m² square-shaped quadrats (fewer than would be used in a formal inventory), arranged in a grid of three columns and five rows, spaced approximately ten meters apart.

Table 3.1 • Using a Quadrat Sample to Describe the Composition and Structure of a Remnant

Muralt Bluff Prairie   6/5/2006

Quadrat	Top Row					2nd Row					3rd Row					Freq.
	1	**2**	**3**	**4**	**5**	**6**	**7**	**8**	**9**	**10**	**11**	**12**	**13**	**14**	**15**	
Forbs																
Amorpha canescens	1											1				2
Anemone cylindrica														1		1
Antennaria neglecta		1	1				1	1		1						5
Asclepias verticillata		1		1		1	1	1	1							6
Aster ericoides	1								1						1	3
Aster laevis			1					1			1			1		4
Aster oolentangiensis				1	1				1						1	4
Aster sericeus									1							1
Cirsium hillii										1		1				2
Dodecatheon meadia														1		1
Erigeron strigosus					1											1
Euphorbia corollata											1					1
Helianthus occidentalis				1												1
Liatris aspera	1	1	1	1	1	1	1	1	1	1	1	1		1	1	14
Liatris cylindracea			1			1	1									3
Linaria canadensis				1											1	2
Lobelia spicata		1		1	1	1		1	1		1					7
Monarda fistulosa											1					1
Pedicularis canadensis				1										1		2
Physalis sp.	1															1
Ratibida pinnata	1															1
Rosa sp.	1	1								1				1	1	5
Rudbeckia hirta	1			1						1						3
Scutellaria parvula			1													1
Sisyrinchium campestre		1	1	1	1	1	1	1	1	1	1	1		1		12
Solidago rigida		1							1							2
Viola pedata	1	1	1		1										1	5
Grasses and Sedges																
Andropogon gerardii		1			1	1	1	1	1	1	1	1	1	1		11
Bouteloua curtipendula			1				1	1	1					1		5
Bouteloua hirsuta				1												1
Carex sp.		1		1	1	1		1	1		1				1	8
Panicum sp.	1			1						1				1		4
Poa compressa		1		1	1	1	1	1	1		1	1	1	1		11
Schizachyrium scoparium	1		1		1	1	1	1			1	1			1	9
Sorghastrum nutans			1						1							2
Sporobolus heterolepis		1	1	1	1		1			1	1					7
Stipa spartea	1		1	1			1				1	1	1			7
Shrubs																
Prunus americana	1															1
Rubus allegheniensis									1							1
Totals	12	9	10	13	17	9	11	12	13	9	12	8	4	11	8	

Sidebar | Quadrat Sampling

The data presented in Table 3.1 indicate the presence (1) or absence of each species in each of 15 quadrats, measuring 0.25-m². The sampling points were laid out in a grid, spaced 10 m apart. The grid cells are numbered sequentially, moving east to west; the top row cells are numbered 1–5. The second row begins 10 m south of the first row, and the cells are numbered 6–10. The third row begins with number 11, 10 m south of the second row.

Using this information, you can describe the spatial pattern of the different species.

For example:

1. What is the minimum number of forbs (non-grasslike flowering plants), grasses/sedges, shrubs, found in Muralt Bluff Prairie?

2. What species are the most and least common (frequently found) at this site?

3. How many species are found in an area of 0.25 m²? (Specify: average number, minimum, maximum, variation by life form.)

Are some species found throughout the site? Are others clustered in the north? The south? Explain.

FOOD FOR THOUGHT _____

Now, think about the following:

1. If you know there is a physical gradient on a site, running from the top to the bottom of a hill, how would you distribute quadrats? Explain.

2. What information is missing from the presence/absence data used in the Muralt Prairie example?

3. Which might change more, the list of rare species or the list of common species, if you increase the number of sampling points?

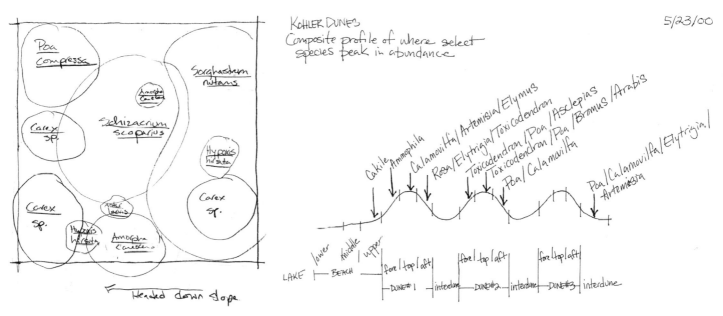

FIGURE 3.4 Quadrat Mapping. These are maps of a prairie quadrat (left) and the elevation changes in a dunal system (right). Each of these represents a distribution of species. The map on the left illustrates the horizontal patterns of prairie species within a 1-m² quadrat. This map shows the location of individual wildflower stems and the extent of several grass clumps. The map on the right shows the distribution of dunal species in relation to topographic changes along a gradient stretching from Lake Michgan inland. The basins in the dunes create an environment protected from winds off the lake and blowing sands. (Student work from Landscape Architecture 666: Restoration Ecology, University of Wisconsin–Madison.)

Quadrat mapping. Another use of quadrats is for understanding the growth patterns of species. This technique provides an even more detailed picture of the horizontal pattern of the community at a micro scale. For example, individual quadrats can be mapped by sketching the placement of plants on graph paper and/or through the use of photographs (Figure 3.4).

Such maps are especially good at helping visualize how plants spread and intermingle with one another. The maps illustrate which species' populations are found in clumps, with individual stems growing close together, or which ones are found as single, widely spaced plants. They are also helpful for visualizing how species mix together in space. By using nested quadrats—that is, by including quadrats of different sizes (1 m², 10 m², 100 m²), one within the other—you can uncover patterns at different spatial scales. Sketches, and especially photographs, can also give an indication of layering (vertical structure).

By increasing the quadrat size, different life forms can be mapped. In forests, for example, trees can be mapped using 0.01-ha square quadrats and the herbaceous layer using quadrats that are 1 m² or less in size (see Sidebar; Figure 3.5). If the distances on the maps are represented with some accuracy, such as by using graph paper, the maps can also aid in

Sidebar | Mapping Trees in a Forest

To quickly create a scaled map of trees in a forest, construct two sides of a 0.01-ha square quadrat to represent *x* and *y* axes using 10-m tapes placed perpendicular to one another along the ground. Then by sighting the trees from both baselines, you can record *x* and *y* distances (see Figure 3.5). By converting the recorded distances to the scale of your graph paper, you can get a good representation of the positions of the trees relative to one another.

If you have access to a GPS unit and a geographic information systems computer program, you can develop tree maps even more rapidly by recording the geographic coordinates of each trunk and using these data to make digital maps.

By identifying the species and measuring the diameter of each tree trunk, you can determine the species diversity of this section of the forest and estimate the densities and size distributions of the trees collectively and by species.

FIGURE 3.5 Mapping Trees in a Forest. A team of students is using two 10 × 10-m measuring tapes to represent the *x* and *y* axes of a map. By recording the locations along both axes of the center of each tree, they can map the horizontal patterns of the trunks. By measuring the diameter of the trunks and recording the species of each tree, they can represent the horizontal structure of the canopy layer. (Photo reprinted by permission from Evelyn Howell.)

understanding densities—how many plants/species are found in a particular area—and in spacing—the distances between stems or clumps.

Site Visit Approach: Targeted Observations

Some restorations are designed with very explicit goals, such as linking and/or enlarging habitats for particular, often threatened or endangered, plant or animal species; achieving a certain rate of carbon sequestration; or filtering stormwater runoff. Remnants provide what is sometimes referred to as "baseline information"—a "natural" standard against which the performance of a restoration can be compared. Because of the spatial and temporal variability of natural systems, as well as the fact that remnants may not be representative of the natural landscapes of which they were once a part, it is important to be cautious in making exact comparisons. Nevertheless, remnants remain one of the best sources of information about "natural" conditions.

For these purposes, studies of remnants take a more focused approach. Instead of making general observations, you formulate explicit questions, and choose tools and approaches to match the question. For example, to determine the habitat components required for maintaining a population of a certain squirrel species, you could visit a community in which the squirrels are found and look to see what foods they eat, what preys upon them, where they nest, where they shelter, and whether the answers vary with the seasons.

3.4.2 Investigations of Historical Communities

As we discussed in Chapter 1, the modern profession of restoration ecology is forward-looking and future-oriented. Unfortunately, the name of the field sometimes causes confusion. The term "restoration" implies that we are putting something back, but what practitioners create is not a literal copy of something that existed in the past. Rather, we are hoping to establish settings in which native plants and animals can continue to coexist, interact, and change with minimal disturbance pressures from modern human society. Of what use, then, are historical communities?

First, by learning about plant and animal groupings that were once found in the region of a restoration site, you can establish a historic range of variation—a catalog of the many different combinations of species, structures, disturbance patterns, successional pathways, and the like—that have existed in the past. This information helps to ground community/ecosystem models, and it leads to a better understanding of the complexities of natural vegetation than an investigation of existing remnants alone would yield. Remnants often do not represent accurate samples of the historic communities. In addition, according to ideas about global climate change, the environment of the future may not be what it is at the moment, let alone what it was historically. Therefore, the future range of variation will likely depart from that of the past. Nevertheless, an understanding of the historic range of variation provides an estimate of the limits within which a community/ecosystem may be successful.

Second, communities are historical and sensitive to initial conditions (see Chapter 2). Therefore, learning as much as possible about the history of a restoration site is one of the keys to success. Site history is thus an important step in the process of site inventory and analysis.

There are several different sources of information about historic biological communities. Some of these can be useful in conducting a site inventory and analysis of a specific site (discussed in detail in Chapter 4). All can contribute to understanding the historic range of variation within a region. For the best results, it is helpful to follow a few guidelines when using historical data:

1. Whenever possible, use three or four different sources of information, and cross-check them in order to confirm your conclusions.

2. Document the source of the information and, if possible, the original data collection methods, and consider potential sources of error in preparation or transmission (e.g., errors introduced by inaccurate mapping tools, by author bias, or by a need to translate original species or community identifications into modern nomenclature).

3. Describe the methods you use to interpret the historic information. Pay attention to the scale and degree of accuracy of the original data. Use the information appropriately, being careful not to draw detailed conclusions using general information.

4. Choose historic time periods that will provide information consistent with contemporary conditions and the restoration goals.

In other words, as with anything related to restoration ecology, remember to document assumptions, methods, and the reasoning that underlies the decisions.

Historic Maps

Maps have been produced by many cultures all around the world for hundreds of years. They can be found in government archives, in public libraries and museums, in historical societies, in private collections, and now increasingly on the Internet. Maps display spatial information and have a variety of themes—topography, soils, property boundaries, places of interest to visitors, and transportation routes, for example. Vegetation and land cover maps are particularly useful in restoration.

Historic maps use various vegetation classification systems, many of which divide community types according to their dominant life form, such as forests, deserts, or grasslands. The maps cover different land areas, up to the scale of continents, and are based on data sources, including explorers' notes, on systematic ground surveys, and, in more recent times, on the interpretation of aerial photographs or satellite imagery.

Historic Photographs

Photographs have been available since the early part of the nineteenth century. They can be very useful tools for figuring out the overall vegetation structure of an area, and in some cases, it is possible to identify individual plant species, if the images are sharp. Photographs of wildlife can also be helpful in placing species in a region at a particular place and time, if the photograph has been so labeled. Some images show areas after windstorms or fires, thus helping explain what impacts these disturbances had on the vegetation, as well as documenting historic occurrences.

Aerial photographs, taken from relatively high altitudes, date in the United States from the 1930s. Depending on the resolution and distortion caused by the angles from which they were taken, these bird's-eye views are especially helpful in revealing landscape patterns. In some cases, moving pictures also capture images of the vegetation of the past. A good place to look for historic photographic images is in the archives of the local historical society. Historic air photos can be obtained for most of the United States from the U.S. Geological Survey.

CASE STUDY   The General Land Office Public Land Survey

One historic mapping project that has proven to be particularly useful to restorationists in the United States is the General Land Office Public Land Survey System (PLSS), which was initiated by the Land Ordinance of 1785. The ordinance established a rectangular geographic coordinate grid system for identifying the boundaries of public land properties to be opened up for settlement. The grid is divided into "townships," each of which is 6 miles long and 6 miles wide. Each township is further divided into 36 squares called "sections," each of which measures 1 mi² or 640 acres. The location of any site within the area covered by the survey—some 30 midwestern, southern, and western states—can be easily identified by this system.

To establish the grid, teams of surveyors walked the township and section lines, taking notes on the vegetation, and marking the locations of the grid corners by designating "witness trees." The surveyors identified the species of the witness trees, measured their diameters, and recorded the distances and bearings of the trees from the grid corners. Using this information, as well as the field sketches required of the surveyors after completing each unit and the notes about the nature of the land they passed through (especially relating to farming and timber potential), you can:

1. Construct a map representing a snapshot of the vegetation of a region during that moment in time (see Figure 4.14).
2. Identify some of the prominent species that were part of historic communities.
3. Identify and partly describe historic disturbance processes.

The notes are particularly useful for describing the nature and extent of community/ecosystem disturbances (see Chapter 2). The surveyors would describe the aftermath of wind throws, recently burned areas, and floods, and if the disturbance patch was located on a township or section line, it is possible to estimate the extent, at least in one direction. ■

Herbaria, Museums, and Other Collections

People love to collect things, for pleasure as well as for scientific study. Collecting natural history specimens of all kinds was a favorite pastime for many during the nineteenth century in Europe and America. Collectors traveled the world hunting for unusual organisms, both living and dead. Many of these private collections eventually became the core of public natural history collections.

Herbaria and natural history museum specimens provide a wealth of information about the historic locations of plants and animals. Herbaria are collections of dried plant materials, each of which is identified by species, and generally also labeled with information about when and where it was collected. Similarly, natural history museums contain skeletons or preserved and mounted specimens of animals, insects, and other invertebrates, also with collection information.

This information can be useful in determining what kinds of species were found in a region historically—at least as far back as the known collection dates. Knowing the species that were present historically, you can infer the extent and presence of the communities that likely supported them. The level of detail available is, of course, dependent on the data provided by the labels. Sometimes plant collectors, for example, went so far as to describe the microclimates and soils of the areas from which the collections were made. But more often, this level of detail is missing. The specificities of the locations also vary. Much of the information about what is housed in various public collections is now available on the web.

Descriptive Narratives, Naturalist Field Notes, and Oral Histories

Written descriptions of historic plant and animal communities can also provide a wealth of information. And these narratives are often great fun to read! They can be in the form of published or unpublished diaries, letters, autobiographies, memoirs, news articles, and oral histories, to name a few. Here is an example, concerning southern Wisconsin, from naturalist John Muir's autobiography:

> It was a great memorable day when the first flock of passenger pigeons came to our farm, calling to mind the story we had read about them when we were at school in Scotland. Of all God's feathered people that sailed the Wisconsin sky, no other bird seemed to us so wonderful. The beautiful wanderers flew like the winds in flocks of millions from climate to climate in accord with the weather, finding their food—in fields and forests thousands of miles apart. I have seen flocks streaming south in the fall so large that they were flowing over from horizon to horizon in an almost continuous stream all day long, at the rate of forty or fifty miles an hour, like a mighty river in the sky, widening, contracting, descending like falls and cataracts, and rising suddenly here and there in huge ragged masses like high-plashing spray. (1965, 128)

This memory from the early 1800s is particularly striking because the passenger pigeon (*Ectopistes migratorius*) that Muir writes about was extinct by 1914.

Other rich sources of information can be found through conversations with long-term residents of a region, and in the cultural traditions and stories that have been passed along through many generations. People have been part of and have inhabited and influenced "natural" communities and ecosystems for thousands of years. Oral histories can provide clues about the historic range of variation of the communities in a region, and understanding how indigenous cultures used (and, in many cases, still use) the plants and animals of a region adds complexity to ecosystem models.

As professional historians are quick to point out, narratives such as these need to be interpreted carefully. The language and tone of the writing may be colored by the selective memory and nostalgia that often attend the passage of time, or by the purpose of the piece (e.g., to entice people to settle the land). Nonetheless, narratives do help you visualize the dynamics and beauty of former landscapes in a way that maps and preserved specimens may not. Interviews with landowners or others familiar with an area can also provide you with insights about the more recent past.

Reading the Landscape to Learn About the Past from the Present

Contemporary remnants are extremely valuable resources for understanding the composition, structure, and ecosystem processes that characterize natural plant and animal communities. They can also serve as windows into the past, if you know what to look for. "Reading the landscape" is a lot like following the clues in a detective story or a treasure hunt—in this case, looking for physical traces left behind by past events. There are many kinds of biological and physical evidence in the landscape that provide clues to regional landscape history. Examples include results from investigations of preserved animal food caches, studying phytoliths (life form–specific microscopic hydrated silica bodies formed within the cells of plants and preserved in soil after death), and looking at soil charcoal layers for evidence of species involved in past fires.

In areas where trees are common, two additional techniques are commonly used. The first involves the growth form of trees. Many trees have wide-spreading branches when growing in the open and a more vertical structure when they are clumped together, as in a forest. Finding a few trees with evidence of having had spreading lower limbs among many others with straight trunks may indicate that the area had once been a savanna (Figure 3.6).

In regions where trees produce annual growth rings in the wood of their trunks, dendrochronology, the science of tree ring dating, is a second technique for estimating both the age of individual trees and the pattern of growing conditions from year to year (see Sidebar). The annual rings are formed by trees growing in seasonal climates because the trees produce visibly different kinds of wood early in the growing season than they do later. The combination of the two kinds of wood together represents one growth cycle (often one year), and when looked at in cross-section, as, for example, on the surface of a stump after a tree is felled, the layers of wood appear to be rings. The number of rings gives an idea of how old a tree is, and, because the width of the rings is positively correlated with growth rate, the frequency of good growing seasons over the life of a tree can be estimated. By sampling several trees of different species and sizes, we can determine the age structure of a site. By looking at the patterns of the rings, we can detect interruptions of growth, as might be caused by drought or other adverse environmental changes.

Dendrochronology is also useful in discovering past disturbance histories. As we discussed in Chapter 2, agents such as floods, fire, windstorms, and even insect outbreaks are important drivers of community/ecosystems dynamics. Understanding the historic spatial and temporal frequency of these events is fundamental to creating conceptual community/ecosystem models and developing management strategies. When these events leave

FIGURE 3.6 A Former Savanna Tree. This formerly open grown tree displays remnants of limbs previously shaded out by rapidly growing saplings that established once fire and browsers were removed from this landscape. (Photo reprinted by permission from John Harrington.)

Sidebar | Collecting and Analyzing Tree Rings

In areas with seasonal climates, you can often estimate the age of a tree after it has been logged by looking at the surface of the cut stump and counting the rings. Dendrochronologists use this method. For example, they might look at logs used to construct a building and date the rings from the time the building was constructed (and presumably the tree was cut). To obtain information from living trees, they have developed an alternative, "no kill" approach that employs a small-diameter (0.5 cm) metal increment borer.

To study the growth history of a tree, using the increment borer, drill into the center of the tree and extract one, usually at least two, pencil-thin rods (cores) of wood. Next, transfer the cores to a lab (using a drinking straw), dry them, glue them to a surface to stabilize them, and then polish them with sandpaper to better reveal the rings. The rings are best viewed through a low-power microscope. You can date the cores by working backward in time from the outermost ring (which represents the present) back to the core.

FOOD FOR THOUGHT _____

The increment borer removes a very thin cylinder of wood from the trunk of a tree. What are some of the potential problems with using this technique, in terms of the health of the tree, as well as in using the information to understand the age structure of a forest or patterns of climate?

traces in tree rings, we can track them through time. For example, in many types of temperate zone forests, ground fires will wound trees without killing them. The wood grows around the scar, leaving characteristic interruptions in the growth rings. Using this historical record, preserved in the wood of living (and dead) trees, we can discover information about disturbance size, frequency, intensity, and even seasonality. We can also correlate the timing of disturbances with information about historic climate patterns.

3.4.3 Establishing Scientific Experiments

A full discussion of how best to conduct science-based research projects is beyond the scope of this text. The purpose here is to introduce you to the basic logic behind different research approaches, and to suggest and highlight some of the ways you can incorporate experiments into a restoration project with fruitful, helpful, and ultimately time-saving results.

Exploring ecological theory, observing remnants, and investigating historic communities can greatly enhance your understanding of natural communities and ecosystems. To fully adopt an adaptive approach to restoration, it is helpful to generate entirely new data by embedding two additional research strategies into a project as implementation proceeds: (1) research experiments, including field pilot studies and those conducted under more controlled conditions in a lab or greenhouse; and (2) demonstration field trials. These kinds of investigations often concern different restoration strategies—how to add or remove vegetation and/or wildlife, ameliorate microclimate conditions, facilitate ecosystem functions, and so forth. The experimental approach, in particular, when used in the course of a restoration project, not only can save time and money, but also lead to advances in ecological theory and therefore our understanding of the nature of communities and ecosystems.

Classical experiments rely on systematic observations in the form of measurable, usually quantitative, data, and they investigate cause-and-effect relationships: "If X happens, then Y will follow." The research is based on the familiar scientific method:

1. Ask a measurable question.
2. Construct a hypothesis. (Predict what the causal relationship will be, based on current theory.)
3. Design an experiment to test the prediction.
4. Examine the outcome. (Did your prediction hold? How confident can you be that the results are accurate?)

The idea is to produce a change in a situation (apply a treatment, initiate a behavior) and examine what, if anything, happens (measure the result). Experimental restoration research differs from the trial-and-error approach in these important ways: (1) Research not only follows established procedures designed to minimize the potential for preconceived notions if the investigator might have to affect the outcome, (2) it ultimately seeks to establish that a particular procedure works but also to explain how and or why it works, and (3) it is based on established ecological (or, in some cases, social or psychological) theory.

Experiments attempt to uncover cause-and-effect relationships by keeping all of the factors (or at least all known factors) that might influence an outcome constant except for the one being tested. As a result, if a change does occur (so the thinking goes), it is most likely due to the relationship being tested.

Phosphorus as a Limiting Factor

Suppose you are having trouble establishing herbaceous species into the ground layer of a degraded forest. The desired species do germinate from seeds and survive for a while, but eventually most of the plants weaken and die. A clue about what might be going on is that remnant forests in the area have higher soil phosphorous levels than the soils of the restoration site. Furthermore, the literature confirms a link between the vigor of the herbaceous ground layer and soil nutrient levels. Based on this, you might establish an experiment to test the survival rates of plantings done with and without added phosphorus.

In this example, the phosphorus would be an independent variable—the experimenter, acting as an external force on the system, manipulates its value. By varying the independent variable, you are said to be applying a treatment. The survival rate of the plantings would be the dependent variable—its value changes when the independent variable changes.

In order to isolate the effect of phosphorous levels, other known factors that might influence the survival of plantings (such as light, temperature and moisture levels, the availability of nitrogen and other mineral nutrients, the presence or absence of soil microbes) must be held constant during the course of the experiment, or at least be made to vary in the same way in all of the plantings (those with and without added phosphorus). These factors are called the control variables (or simply the controls). By managing the experiment in this way, if more plants survive with phosphorous additions than without, you will have evidence to support the idea that phosphorous levels and survival are related.

Lab and Greenhouse Studies and Principles of Classical Experimental Design

One of the best ways to focus on the relationship between independent and dependent variables is to conduct the experiment in a laboratory or controlled environment greenhouse. These settings allow you to keep careful track of all the variables involved in the experiment—independent variables, dependent variables, and especially the controls. In a lab, you can keep the control variables at known levels, apply treatments, and then observe and measure results with precision. Lab and greenhouse studies are especially helpful for understanding how and why something works, and therefore have been very useful in establishing general scientific principles and theory. For restoration, such controlled environments help elucidate abiotic and biotic germination, pollination and growth requirements of plant species, how fluctuating water levels or fire regimes affect survival, and the mechanisms by which herbicides harm target species, among other things.

In the case of the phosphorous addition example, the experimental protocol might include the following steps:

1. Plant a number of herbaceous understory plants in twelve pots in a greenhouse, and add a set amount of available phosphorus (a form known to be usable by plants) to six of the pots (the treatments) and not to the others (the controls).

2. In order to establish the experimental controls, manage the greenhouse environment so that all of the pots experience the same daily light, moisture, and temperature levels and have an identical planting medium—perhaps even soil collected from the restoration site.

3. Determine the results after a set period of time (perhaps the length of a growing season) by counting the number of surviving plants in each pot and comparing these results between the treatment pots and the controls.

Notice that this hypothetical experimental design uses several pots each for the treatments and controls. To ensure accuracy, a number of replicates are needed—several treatment pots that receive the phosphorus and several others (controls) that do not. The practice of using replicates and averaging the results reduces the likelihood that any variability among individual plants with regard to phosphorous requirements might influence the results. In this example, the results would be reported as a comparison of the average number of plants per pot in the control group versus the average number per pot in the group treated with phosphorus.

In addition, in order to account for any unforeseen variation in the microenvironment of the lab (such as "hot" or "cold" spots), it is a good idea to randomize the location of the pots. Rather than placing all the control pots on one lab bench and all the treatment pots on another, the placements are mixed up. In this way, the chance of any pot landing in a hot or cold spot would be the same, regardless of treatment. Randomization is an important tool by which bias and the effects of unforeseen chance variation can be reduced in an experiment.

Variation in environmental factors can sometimes be foreseen. For example, there might be a known microclimate pattern within the lab—one bench receives more light than another. In that case, a randomized block design could be used. A block design creates sets, each of which contains treatments and controls. The goal is that the microclimate pattern within each block will be relatively homogeneous, or at least less variable than the variation between the benches. Blocks (sets) are placed on each bench, and the location of the treatment and control plots within each block is randomized. This arrangement is also a good idea when the potential for environmental variation exists, even if a problem has not been specifically identified. Both the complete randomization and the randomized block design layouts are used to minimize the influence of factors external to the experiment.

The last step in the experiment is establishing the likelihood that the results are "real." If there is a difference in survival rates between the treated and control plants, how likely is it that the change was due to the addition of phosphorus? To answer this question, a statistical significance test can be used. Statistical significance tests are quantitative techniques that present evidence of the likelihood that any difference noted between treatments and controls was due to the treatment, rather than to chance.

There are many kinds of statistical tests available, and it is helpful to ask the advice of a statistician in selecting a test that best fits the parameters (e.g., number of replicates, degree of individual plant variability) of a particular experimental design. The test will estimate the confidence level of the results—the probability that the results were not due to chance. For example, a 99% confidence level indicates that there is only a 1% likelihood that the results of an experiment could not have occurred by chance (emerged randomly). The confidence level that must be reached in order to accept the results of an experiment is established in advance. The 95% confidence level is a common standard in restoration research. If the test indicates that the experimental results meet the confidence level criteria, the results are said to be significant, and accepted as being due to the relationship being tested. Using our example, suppose the experiment shows a "significant increase" in survival of the plantings with added phosphorus, meaning that the results meet the chosen confidence level criterion (95%). Then the results would support a decision to go to the expense of amending the soils of the restoration prior to planting.

However, suppose the findings indicate no "significant difference" between the treatment and control plants. Perhaps the results do not meet the established confidence level and are significant at the 80% level. Then it is time to rethink the situation. Then you would follow these steps:

1. Decide to see whether changing the amount or type of phosphorus used in the experiment will change the results. Set up a new experiment to test multiple treatment levels.

2. Accept that the simple addition of phosphorus is not the key to the successful establishment of the herbaceous layer plantings in this situation. Propose, and then identify and set up, experiments to test additional factors thought to, in principle, influence survival (e.g., herbivory, depth of litter, lack of soil microbes, lack of available nitrogen or other nutrients).

3. Decide to test whether several factors acting together are involved (e.g., levels of phosphorus and the presence of mycorrhizal microbes).

4. Examine the experimental setup for any event that might have altered the results, such as mistakes or anomalies in conditions.

The important point here is that both "positive" and "negative" experimental results, assuming the experimental design and execution were sound, benefit both the practice and the theoretical understanding of restoration.

In making decisions while conducting experiments, it is critical to remember to distinguish between (1) the importance of the magnitude of a difference between the controls and the treatments, and (2) the statistical significance of this difference. For instance, it is possible that the mean number of plants per pot of the phosphorous treatment group was 10.5 and that of the controls was 9, and that the results met the preestablished confidence level. Although this result would be considered statistically significant (a difference of 1.5 plants), you have to decide whether the result would justify the expense of adding phosphorus to the soil of the restoration site.

3.4.4 Establishing Field Trials

A fundamental drawback of lab or greenhouse experiments is that the controlled environments of even the most complex and sophisticated laboratory simulations do not match the environmental complexities and unpredictable events of nature. Furthermore, it is possible that the values to which the control variables are set (light, temperature, soil moisture) may not reflect conditions in all sites. Relationships demonstrated in a lab setting may behave differently amidst the complex interactions occurring in natural ecosystems. In some facilities, it might be feasible to mimic the composition and structure of the soils of a degraded forest, and set light and soil moisture levels to match those found in a natural setting. However, much about the interactions in nature remain unknown, and therefore cannot be simulated in a lab. Because the solutions tested in the lab will ultimately have to be applied in natural settings, many questions related to restoration—for example, optimal planting densities, planting techniques, plant removal strategies, and even soil nutrient additions—are often best addressed in the field. Moreover, many restorationists do not have ready access to lab facilities. For both of these reasons, much restoration research is carried out in the field.

The steps in field experiments are similar to those used in the laboratory setting: You pose a question, develop a hypothesis about a cause-and-effect relationship, apply a treatment (or a set of several different treatments), and compare any measurable changes that occur in the dependent variable(s) to the changes, if any, occurring in a control. The main difference is that most of the so-called control variables are essentially out of your control and will fluctuate over time. They may interact with the treatment in unpredictable ways, and are often not uniformly distributed across the study site.

In the case of the phosphorous addition experiment, light, temperature, and moisture levels, the availability of nitrogen and other mineral nutrients, and the presence or absence of soil microbes are more likely than not to vary across a site and are difficult to manipulate in the field for experimental (or restoration) purposes.

To account for the natural variation of the control variables through space and time, once again, the experimental design includes replicates and the randomization of the locations of treatments and controls. Block designs are also commonly used in field research experiments. The blocks (containing sets of treatments and controls) are placed in different locations throughout the site to account for the environmental variation that exists.

In spite of the advantages of incorporating research projects into a restoration, many barriers exist, not the least of which are having access to the requisite facilities and supplies, and being able to convince clients to support the inclusion of such experiments into a restoration's schedule and/or resource budget. Many public and private clients are impatient to get on with what they see as the real work of a project, and do not want to wait or spend resources on experiments, which they view as having theoretical, rather than practical, consequences. They want immediate "results, not information." However, this situation is changing, as more evidence accumulates supporting the value of spending time conducting research at the beginning of a project, to benefit the long-term success of restorations.

In some cases, restorationists are turning to volunteers not only to help implement and manage restorations, but also to participate in maintaining, collecting, and analyzing research data. To the extent that willing volunteers skilled in research are available and/or training is possible, such situations have the potential to greatly increase our knowledge.

3.4.5 Demonstration Field Trials

Just as the name implies, demonstration field trials evaluate one or more implementation or management techniques on a small portion of a restoration site. Usually these trials are established at the beginning of the implementation phase, and they have two purposes:

1. To compare the onsite performance of alternative strategies before choosing one to implement over the whole site.
2. To determine the best way to use or adapt a particular tool to the unique conditions of a site.

Demonstration field trials can be thought of as being a step down from experimental research and a step up from the trial-and-error method, in terms of their value to restoration ecology. They lack the rigor of field experiments. Randomization is not considered, and controls are not established; replicates are few. Therefore, the results can be misleading, and it is difficult to use them to develop sound theory. However, the results of demonstration field trials are very helpful in generating experimental hypotheses.

Especially when the results of demonstration field trials are dramatic—if, for example, an onsite comparison of planting techniques produces rates of establishment four times higher for one method than for the others—they can be very valuable in increasing the success of a particular restoration. This is especially the case in a situation in which the alternative is to proceed by trial and error. It is better to spend some time at the beginning perfecting techniques than having to start over after having made a poor choice at the outset!

| CASE STUDY | A Demonstration Field Trial: Erosion Control |

The Lakeshore Nature Preserve on the UW–Madison campus has a number of wooded hillsides that are largely devoid of vegetation due to erosion from foot traffic and stormwater runoff. The restoration goal for these areas includes maintaining an herbaceous ground layer throughout the woods. The implementation plan is to (1) stop the foot traffic and divert the stormwater and (2) plant desired ground layer species on the denuded slopes. In order to secure the plantings, Project Assistant Rebecca Kagle determined that she would have to secure the remaining soil.

There are several kinds of erosion control materials available commercially. Before deciding on an approach that could be used in many areas of the Preserve, Rebecca decided to test two—coir and jute—both of which are commonly used in ecological restoration projects that emphasize the use of biodegradable materials. Coir is a semipermanent mat made from coconut fiber. Jute is an organic, 100% biodegradable brown, open weave, light-duty erosion control matting. Coir is three times as expensive and lasts three times as long (4–6 years, as opposed to 1–2 years) as jute. Because of time and space limitations, Rebecca set up a demonstration field trial rather than a research project to test the performance of the two materials. The trial included the planting of two bare slope areas: one with coir, the other with jute.

After 2 years, fewer plants took hold on the slope that had the jute on it, whereas the slope covered in coir had a much denser plant covering. While the slope with the jute did not collapse and it does have more vegetation than it did prior to the restoration project, there is currently no remaining fabric present to hold water, or to slow soil and debris movement. The plants Rebecca added have had low survival rates and needed more watering than the corresponding slope that had coir added to it. Based on this trial, Rebecca recommended that the longer-lasting coir seems to aid vegetation establishment, thereby making it a worthy investment. ■

Because demonstration field trials usually take little space and can be accomplished in relatively short order, they are often easier to justify than complete field experiments to action-oriented clients and funding sources. The trials are seen as leading to immediate practical results, directly tied to the success of the project.

3.4.6 The Role of Systematic Observations in Research

Most restorations do not incorporate formal experiments or field trials. Instead, the restoration planners choose implementation or management techniques based on their previous experiences, on information gleaned from publications or conference presentations, and/or by trail and error. With regard to some communities—for example, the grasslands of the American Midwest—there is a great deal of available information about what works under a number of different site conditions and land use histories. In addition, on any given site, it appears that several approaches can work with grassland restorations. As a result, the risk in making a wrong choice is lessened. However, there is no way of knowing if a different choice might have led to faster, less expensive, or more long-lasting results!

If such projects include the observation and documentation techniques discussed earlier, they can still help increase our knowledge of restoration theory and practice. By looking for patterns in the observations, we can refine techniques and identify questions for further study, thus benefiting both practice and theory.

A comparison of recent grassland restorations in the American Midwest, for instance, indicates that those planted with seed in the fall have resulted in communities with a greater presence and abundance of forbs relative to the grasses than is the case with spring-seeded plantings. These observations have led many practitioners to switch to fall plantings. However, unless we understand why the timing of the plantings matters, it is difficult to be sure that the observations did, in fact, reflect a seasonal effect rather than some other factor, such as a coincidence of seed lot quality. Moreover, it is not certain that the results will hold for grasslands in other parts of the world; that is, the patterns may not be able to be generalized. Controlled experiments can help uncover the how and why of the observations. In other words, the observations generate hypotheses for experimentation, and the results of the experiments can lead to explanatory theories.

3.4.7 The Contributions of Experiments, Field Trails, and Systematic Observations

Experiments conducted during the course of a restoration benefit both the project at hand, and the profession as a whole. It is for this reason that creating a research plan is included in the outline of the restoration process (see Chapter 1).

Benefits to Individual Projects

Restoration experiments, especially those conducted in the field, are a key component of adaptive restoration. Because each site has a unique history and composition, it is not always certain, even in the case of community types with a long history of restoration, which of a number of possible site preparation or species introduction techniques will work. In some situations, there are few, if any, precedents to follow. In both cases, it is often helpful to start a project by establishing one or more experiments, preferably on site, to check out different approaches. The idea is to test possible techniques on a portion of a site before restoration begins in earnest, and use the results to select the best restoration strategies with which to proceed.

If a restoration can be implemented in stages, a practice that may even be desirable on large sites due to resource limitations, experiments and restoration practices can be refined as time goes on. The results of one experiment can lead to more questions, which in turn lead to the development of more focused restoration techniques, which lead to more experiments, more results, more questions, and so on. Professor Joy Zedler, of University of Wisconsin–Madison, and several collaborators used this iterative approach to test planting strategies while restoring herbaceous cover to a portion of the Tijuana Estuary.

Contributions to Restoration Theory

The results of experimental research can lead to the development of general principles and even to theory (see Chapter 2). An exciting approach to generating restoration theory is performing a meta-analysis: examining the results of many different experiments that explore a common question to see if the results, as a group, provide a common answer. Meta-analysis uses systematic statistical procedures to reach conclusions. The more such experiments are embedded in restoration projects, the more likely it will be that restoration techniques will gain in sophistication and complexity.

Bayesian statistics is a meta-analysis technique that has great potential to advance the field of restoration ecology. This technique enables a researcher to draw conclusions based on lessons learned from past experience, predictions based on theory, the results of field trials, and the assessments of experts based on systematic observations, as well as data from experiments. The implications are that every restoration has the potential to advance the profession, as long as the restoration team shares the results of each project. It also means that the various communication networks could take a leading role in serving the profession by facilitating exchanges between researchers and practitioners in order to suggest procedures for site managers to use in collecting information for researchers, or questions researchers could address to help practitioners.

In addition, the rising interest in restoration ecology and the increasing recognition of the power of meta-analysis have spurred an increase in restoration research in universities and other settings with appropriate facilities. There is a long history of partnerships between people trying to solve practical problems and scholars who are trying to understand how the world works. The questions raised by practitioners become hypotheses investigated by researchers; the theories developed as the result of scholarly research become solutions applied by practitioners. Two long-standing results of such partnerships in the United States are the system of land-grant colleges and universities, established to assist the agricultural industry, and federal funding programs, such as the National Science Foundation, that spend taxpayer dollars on research applied to "national needs." Private firms and public nonprofit organizations, such as the National Audubon Society and the National Wildlife Federation, also provide funding for this kind of "applied" research—restoration problems among them.

KEY CONCEPTS

- The adaptive approach to restoration includes an expectation that projects be flexible and able to switch strategies or desired outcomes in the face of changing conditions or new information.

- Project documentation involves recording notes and observations, compiling work journals and team reports, making photographic records of the site and restoration activities, and keeping records of budgetary expenditures, including personnel hours, equipment, and materials. Once information is collected, it is stored in an archive.

- Restoration research informs and leads to improvements in restoration practice, and contributes to theory and our understanding of community/ecosystem ecology. Research approaches include observations of remnants, investigations of accounts of historic communities, field and laboratory experiments, and performing field trials.

- The evaluation and communication of information generated by research and documentation are essential for advancing the restoration ecology profession.

FOOD FOR THOUGHT _____

1. Which aspects of a restoration project would be best documented using photographs? Written notes? Annotated maps? What are the advantages/disadvantages of each of these media in terms of accuracy, reliability, longevity, and ease of storage?

2. Why is it helpful to understand the range of variation of historic and contemporary natural communities before designing a restoration?

3. How might the idea that communities vary in composition over space and time influence the choice of species for a restoration?

4. What arguments might you use to convince a funding agency to support embedding one or more research projects in a restoration?

5. Design a field experiment to compare two different techniques for planting herbaceous species in the understory of a forest restoration.

Resources for Further Study

1. For a good overview of techniques for investigating historic communities, see:

   Egan, D., and E. A. Howell, eds. 2005. *The Historical Ecology Handbook*. Washington, DC: Island Press.

2. Well-know historian William Cronon maintains a website that contains valuable information on conducting historical research:

   Cronon, W. 2009. "Learning to Do Historical Research: A Primer for Environmental Historians and Others." http://www.williamcronon.net/researching/index.htm

3. Here is a good reference for understanding ecological experiments:

   Scheiner, S. M., and J. Gurevich, eds. 2001. *Design and Analysis of Ecological Experiments*. Cambridge, MA: Cambridge University Press.

4. For an interesting perspective on the value to restoration theory and practice of conducting multiple restoration experiments and/or examining the results of restorations independently conducted on numerous sites, check out the following article. This paper also gives you a glimpse of the potential of Bayesian statistics:

   Holl, K. D., E. E. Crone, and C. B. Schultz. 2003. Landscape restoration: Moving from generalities to methodologies. *BioScience* 53:491–502.

4

Site Inventory and Analysis

LEARNING OBJECTIVES

After reading this chapter, you will be able to:

- Describe each major step of the site inventory and analysis process.

- Apply site inventory and analysis to decision making within the master, site, implementation, and management plans.

- List the attributes that are inventoried during the site inventory, identify the sources for each, and discuss how each informs the restoration process and decisions.

- Differentiate between three general methods of analyzing the inventory data.

Perhaps the most important step in the restoration process is site inventory and analysis, the process by which you gain a thorough understanding of the situation at hand. The site inventory is a catalog of information about the site's features, including present and projected climate, as well as its physical (topography, hydrology, soils, etc.), biological (plants and animals), and social components (e.g., built features, human uses of the site, legal restrictions), and the surroundings. The analysis is an evaluation of the limitations and opportunities presented by these site features at each stage of the restoration process (see Chapter 1).

The general process for conducting a site inventory and analysis consists of the following steps:

1. Determine what data to inventory, map, and analyze.

2. Determine the data requirements (quality, level of precision).

3. Identify the data sources.

4. Determine the methods of data collection.

5. Determine the methods of data analysis.

6. Collect and analyze the data.

7. Plan the restoration based on the analysis results.

The site inventory and analysis begins in conjunction with the master planning stages of the restoration and is linked to the previously discussed background information (see Chapter 2). It continues throughout the site, implementation, and management planning phases, as well as during and after the restoration is implemented. As you begin to investigate the site and discover new information, you will often need to reevaluate the project goals; the new goals may in turn mean that you will need to collect additional site information. This is not unlike remodeling a residence. Housing contractors assess a situation and the remodel plans prior to bidding on a project. They develop implementation plans before beginning the actual implementation. Often, as siding, walls, and so on are being removed, they discover new constraints (such as unconventional framing) or opportunities (such as reusable materials) that require study and reassessment of the next steps or, at times, the entire project. Figure 4.1 illustrates the cyclical nature of the adaptive restoration process and the central role of site inventory and analysis.

The first step is to walk through the site and become acquainted with it, by essentially using the site immersion technique introduced in Chapter 3 in conjunction with the study of remnants. Ask questions such as: What are the general land cover, site condition, and site context? Describe the soils. Are they sandy or clayey? Are there signs of erosion? Does the soil feel hard when walking on it, suggesting it may be compacted? Are there pools of water standing after a rain, or is the soil dry or lacking organic matter? Is the topography rolling and varied, or is it level? Are there unique landforms, such as exposed cliffs, deep valleys, or unique rock formations? What plant and animal species are seen, and do any belong to the community types to be restored? Are there undesirable species? Has the natural hydrology

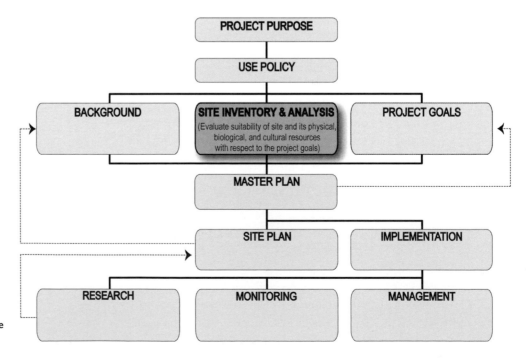

FIGURE 4.1 The Restoration Process: Site Inventory and Analysis.

Sidebar	A Site Inventory for the Merced River Corridor Restoration Plan

A proposal by Stillwater Sciences for the Merced River Corridor Restoration Plan (Vick et al. 1997) lists the following steps to be taken in the site inventory:

1. Identify factors limiting the fall-run Chinook population in the Merced River.

2. Map current channel planform (channel boundaries, floodplain and terrace features, pool and riffle habitats).

3. Update hydrologic analysis.

4. Assess and predict sediment supply and transport in the gravel-bed reach.

5. Assess and map current riparian vegetation and wetlands.

6. Assess constriction of floodplains by levees downstream of the gravel-bed reach.

7. Identify social, institutional, infrastructure, and legislative opportunities and constraints.

8. Identify local, state, and regional stakeholders.

been altered? What are the adjacent land uses and view points? Are there hazards to remedy? Is there evidence that people are using the site, and if so, for what purpose? Through observation alone, you can learn much about a site and how appropriate the proposed goals might be.

The next step is to design a more formal inventory, the nature of which will depend on the stage of the restoration process it supports. Prior to beginning each stage of the inventory, you and your restoration team should determine what it is you need to know about a site at that point in the restoration process, and design the data collection protocols accordingly. There are many examples of projects with collected, but unused, data, a practice that wastes both time and money. The project purpose and goals (see Chapter 6) help guide the initial inventory, as does ecological theory (see Chapter 3). The inventory is specific to a site and it is intertwined with the site analysis. For all of the data you collect, you should know how they are to be analyzed and for what purpose (see Sidebar).

In this chapter, we will introduce the major site resources that are usually included in a site inventory, discuss where to find existing data sources, and describe the most common techniques used to evaluate the results of the inventory.

▶ 4.1 OVERVIEW OF THE SITE INVENTORY AND ANALYSIS PROCESS

The restoration activities occur at the site level, a specific locale with legal descriptions of the property boundaries and any legal constraints that are placed upon it, such as zoning ordinances and easements. The proposed restoration site may be all or only one part of a legal property parcel. Often, the site boundaries are locked; flexibility to extend or modify site boundaries is therefore limited. A first step of the site inventory, then, is to define the site's legal properties, including property boundaries and rights (political boundaries), resource rights assigned to private parties, area land ownerships, land use regulations and incentives (zoning), and applicable regulations, permits, and easement and deed restrictions (particularly those that might limit site access and site activities).

However, the site also has a context that extends beyond its legal boundaries and refers to the world in which the restoration occurs. The nature of this context affects your ability to accomplish a restoration, just as your health, ability to succeed in a job, or raise a family depends on a wide array of factors. Climate, hydrology, vegetation, and wildlife are several forces that interact within a defined site but are not bound by it. These interactions will significantly influence the decisions you make during restoration.

Therefore, the word "site" in site inventory and analysis can be misleading; you should always look beyond the boundaries of the physical site that you plan to restore. A site inventory and analysis process explores all resources that could affect the site, regardless of their spatial locations or temporal origins. This site exploration includes not only physical and biological features, but also socioeconomic influences—past, present, and future. For example, the restoration of a wetland often depends on your ability to control runoff from uplands, mitigate water diversions by adjacent landowners, or work with legislation that restricts changes to wetlands or water levels. Climate change appears to be inevitable, but the impacts it may have on ecosystems are only beginning to be understood. Legal restrictions, zoning, and fire regulations are political aspects that have significant ramifications on restoration activities, but they are not physically on the site. The political, economic, and social contexts are quite different between developed and developing countries, and these distinctions can have major influences on the goals, objectives, and implementation of the restoration.

CASE STUDY   The Walnut Creek (Neal Smith) National Wildlife Refuge

In 1991, some 1,457 ha (3,600 ac) were purchased from Iowa Light and Power to establish the Walnut Creek National Wildlife Refuge (now the Neal Smith National Wildlife Refuge) and Prairie Learning Center in the southwestern corner of Jasper County, Iowa. The project goals were to restore these refuge lands to the natural condition that existed immediately prior to European settlement, and to provide opportunities for research and environmental education. Since this initial purchase, the refuge has grown to approximately 2,266 ha (5,600 ac) with a final target of 3,502 ha (8,654 ac).

A team of planners and ecologists used historic and contemporary ecological data to design and establish long-term restoration objectives for the site. They gained a general understanding of the historic natural landscape character by reviewing General Land Office surveyors notes from the 1860s, soil survey data, local histories, 1800s news articles, and plat map books. They then conducted a series of site inventories to learn what, if any, natural characteristics still persisted on this agricultural landscape. In doing this inventory, they discovered small, isolated prairie and savanna communities and proceeded to evaluate the quality of these remnant natural communities that existed there. The team then used this information to evaluate the site's potential as a landscape-scale grassland restoration. To determine which kinds of prairie and savanna would match the site characteristics, they further examined the soils, existing vegetation, topography, and hydrology of the landscape. They also studied reference prairie, savanna, and sedge meadow communities within a 145-km (90-mi) radius to further define a target natural character and to provide guidelines for the composition and structure of the restored communities. Each factor contributed information for evaluating the opportunities and constraints that affected decisions about how to restore the site—that is, to determine the final land cover and land use goals and objectives for the site, as well as to guide the restoration implementation and management.

Since the plans called for the area to contain a visitor center, the planners collected additional data to determine suitable locations for trails (topography, soil bulk density, view points), roads (soil bulk density, access points with external roads, access points on site, views), and a visitor center building (soils, views, topography). They also acquired information on the demographics of possible visitors. Planning and construction of these amenities happened concurrently with early ecological planning and development, thus creating both limitations and opportunities. ■

4.1.1 The Purpose of Site Inventory and Analysis

One way to understand the purpose of site inventory and analysis is to explore some of the broader questions the process is intended to answer. We consider these for three of the major steps in restoration planning: master/site planning, implementation, and management.

Master/Site Planning

The master plan and site plan are closely related; both depict the desired outcome of the restoration as it will appear on the site. The role of the site inventory is very similar for both efforts, with the major differences being that the inventory for the site planning efforts will be at a more detailed level. Hence, the questions we might ask are similar, and we discuss the two together.

What are the possible target restoration models, given the site conditions? Chapter 2 discusses community/ecosystem models and the importance of having thoroughly identified and studied reference models for composition, structure, processes, and relationships with the biological and physical environment. With this material as background, you will conduct an initial site inventory and analysis to evaluate the existing physical and biological conditions to determine what communities/ecosystems the site can reasonably sustain with minimal human input.

Does the site meet the conditions needed to accommodate the restoration objectives when the community or ecosystem target is predetermined? In some situations the restoration process begins with a specific habitat, plant community, or ecosystem in mind. One example would be attempts to reestablish habitat for an endangered wildlife species. In such cases, you will focus the site analysis to understand whether a site has the physical and biological conditions necessary to support the desired species or community type.

What site opportunities for the restoration are present (existing vegetation, wildlife, soils, access, etc.)? All landscapes have opportunities or circumstances that enhance the likelihood of achieving the restoration objectives. The site inventory and analysis helps discover what these are. Opportunities can include nearby community remnants that serve as "models" or seed sources, plants or animals existing on the site that belong to the communities to be restored, streams and other onsite water sources for implementation requirements, and existing roads that may serve as future firebreaks or provide service access.

What site limitations and constraints to the restoration exist on the site (vegetation, wildlife, soils, access, etc.)? As with opportunities, sites also have limitations (constraints) that, unless removed, will inhibit the restoration. For example, an altered hydrology might mean that a desired historic community type cannot be supported under current conditions. In some semi-arid regions of the world where agriculture is a major land use, water levels have dropped so significantly that the vegetation supported has transitioned to one adapted to a more arid environment. The invasion of foreign plant or animal species often tends to increase under disturbed conditions and adds to these alterations. Restoring this system has several challenges: (1) the politics of obtaining the water needed to restore the system; (2) the economics of such restoration, including perhaps removing salts that may have built up in the soils; and (3) the reality that even if the hydrology is restored, the site may be so altered that returning it to the original state prior to the disturbance is unlikely. In such cases, you may need to change the restoration goals.

What wildlife can be reasonably supported given the site's present and potential future conditions? We have alluded to factors that support vegetation, but creating wildlife habitat and aiding its reintroduction are also important to our restoration efforts. Wildlife needs are often divided into the categories of food, nesting, cover, and water. Take, for example, the regal fritillary butterfly (*Speyeria idalia*). The larvae depend on violets for food, while the adults feed on the nectar produced by a variety of plant species with purple flowers found in the North American tallgrass prairie and Great Plains. The butterflies lay eggs in the fall, and the larvae are thought to overwinter in duff. During moist periods, the adults are frequently found feeding in upland prairie, but during drought periods, they move into lowland wet prairie. In order to restore and sustain habitat for this endangered species, you must consider the vegetation, litter, soil, topographic, and hydrologic conditions required for supporting it throughout the year. As many native wildlife species are similarly dependent on specific habitat factors, it is essential to define which wildlife species are of interest in the project goals, and then inventory for those resources needed to sustain those species.

What functions and processes are present? Which ones are missing or altered? The purpose for restorations varies (community establishment, reintroduction of rare species, water quality improvement, reclamation of brown fields, etc.), and the processes that exist onsite that either contribute to or hinder the purpose need to be recognized. Likewise, you must identify the processes that are necessary for the restoration purpose to be met but that are not currently present on the site. Such processes and functions might include productivity, decomposition, and nutrient cycling; disturbance regimes, including drought, fire, flooding, wind, and human activities; hydrology; and species interactions and movement into and out of the site.

What past, present, and future human activities are compatible with the model and goals? Most landscapes considered for restoration have had, and will continue to support, human activities. The site inventory and analysis will help determine what ongoing activities are compatible with or damaging to the site resources and goals. Your restoration plan must address how to provide for or eliminate these activities as the restoration is implemented. Many activities, such as ecotourism and some agricultural practices, have the potential to benefit a local, regional, or state economy while sustaining the natural character of the land. Other activities have minor impacts on the landscape but substantial benefits for educational activities, such as photography and nature study.

Some uses that are controversial can be compatible and even beneficial if they are monitored and regulated, including hunting and harvesting. For example, U.S. deer populations have risen dramatically over the last 100 years from a density of fewer than 3 per km^2 to more than 39 per km^2 (8–100 mi^2) in some areas. The resulting increase in browsing on herbs and woody seedlings has significantly altered the composition and ecological processes within both forest and grassland systems. Many preserve managers allow regulated hunting to reduce deer populations to levels that were thought to occur prior to settlement, and which the natural system may be able to accommodate.

However, some activities, such as all-terrain vehicle (ATV) use and harvesting organisms from coral reefs for the aquarium trade, are generally incompatible with restoration. Incompatible but long-existing land uses and human behaviors can be difficult to change without education, public awareness of the restoration purpose, and onsite efforts to work with visitors. The site inventory and analysis will help you to predict such user conflicts and opportunities.

Implementation Planning

The site inventory and analysis also provides the information that drives restoration implementation. Implementation planning considers the activities required to physically produce the site objectives and plan, and is influenced by the current social and political context. Site inventory and analysis helps guide implementation planning by examining several questions, three of which are discussed next.

What are the site preparation needs? The results of site inventory and analysis provides information useful to many decisions made during site preparation (see Chapter 8). For example, you may want to know: Does the grade of the land need modification to create a natural contour? Are soils compacted, and is the natural hydrology intact? Have past agricultural practices left chemical residues in the soil that may be detrimental to species establishment? Are there physical structures, including roads, buildings, or dams, to be removed? Is the existing vegetation compatible with restoration goals, or will some or all of it need to be removed? What are the past and current uses of the land that may affect soils, hydrology, vegetation, and future implementation and management activities? For instance, brownfields, or sites that are severely degraded by past industrial or agricultural uses, may require specific tests for toxic chemicals in the soil and/or water, along with expensive or specialized methods for containment or removal that may alter the process of introducing targeted restoration species.

How might physical characteristics of the site, such as topography, hydrology, and soils, influence vegetation establishment? Both the propagule type and the planting technique will vary with site conditions. The appropriate site preparation technique and planting strategy will differ for areas that pond or flood from areas that occur in uplands that rarely, if ever, flood. Sandy soils, often infertile and prone to drought, will require different planting techniques, seeding rates, and planting times than fertile moist loams. Steep slopes are often difficult for machinery to negotiate, and the conditions worsen if the slopes are rocky or have soils prone to erosion and movement. In such situations, manual planting of seedlings and saplings, along with erosion fabric, may be necessary. No single planting technique or seeding rate will work at all sites; therefore, you will have to tailor your method and approach to the specific site conditions.

What existing vegetation and wildlife may be impacted by implementation schemes? Land clearing, chemical use, fire, mowing, and planting are a few of the implementation activities that may interfere with the short-term or long-term survival of some species on the site. These disturbances can also affect the public's support for restoration.

For example, the 350-acre Lakeshore Nature Preserve is located along the shores of Lake Mendota on the University of Wisconsin–Madison campus. The mission for this area is restoration through student research and involvement. The preserve, however, borders residential neighborhoods, and many residents value and use this land for birdwatching. One goal of the restoration is the removal of pest species and the reopening of the forest canopy in areas that were formerly oak savanna. However, because of the potential impact of shrub removal on bird species that use the shrubs, a group of neighbors strongly opposes the restoration goals.

Similarly, the preserve hosts healthy, large mature Norway maple (*Acer platanoides*) and catalpa (*Catalpa speciosa*) trees that were planted when the area was farmed. These species do not belong to the community types associated with the Midwest, yet there is significant disagreement among members of the public about whether older and historic individuals of these species should be replaced with species that are native to the area.

Management Planning

Management is the third category of restoration that benefits from a thorough and strategic inventory and analysis. As you evaluate the site and determine the restoration targets, you must concurrently consider management needs, such as burning, grazing, cutting, flooding, and user oversight to sustain the restoration.

How do the existing site conditions affect management activities? Decisions on how to manage are guided not only by the needs of the restored plant community but also by the site conditions, such as topography, hydrology, and adjacent land uses. For example, erosion may be a concern for specific management techniques on steep slopes. The use of prescribed management burns may be acceptable in sparsely populated areas, but may be prohibited or highly restricted where smoke could travel across roads or into dwelling structures. At this stage of the process, it is not too early to consider what management may be required and, given the site conditions uncovered during the inventory, how it will be accomplished.

How might management affect desired community elements? Management will affect vegetation and wildlife, including soil fauna, either positively or negatively. Fire, for example, can be highly effective in reducing fuels (litter) and is a major tool for constraining unwanted shrubs and saplings. A fire that is too intense, however, may result in the killing of species, even those adapted to fire, or result in soil sterilization, thus opening up new areas for colonization by pest species. An area's climate, topography, and soils will influence how and when you may want to burn in order to control the fire's intensity.

▶ 4.2 RESOURCES TO BE COLLECTED BY INVENTORY

The specific site and resource data you collect for analysis depend on each project's purpose, goals, and landscape setting. However, there are several resource categories to consider in almost any restoration project. These include physiographic information (location, soils, landforms, hydrology, etc.), biological resources (vegetation, wildlife), and socioeconomic information (human population demographics, zoning regulations, historic land uses, etc.). Once you identify the information you need, the next step is to determine where to obtain it (onsite; aerial photography; local, state, or federal/national agencies), how to collect it, and how to analyze and interpret it.

Table 4.1 shows a sample of the kinds of data you might collect. It is often helpful to rank the various types of potentially useful data in terms of how important they are to creating the master, site, implementation, monitoring, and management plans. This helps in setting data collection priorities and also with the evaluation. The matrix within the table provides one example of such assignments.

▶ 4.3 REGION AND CONTEXT

Recognizing and understanding the regional context for a possible restoration site are invaluable in deciding what, how, and where to restore. One of the things to consider, in addition to location or geographic coordinates, is the degree of isolation from other similar sites (habitat patches) or compatible landscapes (Figure 4.2). The concepts of landscape ecology discussed in Chapter 2 are helpful for understanding the benefits and disadvantages of site size and distance for supporting sustainable plant and wildlife populations, and their potential for migration between sites. It is also helpful to note the location of a site with respect to the landforms of the region, water bodies, directional winds, farmland, industry, roads, and remnants that are similar in composition to the desired restoration. For

Table 4.1 • Site Resources Collected During a Site Inventory and Analysis and Their Level of Influence on Selected Resources Pertinent to the Restoration Site, Implementation, and Management Plans

RESOURCES	Site Plans								Implementation Plans				Management Plans		
	Plant Community	Plant Species	Wildlife (Proposed)	Access Locations	Trails	Service Roads	Human Services	Aesthetics	Site Preparation	Planting Method	Timing of Planting	Trail Installation	Management Method	Timing of Management	Frequency of Application
Climate															
Temperature	■	□	□								■		■	□	■
Precipitation	■	■	□				■		■	□	□	□	■	■	■
Wind	□	□	□		□		■		□	■	□	□	■	□	□
Geography															
Size	■	□	■	□	□		■	□	□				■		
Shape	□		□	□				□		□			■		
Isolation/Location	□		■	■		■	■			□			■		■
Physiography															
Slope	■	■	□	■	■	■		■	■	■	□	■	■		■
Aspect	■	□	□		□	□		□	□	■	■	■	■		■
Elevation	■	■	□	□	□	□		□	■	□	■	■	■		
Landform	□			□	□	□		□				□			
Hydrology															
Surface Water	■	■	■	□	■	■			■	■	■	■	■	■	■
Groundwater	■	■	□						□						
Water Quality	■	□	■					□							
Water Quantity	■	■	□						■	□	■		■	■	■
Flow Rates	■	■	■										■		
Water Depth	■	□	■	■	■	■		■	■	■	■		■	■	□
Seasonal Pulses and Flows	■	□	■	■	□				■	■	■	□	■	■	□
Wave Action	■								■	■	■		□	□	
Soils															
Texture/Structure	■	■	□		□	■			■	□		□			□
Depth	■	■					□		■	□					
Fertility/pH	■	■							■				□		□
Bulk Density	■	□	□		□	■			□	□		■	□		
Erosion	□	□	□		■	■			□	■		■	□	□	□
Vegetation															
Existing	■	■	■			□		□	■	■	■		■	■	■
Age Structure	□	■	■					□	□	□			□	□	□
Invasive Species	□	□	□		□			□	■			□	■	■	■
Wildlife (Existing)		■	■					□	■	■	■		■	■	■
Humans															
Existing Land Use	□		□	□		□			■	□	□	■	□	■	□
Adjacent Land Use	□		□	■					□	□			□	□	
Legal Restrictions			■			■	■		■				□	■	
Public Context			□	■		■	■		■	□			■	■	□
Public Safety		□	■	■	■	□	■	□	■	□		■	■		□
Utilities/Structures				□			■	□	■			■	■	□	□
Cultural Resources				□	■			□	■			■	□		
Outstanding Features	□			□	■	■		■	■			■	□		
Resources/Economics	□	■		■	□	■			■	□	■	■	□	■	■
Hazards					■	■	■		■	□		■	■	■	□

■ = Highly influential

□ = Moderately influential

example, a site may be on the leeward side of a mountain and therefore sheltered from wind, or on uplands or along a ridge that has significant exposure to wind and sun.

A site's shape and size have a significant influence on the success of accomplishing restoration goals. Small restorations under the appropriate environmental conditions can contain a high diversity of plant species, yet these would be inadequate in size to accommodate the needs of many large mammals, particularly predators and ungulates. An area of 2,690 km^2 (1,000 mi^2), for example, has been suggested as the minimum range for a stable population of African elephants residing in semi-arid environments of Africa. However, for some invertebrates, 2,000 m^2 (0.5 ac) may be adequate as long as populations of the host plants are sustainable. A small site's size can sometimes be mitigated for mobile species or organisms if it is possible to cluster several sites within a region, and such that organisms can move between them. Of course, this works only if the organisms are able and willing to cross the land cover types and land uses that fill the gaps between sites.

0   250   500   1,000   1,500 Meters

2005 Prairie Remnants
Region Landscape Extent

FIGURE 4.2 **Spatial Distribution of Largely Degraded Prairie Remnants Within an Agricultural Landscape.** Mechanisms to restore the biology of these remnants in southwestern Wisconsin, and also restore their connections, depend on many variables, including the present local land use; physical features, such as slopes, aspects, and soils; and biological features, such as existing vegetation and wildlife that occur within and between remnants. (Adopted from Read, Carrie 2008 and Wisconsin Department of Natural Resources).

4.3.1 Implications for Restoration

How can knowledge of site context aid the inventory and analysis? If the landscape surrounding the restoration site contains potential colonists of either desired or undesired species, this information is valuable in terms of implementation and management. If the remnants are too small to serve ecosystem functions on their own, you might want to know whether they would likely benefit from nearby restorations.

On the other hand, if there are no remnant patches nearby, it is important to determine whether the proposed restoration will be large enough to function and maintain the desired species composition. If the restoration site is not large enough, you can then consider if there are future opportunities to extend the restoration area. Perhaps the adjacent land, even if not in a natural state, can still support the restoration and its purpose. For example, many grassland birds require open stands of grasses with erect stems for perching, but the plant species do not have to be native for all of these species. A stable matrix of grassland remnants, or restorations among pastures or open landscapes, may suffice to meet their needs.

In situations where structure and function on adjacent lands are quite different, you must consider how to deal with potential unwanted inputs. Areas adjacent to croplands, for example, are often exposed to inputs such as herbicides, sediments from erosion, and stormwater and nutrient runoff. The future context in which a restoration will occur is also important to consider. Planted in the 1930s within an agricultural landscape, the University of Wisconsin–Madison Arboretum's Greene Prairie was an outstanding restoration. In recent years, silt and nutrient-loaded storm water from ongoing development to the west and south have significantly reduced the quality of approximately 18% of this restoration. The native warm season grasses throughout a considerable area of the project have been largely replaced by an exotic species, reed canary grass (*Phalaris arundinacea*). In addition, several rare species that were found here, including the prairie white-fringed orchid (*Platanthera leucophaea*), are threatened. The constant flooding this site now endures brings into question the practicality of further restoration without substantial engineering and cooperation from surrounding municipalities. (See Chapter 14 for more information on Greene Prairie.)

Adjacent land uses can also restrict the management of restored properties. Roads, and institutional and residential land uses that require noisy equipment or the use of fire, often conflict with land management. A detailed management plan can generally resolve such issues, but without a site analysis that intentionally looks for such conflicts, these issues may arise at inopportune times.

4.3.2 Data Sources

Satellite imagery, aerial photographs, and resource maps are excellent sources of information about the landscape context of a site. These sources are available worldwide from government agencies that include the U.S. Geological Survey (USGS), the USGS Mid-Continent Geographic Science Center, and Geoscience Australia (Australian Department of Resources, Energy, and Tourism).

On their websites, these organizations provide catalogs describing the availability of, and means of access to, digital remote sensing imagery, air photos (past and present), location maps (i.e., USGS quadrangle maps or 7.5′ maps), and onsite mapping of land features, as well as historic land records from books, maps, and serials, and a wealth of information about the Earth's features and dynamics. At a bare minimum, these data sources can help you determine the location and context of a site, and find additional sites for possible restoration. Beyond this, agency reports, bulletins, and websites can help you understand past and possible future land use and land cover changes in relation to both natural and human-caused events within your site's regional location.

A nongovernmental source that is readily accessible via the Internet is Google Earth. This website is a geographic information center comprising satellite imagery and aerial photography for the entire world surface. Depending on the source, resolution can be quite high or very low, and many images are available in three-dimensional format. Google Earth can provide an excellent overview of many landscape features, including streets, structures, nearby population centers, terrain, water bodies, and land cover.

▶ 4.4 CLIMATE

Regional and local precipitation and temperature patterns are important features of most inventories. Ecoregions and vegetation types often correspond to the global gradients of temperature and precipitation of these patterns (Figure 4.3). For example, hardwood forests typically develop in temperate regions where precipitation is greater than 84 cm (33 in.) annually, whereas grassland and savanna communities are common to interiors of large landmasses often in semi-arid and arid regions. Tropical rainforests occur within 10°–25° north and south of the equator at elevations of up to 3,000 m, and have a normal annual rainfall of 150–400 cm (60–160 in.) or more per year, with a narrow range of temperature variation averaging 22°C–34°C (72°F–93°F) year-round. On the other extreme, rainfall in desert environments averages less than 25.4 cm (10 in.) per year, and temperatures exceed 38°C (100°F).

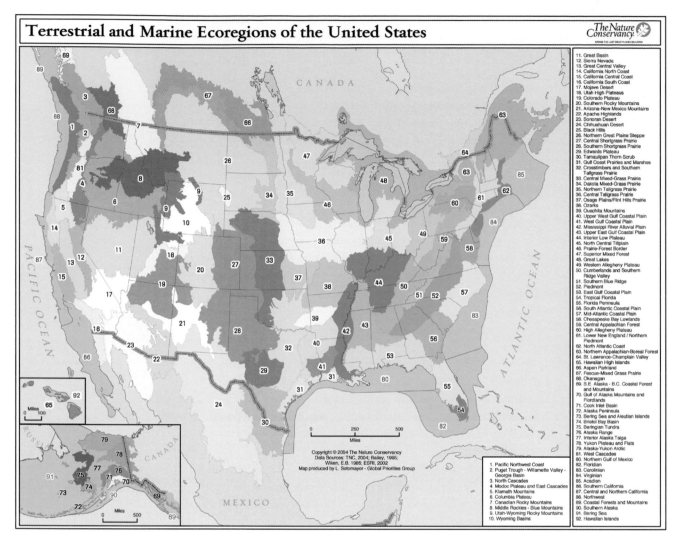

FIGURE 4.3 Terrestrial and Marine Ecoregions of the United States. The World Wide Fund for Nature defines ecoregion as a large area of land or water containing a geographically distinct assemblage of natural communities that share a large majority of their species and ecological dynamics, share similar environmental conditions, and interact ecologically in ways that are critical for their long-term persistence. Each unique color and number in the above map represents a different ecoregion. (Graphic reprinted by permission and under copyright of The Nature Conservancy 2004. Sources: Bailey, R. G. 1995; Wiken, E. B. 1986; ESRI 2002. Map produced by L. Sotomayor.)

4.4.1 Implications for Restoration

Climate has significant and multiple implications for restoration site, implementation, and management planning. They can be grouped into the following categories:

- Vegetation and animal communities supported regionally and locally.
- Methods and timing of implementation and management.
- User activities permitted and related infrastructure.

Although you will likely be familiar with the general climate of a region, climatic conditions specific to the restoration site will need further investigation. Within an ecoregion, climate can be sufficiently modified by local landforms, vegetation, and soils to allow for a diversity of community types. Topography, in particular, is highly effective at modifying onsite climates. Here are some of the climatic factors that are important to know for a site:

- *Precipitation*: Seasonal and annual rain and snow amounts, including average, maximum, minimum; frequency and timing of weather extremes, such as droughts and storm events.
- *Temperature and relative humidity*: Influenced by solar radiation (amount, duration, and timing) and air mass (body of air with horizontal layers of uniform temperature, humidity, and pressure), and locally influenced by topography (slope steepness and aspect), vegetation, and soil surfaces.
- *Wind*: Movement of air masses, speed, seasonal direction, frontal activity.
- *Frequency of extreme events*: Drought, storms, fires, and catastrophic events, which are often influenced by the other categories above.

Climate change will be a factor in the future for restoration planning. The effects of climate change will not be uniform around the world, and predictive models of climate change are still under development and revision as our understanding of this phenomenon grows.

4.4.2 Data Sources

Climate data for most regions of the world are now available online. In the United States, the websites of the U.S. National Aeronautics and Space Administration (NASA) and the U.S. National Oceanic and Atmospheric Administration (NOAA) are particularly useful sources for regional and national data. Both of these agencies also cooperate with other countries to provide data on global climate history and current weather events and trends. Similarly, the Australian Bureau of Meteorology provides climatic, hydrological, and oceanic services for the Australian and Antarctic regions.

Using these sources, you can find annual climate data and seasonal patterns, as well as year-to-year patterns of variation, such as drought cycles. The data include average precipitation, temperature, and relative humidity values, as well as historic highs and lows.

▶ 4.5 TOPOGRAPHY

Topography, or the landform of the site, has a significant influence on the local climate; it influences soil properties, hydrology, and vegetation composition and structure, as well as some disturbance events caused by extreme winds or flooding. The three topographic elements that are included in most site inventories are slope steepness, slope aspect, and elevation. These are all easily displayed using mapped contour lines (Figure 4.4).

We usually measure slope steepness in degrees or percent. You can barely perceive a slope of less than 3%, but will certainly notice a 6% slope when walking. Slopes that are greater than 12% are considered steep. Aspect is the compass direction a slope faces. In the Northern Hemisphere, as the steep north-facing slopes increase, the degree to which they are exposed to the sun declines rapidly, whereas south-facing slopes in most situations receive high exposure to solar radiation. Elevation or the vertical landscape position of a landmass is relative to the lands surrounding it.

The amount of solar radiation and moisture received by a landscape is influenced by slope steepness, aspect, and elevation. Temperature differences between slopes of different

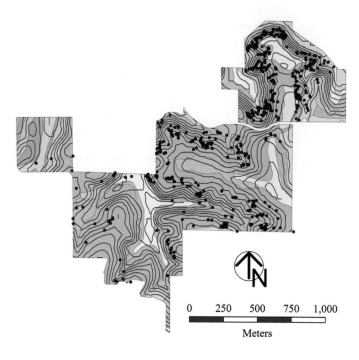

FIGURE 4.4 **Soils and Topography of the Cross Plains Ice Age National Scientific Reserve.** Note the relative location of large oaks (depicted by the black dots) to soils and topography. (From Harrington 2008.)

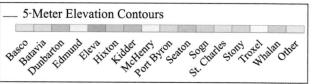

aspects can be significant, particularly as the slope angle increases and where slope exposure becomes perpendicular to the sun as the day progresses. Level sites that are exposed to the sun have their greatest sun exposure and temperature increases near noon when the sun is at its maximum angle to the Earth. Soils on slopes that receive limited to no direct sun dry more slowly than slopes exposed to the sun after precipitation. In cold climate regions, snow will often persist for much longer periods on shaded slopes, slowly supplying moisture to soils and protecting them from drying out (Figure 4.5).

Since shaded slopes remain cooler as well as moister than those exposed to sun, they tend to be better at supporting woody vegetation in dry regions. As vegetation grows and shades the soil surface, the soil remains cooler and moister for longer periods, thereby supporting greater canopies. These canopies supply organic matter to the soil, which in turn builds soil moisture-holding capacity and fertility.

Aspect also shelters a landscape from, or exposes it to, strong and sometimes harsh winds. Winds influence surface soil moisture, erosion, and foliage moisture, as well as plant breakage or stability. Some wildlife species, such as many butterflies, find that strong winds hinder their mobility, and they are often restricted to sheltered or leeward sides of hills and mountains.

For plants, the importance of aspect increases with slope steepness, as the two interact to influence how much exposure to the sun a piece of land receives. Where topographic extremes are present, such as mountains and deep canyons and gorges, the type of vegetation and wildlife supported changes rapidly with elevation, aspect, and the subsequent changes in temperature, moisture, and light/shade. Even with gentle topographic relief, elevational changes result in a heterogeneity that brings subtle changes and diversity to a landscape.

FIGURE 4.5 **General influence of aspect on vegetation and local environment.** Baraboo Hills, Wisconsin, USA. (Photograph reprinted by permission from Jeb Barzen)

South Facing Aspect	**North Facing Aspect**
Lowest moisture levels	Highest moisture levels
Grasses dominate	Trees and shrubs dominate
Direct solar gain	No direct solar gain
Highest average temperature	Lowest average temperature
Early snowmelt	Later snowmelt
Early spring greening	Later greening in spring
Light tolerant vegetation	Shade tolerant vegetation

4.5.1 Implications for Restoration

Topography has restoration implications for the vegetation supported, landscape continuity, wildlife presence and activities, site preparation, planting schemes, management techniques and timing, and user activities in the form of trails, services, infrastructure, views, and aesthetics. Usually the first step in a topographic inventory is looking at the overall landform. The general landform—whether the site is flat, for example, or consists of ridges and valleys—influences the types of communities it can support, and which implementation techniques may be required to restore them.

Slope, aspect, and elevation help determine the specifics of structure and composition, as well as the timing and methods of restoration implementation. In some cases, slopes may be sufficiently steep that you can only use equipment with an extremely low center of gravity. As slopes become steeper, even traveling on foot can be difficult, and the use of chain saws and other types of mechanical equipment become more precarious. In addition, steeper slopes do not allow the same type of site preparation as level slopes for minimizing surface damage, such as erosion. As noted previously, aspect and elevation influence moisture levels and soil temperatures, thereby influencing planting dates. Finally, slope and aspect have an impact on the location and types of trails and activities permitted on site.

Slope steepness and aspect also influence the types of management and associated tools that can be used on a site. Fire, for example, is used to help in the establishment and long-

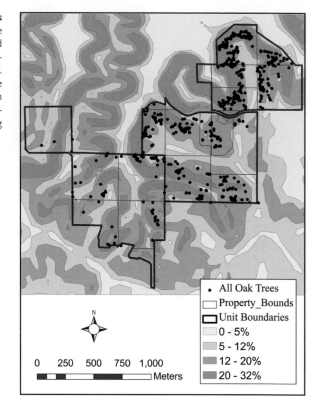

FIGURE 4.6 Slope Patterns of the Cross Plains Ice Age National Scientific Reserve. The slopes are classified by percentage ranges and overlaid with existing oaks that have a diameter at breast height (dbh) of 40 in. or greater. The majority of larger oaks exist on or near the steepest slopes. Using only this image, it can be difficult to tell which of the 0–5% classified slopes are in valleys and which are along ridgetops. (From Harrington 2008.)

term management of many types of restorations. However, in colder climates where snow exists, managers can often burn earlier on slopes that receive direct exposure to the sun than on shaded slopes. It is not unusual to observe shaded, snow-covered slopes in conjunction with exposed slopes that are dry with renewed vegetative growth. Where a mix of slopes and aspects come together, such as where several ridges meet, winds can be erratic, and fire behavior becomes less predictable.

4.5.2 Data Sources

You can find information about the topography of North America on the websites of the USGS, NASA, and the United Nations Environment Programme (UNEP) GRID-Arendal. Many private agencies provide topographic data and services for a fee, including GeoEye, which operates the IKONOS Earth observation satellite imagery at resolutions of 1–4 m. These data include aerial photographs and satellite imagery for producing topographic contour maps, such as the USGS quadrangles and digital elevation models (DEM, or three-dimensional relief maps), for use in computerized geographic information systems. You can calculate and classify slope steepness manually or with software from either topographic contours or DEM (Figure 4.6).

▶ 4.6 HYDROLOGY

Hydrology is the study of water systems. A restoration site inventory planner is particularly concerned with the characteristics of the surface and groundwater resources that influence what plant and animal communities a site can support. Additional concerns are the restoration site preparation requirements, the propagules used for planting, planting methods and timing, and management techniques and schedules.

Some of the major features to inventory are water origin (surface and ground sources); characteristics of surface water features, such as volume and depth, rate of flow, wave action, and quality (oxygen levels, pH, and mineral, nutrient, and sediment content); seasonal flow patterns; and the frequency of extreme events (flooding, ponding, and drought).

Surface waters contribute to fens, streams, lakes, rivers, marshes, bogs, and other water bodies. Storm water enters these features directly from precipitation and through surface runoff and groundwater discharge. The ratio of permeable to impermeable surfaces has significant effects on the volume, rate, and quality of stormwater runoff. When surface runoff flows across agricultural lands, it can pick up high concentrations of fertilizer, nutrients, pesticides, herbicides, and/or sediments that then influence the water quality entering the surface waters listed above.

Groundwater, or water that is located beneath the Earth's surface, is stored in the spaces between soil particles and the pores and fractures of rock and unconsolidated deposits. Groundwater is recharged or replenished by precipitation, snowmelt, and water that leaks from the basins of lakes, rivers, streams, and other water bodies. The water chemistry is influenced by both the source of the water and the bedrock or aquifer through which it flows. Flow rate depends on the composition and the degree to which the pore structure of the rock in the aquifer is connected.

Groundwater depth is somewhat constant, but it will vary with extreme periods of precipitation or drought. Depth to groundwater may increase where human populations depend on it as a major source of water. Pollutants from septic systems, agricultural applications, and industry impact groundwater quality. Depth to water table is one of several determinants for wetland soils. Where groundwater is near or above the soil surface (surface discharge), soils remain saturated, decomposition slows, organic matter builds, and peat substrates form. In situations where groundwater flows through softer rock, minerals are often picked up, and the resulting water chemistry can have a tremendous impact on the plant species composition when it discharges across a soil's surface.

4.6.1 Implications for Restoration

The existing site hydrology (water quantity, depth, and duration) will influence what plant communities can be restored. For example, sedge meadows occur in soils that are saturated year-round; wet prairies and meadows occur where soils flood early in the growing season but drain and dry by the middle of the growing season. Floodplain forests tolerate and require seasonal periods of flooding but rarely tolerate permanent flooding; swamps often remain flooded year-round.

Water quality also affects which plant and animal communities can be established. Some plants, for instance, depend on high concentrations of minerals occurring in groundwater that flows through soft rock such as limestone. Underground springs bring oxygen to the roots of many plant species. Calcareous fens are one example of a plant community that relies heavily on subsurface water flow that is high in both dissolved oxygen and calcium.

The presence, depth, and duration of surface water, as well as the availability of groundwater to plants, will influence not only the restoration goals and which species will occupy a site, but the method and timing of both site implementation and management. For example, if you are restoring a site that experiences occasional to frequent flooding, you need to choose your planting methods according to whether the land to be planted is likely to be dry, saturated, or flooded. Seasonal rains and flooding will also influence the type of management, as well as when it can be applied.

Alterations to the regional landscape and watershed in which a site occurs have significant impacts on what you can successfully restore. Site drainage for agricultural purposes is often an obstruction to restoration. For example, the Shea Prairie Preserve contains a former fen, which the property owner wishes to restore. The source of water for this fen, however, is located on private property on the hillside above. That land was previously tiled, which in effect lowers the water table, draining water away from the former fen. Any restoration efforts would have to recognize the difficulties in restoring the fen under these circumstances. In the best scenario, breaking or plugging the tile would restore the hydrology. These efforts are not always successful, because tiles that are cracked but not shattered can continue to drain water. In some instances, complete tile removal is necessary. However, any of these actions could also result in higher and undesired moisture levels on the neighbor's property, thus limiting its use for agriculture. For restoration of the fen to be pursued, the site planners would need to work with the neighbor and, if given permission, incorporate the neighboring property into the site planning process. In this case, the property in question was purchased through funds for wetland mitigation.

A goal for some river restorations is the improvement or reestablishment of spawning beds. These gravel beds in the river remain free of sediment and sand due to periodic floods. River systems altered by dams, for example, can see this scouring process disrupted. The gravel beds become covered in sediment and sand and no longer meet the spawning needs of fish.

4.6.2 Data Sources

To prepare for an onsite visit, you can explore online data sources. The websites of the U.S. Army Corps of Engineers and the USGS provide data for major lakes and rivers, including seasonal depth, stream flow rates, and water quality. Records of wetland presence, some dating to the 1920s, are also available. Graphics include maps for groundwater elevation, geology, watersheds, flood zones, and permanent and intermittent streams. UNESCO's International Hydrological Programme is an additional source of global information on hydrologic regimes. The Natural Resources Conservation Service (NRCS) Soil Data Mart, of the U.S. Department of Agriculture (USDA), provides data on wetland soils.

▶ 4.7 SOILS

Over thousands of years, the interaction of climate, geology, and hydrology has had a tremendous influence on soil development. The resulting soils greatly influence the type of vegetation that grows on a given site and that can be restored. Some soil characteristics that are particularly useful when considering restoration decisions are texture, porosity, drainage, organic matter, depth, fertility, pH, and soil organisms, including mycorrhizae, nitrogen-fixing bacteria, arthropods, nematodes, and termites. The importance of soil fauna to native plant establishment and growth is an expanding area of scientific study that should be continuously monitored.

4.7.1 Implications for Restoration

The characteristics of a site's soils will affect which plant community types to target, the site preparation (including the possible need for soil amendments), planting dates, and plant sensitivity to wind and water erosion, as well as compaction, susceptibility to invasion by opportunistic species, and constraints for infrastructure.

Soil texture, defined by the size (diameter) of mineral particles, influences soil aeration, water infiltration and movement, cation exchange capacity, nutrient availability, and sensitivity to erosion and compaction. Soils that are composed of all three particle sizes—sand (0.05–2.0 mm), silt (0.002–0.05 mm), and clay (<0.002 mm)—are typically referred to as a loam. They are described as sandy, silty, or clayey, depending on the proportions of particle sizes (Figure 4.7). Organic matter is of significant importance to building soil structure and, along with clay content, influences water and nutrient exchange and retention. The mineral particles and organic matter group together to form aggregates and create the framework of a porous soil matrix. Plant roots receive both water and oxygen from pores, and it is through these pores that water flows to underground aquifers after storms. Soils with a high percentage of fine particles or clays are easier to compact; the pore space reduces more than in soils with larger amounts of coarse sands or silt particles.

Depth to bedrock and soil depth are additional indicators of the community types that can be restored. Shallow upland soils typically do not have the capacity to hold more than limited amounts of water and nutrients, so the vegetation communities that occur on these soils are often adapted to dry, infertile environments. Exceptions can occur where water constantly seeps through the rock, creating a moist but thin soil bed where select groups of plant species can establish. Another exception is regions of high rainfall, such as the tropics and some coastal areas. Bedrock chemistry also contributes to soil pH and fertility.

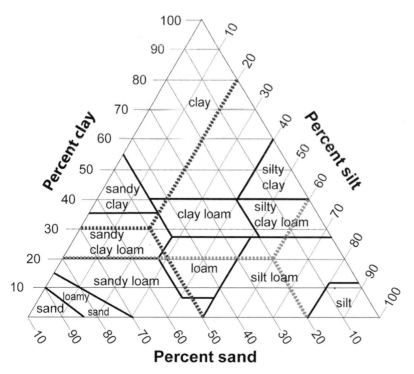

FIGURE 4.7 **The Soil Texture Triangle.** A soil with 20% sand, 20% clay, and 60% silt would be typed a silt loam. A soil with 50% sand, 30% clay, and 20% silt would be a sandy clay loam. (Adapted from U.S. Department of Agriculture soil texture triangle.)

Soil fertility is characterized by both macronutrients (nitrogen, phosphorus, potassium, calcium, magnesium, and sulfur) and micronutrients (boron, copper, iron, chloride, manganese, molybdenum, and zinc), thirteen mineral nutrients that are considered essential for plant growth. Fertility is often considered in the context of the vegetation to be supported, and nutrient availability is influenced by soil texture and pH. Cation exchange capacity, one measure of potential fertility and nutrient retention, increases with the proportion of clay and organic matter (humus) a soil possesses. Soil pH influences the solubility of nutrients in the soil and consequently impacts their availability to plants (Figure 4.8).

Agriculture has resulted in large quantities of chemicals being applied to the land in the form of herbicides, pesticides, and fertilizers. Many of these substances break down into products that are not known to impact future planting. However, some chemicals have fairly long lives and can affect seed germination or plant establishment for several years after agricultural activities cease. Organic matter and clay particles will absorb many chemicals; these same chemicals will leach through soils that are low in organic matter. Along roads in northern climates where snow and ice are issues, salt or sodium buildup can alter soil moisture and nutrient relationships with plants. Typically a bioassay of soils is requested if there is any suspicion that chemicals may have been previously applied to the land.

Soil organisms, including soil bacteria, fungi, and invertebrates, have an important influence on the germination and establishment of plants. Mycorrhizal fungi, a group of fungi that form symbiotic relationships with plants by colonizing their roots, have a significant effect on plant community structure and dynamics. These fungi have been shown to facilitate nutrient exchange, particularly phosphorous uptake, and occur in at least 70% of known plants (with the exception of sedges); however, their influence on plant establishment, disease resistance, and soil development is only beginning to be understood. In addition to the acquisition of soil nutrients, mycorrhizal fungi influence species composition,

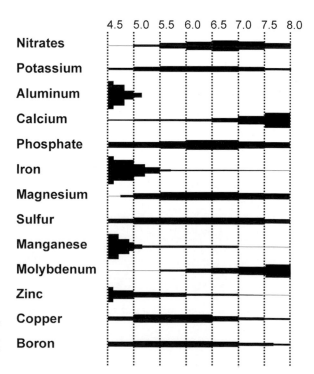

FIGURE 4.8 The Relative Availability of Soil Nutrients in Soils of pH 4.5–8.0. (Adapted from University of Kentucky Cooperative Extension Service 1961.)

diversity, and succession by enhancing water relationships and disease resistance. *Rhizobia* bacteria reside in root nodules where atmospheric nitrogen (N_2) is reduced to ammonia (NH_3) for plant uptake. The direction of these influences is often context-dependent. Plants that form symbiotic relationships with mycorrhiza or *Rhizobia*, for example, may benefit most in low-fertility soils.

Soil compaction and erosion hinder site preparation, planting, and plant establishment, while also damaging the environment and the economy of a region. The tendency of soils to compact when under heavy pressure from equipment, grazing animals, human traffic, and even raindrops increases with clay content (see Figure 4.7). Compacted soils are sensitive to erosive forces; they tend to shed instead of absorb water, have low pore space for air (oxygen), and have less insulation capacity in climates where temperatures change rapidly. Bulk density, or the weight of dry soil per unit volume, is one method for determining compaction (see Chapter 5).

Soil texture and composition are factors in soil-bearing capacity—the ability of soils to support buildings, roads, bridges, and other structures. In many instances, particularly at arboreta and nature centers with an educational mission, such infrastructure is required for human access and participation.

In addition to soil type, erosion is a factor of rainfall intensity and frequency, wind exposure, slope steepness and length, and land cover. How susceptible a site is to erosion is one factor that determines the methods of site preparation, planting, and management. Trail placement and construction are also restricted by the sensitivity of soil to erosion and compaction.

4.7.2 Data Sources

The NRCS Soil Data Mart soil surveys are available by county via electronic files for the entire United States from the Soil Survey Geographic (SSURGO) Database and U.S. General Soil Map (STATSGO2) description. SSURGO files include soil maps for U.S. lands and provide data and interpretation of physical and chemical properties, water-holding capacity, salinity, flooding, water table levels, depth to bedrock, land cover, and development. STATSGO2 files are based on general soils association units and are at a less detailed scale than the SSURGO files. Soil units are mapped as polygons formatted for use with geographic information systems software. A linked database describes each of the soils found within the polygons. These sources are extremely helpful, but their degree of precision may be inadequate for the level of site detail needed for many projects. Onsite spot sampling of soils is invaluable for determining the texture, fertility, pH, and residual chemicals (see Chapter 5). EUSoils (European Soil Database Map Library) is a second source for soil maps that cover Europe, Africa, Asia, and North and South America.

▶ 4.8 FAUNA (MAMMALS, BIRDS, REPTILES, AMPHIBIANS, INVERTEBRATES, SOIL ORGANISMS)

Restorations are often conducted to restore or enhance habitat for endangered or threatened wildlife species, and knowing the food, nesting, cover, and water requirements for species of interest is critical. Habitat is organism-specific; it relates to the species' ability to successfully interact with the site's physical and biological components, as well as those of the surrounding region. The extent of that ability depends on the species. Several spatial factors are important to recognize: minimum site size or scale needed to support the species, distance to the nearest known populations of the species, the distance the species can travel between habitats, and the potential for pronounced edge effects. To preserve a species, ecologists develop a habitat suitability model that describes the combination of these factors that are necessary for the species' survival.

When planning a restoration for wildlife habitat, the first step is to create a habitat suitability model and then determine which, if any, of the components of the model are present on the site proposed for restoration. Of course, knowing whether a species occurs onsite or in the area surrounding the site is important for determining whether it needs to be actively reintroduced, or whether restoring the site's physical and vegetative conditions might be sufficient for species colonization and population growth.

Such a study can be a lengthy process, because recording the presence of species' habitat needs at a single point in time or a single spatial scale can be misleading. In other words, a species' habitat requirements vary over both time and space. For many species, fluctuations in population size, as well as presence, vary by season, year, or climatic periods. Temporal variations in habitat brought about by drought, flooding, windfalls, and other natural disturbances are common, and wildlife populations quickly react to them.

The timing and frequency of inventories to identify wildlife populations are based on the life cycle, mobility, and food needs of a species. For example, many insects have relatively short life cycles, consisting of several stages, in their growth from egg to larva to adult. Feeding requirements can vary at each stage, placing individuals in different locations in a habitat at different times. The adult period for some insects may last days, weeks, months, or years. Various neotropical migrant bird species have specific times during which they are present in a geographic region, and summer and winter habitats, as well as their behaviors in each, can be quite different. Large mammals may move in and out of a region following prey or food sources. Recognizing yearly population fluctuations can help in estimating how often, or over what time periods, preplanning inventories must be carried out.

Many species of insects, birds, mammals, and fish move in and out of regions. Seldom will a single restoration site be of significant size to provide all the requirements for wideranging mammals, such as grazers, in all seasons. Populations of some species may occupy several isolated patches of habitat within a region; the total is often called a metapopulation. The intervening habitat, if compatible, may be used by individual animals to travel from one site to another, providing a limited exchange of individuals between population units.

Thus, inventories conducted to determine a site's ability to be restored to a habitat to support selected fauna include, but are not limited to, the presence of the species onsite or within the region; the size and scale of the site to be restored; the scale, context, and connectivity of surrounding lands or aquatic systems; temporal scales and variations in habitat; species behaviors; vegetation cover (structure to composition); and the presence of other animal species that might interact with the species of interest.

4.8.1 Implications for Restoration

Whether or not the fauna that are indigenous to the ecosystem or landscape proposed for restoration are present in the region of the site has major implications to the restoration planning, implementation, and management. Here are several initial questions: What fauna would typically be present, and what roles might they play in maintaining or enhancing the community or ecosystem? Are the pollinators needed to maintain plant species populations available or nearby? What is the relationship between grazing or browsing and plant composition and structure? Do the fauna support soil disturbance, decomposition processes, or seed dispersal? Are species of interest isolated to one population, or do several populations of a species exist, and do these populations have the ability to interact?

Not all fauna may be desired—during the restoration or in the restored habitat. Wildlife can sometimes interfere with the restoration process. For example, Canada geese (*Branta canadensis*) are problematic for salt marsh restorations along the eastern U.S. coast, where they will pull out newly planted plugs of cordgrass for food. Whitetail deer (*Odocoileus virginianus*) browse young woody seedlings as well as herbaceous plants; in locations where deer densities are high, this can have a significant impact on plant establishment, vegetation composition, and restoration success. Squirrels and other rodents are very efficient at removing acorns that may have been planted only hours earlier during an oak woodland restoration, but they are also effective at distributing these seeds. Globally, feral pigs (*Sus scrofa*) disrupt soil while foraging, resulting in the decline of native species and an increase in non-native vegetation. Birds distribute seeds throughout landscapes, and in some cases, these seeds belong to invasive or pest species. Asian carp are rapidly altering the trophic structure of native aquatic species in the Mississippi River Basin, and as they move upstream, pose future threats to the Great Lakes aquatic ecosystem. Such problems need to be recognized and addressed prior to initiating restoration.

Some faunal groups have aesthetic or human values that require careful consideration in order to broaden support for the project. Enhancing habitat for some species is rewarding to many people but problematic for others. Conflicts between human interests and wildlife are common in habitat restoration efforts for collective groups such as fisheries, as well as for individual species, including wolves (livestock and game predation), bears (nuisance, property damage, human safety), numerous felines (human safety and predation), and sandhill cranes (crop damage) (Figure 4.9).

FIGURE 4.9 Sandhill Cranes (*Grus canadensis*) Feeding in Cornfields Along the Platte River, Nebraska. The Platte River hosts more than 500,000 sandhill cranes along with hundred of thousands of waterfowl, during spring migration. The cranes' diet largely depends on corn residue in fields and macroinvertebrates from native grasslands. During the 1970s, technology allowed farmers to become more efficient at harvesting corn, and its availability to cranes diminished. At the same time, native grasslands were being converted to cornfields. In the past decade or more, efforts by area farmers and environmental organizations have expanded the availability of native grasslands and cornfields to cranes. The Platte River water levels declined significantly starting in the 1940s, changing the river dynamics and resulting in more exposed sandbars, trees, and shrubs, thus effectively reducing the habitat for cranes. Power lines crossing the river create lethal obstacles to cranes as they enter and leave the riverbed. Efforts to restore the waters and the openness of the Platte River in areas of migration continue. (Photo reprinted by permission from John Harrington.)

4.8.2 Data Sources

Specific onsite inventory methods have been developed for most animal groups, and these are documented in various texts and journal articles (see Chapter 5). In addition to onsite inventories, sources of information about wildlife can be found in the technical bulletins of government conservation agencies, in reference books, and in nonprofit land trust organization publications. The website for the U.S. Fish and Wildlife Service is one source for wildlife information. The World Wide Fund for Nature (WWF) and The Nature Conservancy (TNC) websites provide broad information on wildlife and habitat needs, and showcase the work of nonprofit agencies.

▶ 4.9 VEGETATION

A site inventory seeks to determine the existing vegetation composition (species presence) and structure (species arrangement and abundance) on the site. The intent of the vegetation inventory is to provide a framework for restoration planners to identify the plant and animal communities or ecosystems the site could support, the species for initial and future planting, any existing species that need protection, site preparation and implementation methods, and management needs and methods.

Here are some of the many items to include in a vegetation inventory: landscape cover pattern; community types found onsite; composition of communities onsite and on lands adjacent to the site; vegetation structure, including age and size distribution; life-form layers (trees, shrubs, ground covers); stem densities among age classes; specimen trees or plants with unique characteristics, such as size, age, and form; the biology for species of interest, such as those that are threatened or endangered; seed bank composition; colonization sources from surrounding sites; onsite and area pollinators; and mechanisms for seed dispersal.

We will discuss techniques for collecting information about vegetation in Chapters 5 and 9. The presence of invasive (pest) species, including estimates of their abundance and potential for damaging effects, is also very important part of the inventory (see Chapter 11).

4.9.1 Implications for Restoration

One of the first questions to consider is whether any of the desired species, determined from your ecosystem and reference models, currently occupy the site and, if so, in what amounts. You then use this information to decide which species require planting. Besides identifying which species are dominant on the site, it is important to collect data to see whether any endangered or threatened species exist and at what population levels.

Knowing what species are present in the soil seed bank will assist decision making regarding site preparation, planting mixes, and future management. In many cases, the species currently growing above ground are not represented by the seeds found in the soil. Instead, species represented in the seed bank are often opportunistic and weedy.

The occurrence of non-native plants and native opportunistic species are often associated with human activities that have disturbed the site in a manner that would not have occurred naturally—for instance, water use, logging, intense grazing, and cropping (see Chapter 12). Each activity impacts both the physical and the biological attributes of the system, and their interrelationships. The site inventory and analysis can establish what, if any, undesired species are onsite; if their population is likely to increase, remain stable, or decrease under the restoration activities being considered; and how their presence may interfere with the restoration target goals.

4.9.2 Data Sources

Onsite inventories are necessary for understanding the vegetation that occurs on a site (see Chapters 5 and 9). But you can determine general land cover (wooded, open, pastured, crops, wetlands, etc.) with air photographs and satellite imagery. Regional vegetation maps are available for many regions of the world. The USGS Land Cover Institute's website is dedicated to providing land cover information in the form of digital maps and data for much of the globe, and has links to the European Environmental Agency, the Global Land Cover Facility, and the International Geosphere-Biosphere Programme. Natureserve.org is another website that contains a database for plant and animal species, as well as ecosystems, occurring in the U.S. and Canada; it also provides assessments of organisms for Latin America and other countries. Natureserve is a nonprofit organization whose mission is to provide the scientific basis for effective conservation action.

▶ 4.10 INVENTORY AND ANALYSIS OF CULTURAL FEATURES, SOCIAL USES, AND PERCEPTIONS

Although the site inventory activities often focus on the physical and biological components of the project site, it is the socioeconomic component that may have the greatest impact on whether the restoration is implemented and maintained. Citizen and neighborhood support and government cooperation are, in many cases, essential to moving beyond the planning stages. Gaining that support requires knowing with whom to communicate and what needs to be communicated, what issues are of interest or controversial among interested parties, and what government oversight or involvement may be required.

4.10.1 Implications for Restoration

Former site uses can impact the planning, implementation, and management of a restoration. Formerly grazed landscapes often have compacted soils and a suite of existing vegetation that is inappropriate to the restoration and sometimes problematic, as well as an inappropriate seed bank that can be stimulated into germinating during site preparation processes. Intense grazing often results in altered nutrient cycles and soil microbial activities that may be important to a community's restoration and stability. Overgrazing, overcutting of woody fuels, intensive agriculture, and overirrigation have been linked to desertification of semi-arid and arid lands, accompanied by increased salinity, alkalinity, and changes in plant composition often toward non-native species. The past use of herbicides, particularly those that have a long half-life or that break down over a period of years, inhibits the establishment of many species.

In cases where lands were previously used for recreational activities that are not compatible with the future site goals, significant outreach may be needed to increase acceptance among past users. If an area of land has historically been used for such purposes as hunting, dog walking, ATV enjoyment, or birdwatching, restoration planners must carefully assess whether these uses will or will not impact the restoration and project goals. For aquatic systems used as fisheries, for example, you and your team will have to balance the economic impact of any proposed changes with the ecological goals of the restoration.

Future road expansion, subdivision development, crop management activities, tree plantings, and new infrastructure that could influence the restoration planning and results must be noted. The scale of such activities transcends individual properties, because land use policies, such as zoning, are often decided at higher government levels. A region's demographics can indicate the potential for local support, as well as pressures on the land once the restoration is established.

Existing infrastructure, such as utilities or abandoned buildings, influences design, implementation, and management decisions. Restoration activities must avoid conflicts with public service utilities by either altering plans or resolving conflicts with the appropriate agencies. Aboveground utilities, such as transmission lines, may prohibit plantings beneath them or restrict plant height. Often utility companies will cut, mow, or apply herbicides to keep vegetation within an acceptable height. This practice can result in disturbances that encourage invasive species and the degradation of the desired community. Aboveground lines and fences also provide perching opportunities for birds that could bring in undesired plant species by passing seed through their digestive systems. The location of underground utilities, such as pipes and power lines, has to be determined to avoid disrupting services to the public during restoration implementation. Other types of infrastructure may be abandoned or, if no longer needed, removed. Earthen dams for retaining water, and ditches and tile systems for draining lands for agriculture, are examples. For wetland communities, these drainage systems may require dismantling in order to restore the natural hydrology of a landscape.

Planning for potential access to the site and for onsite circulation requires clear objectives about the site's use, and a complete inventory and analysis of site features and characteristics. Soils, slope percentages, and aspects are important to trail and road placement, but so are attractants such as views, geologic features, water bodies, plants, and wildlife. In some instances, sites have sensitive areas, including erosive soils or endangered species habitat, that may be inappropriate for trails or human use.

Trail design also has to consider low-lying areas that flood, and shaded slopes where snow and ice remain concerns throughout much of the year. Safety, as well as preventing the development of ad hoc trails, must be addressed, and this is easiest to resolve if a site analysis has identified those areas where such issues are most likely to exist. For example, in regions where snow and ice are present, trail placement on steep slopes that do not receive direct sun is problematic. In the spring, trails on sunlit aspects are dry and open, but the same trails can be treacherous in areas where the aspect remains shaded and ice persists (see Figure 4.5).

Access and circulation, of course, influence more than visitation. The ability to implement and manage or service a restoration will also require trails. Multiuse trails can be built to accommodate larger equipment for management, or to serve as dividers between management units. For sites surrounded by privately owned lands, easements or the lack of easements may restrict or limit access and management activities.

Cultural features such as those left by historic indigenous societies are found in many parts of the world. Effigy mounds, burial mounds, and building foundations are just a few of the artifacts that occur throughout the landscape. Archaeological surveys may be required when a landscape is believed to contain possible relicts from historic cultures.

Finally, current and potential adjacent site uses that would impact the planning and success of a restoration should be identified. Neighboring pine plantations or woodlots that cast shade into the restored setting may limit the use of shade-intolerant species or communities. On the other hand, nearby vegetation may blend well with the restoration goals. Some agricultural practices next to a restored landscape are problematic. In addition to herbicide runoff and soil erosion, which can occur with tilling, agricultural fields serve as weed sources. However, farmers may also feel that plants and wildlife found in the restoration are causing problems for them.

The social context of a restoration site has numerous implications to the site plan and its implementation and management. These include the need for infrastructure, such as roads, trails, utilities, visitor centers, restrooms, and informational signs; government program support and permitting; public support for the restoration, as well as the implementation and management strategies; nearby resources, including a volunteer base; and solutions for mitigating the effects of adjacent land conditions and uses, along with the ability to limit any harm to neighboring properties caused by the restoration.

4.10.2 Data Sources

Data for cultural features, social uses, and human perceptions are extensive and obtainable from many sources. Air photos can provide imagery showing the location of structures, housing density, and roads, as well as changes in canopy, land uses, landforms such as gullies and stream channels, and ditching, to name a few. Land ownership and property boundaries can be derived from land record offices, where they exist. Information about land use regulations, zoning, deed restrictions, and the need for construction permits can be secured from appropriate government authorities.

Landowner interviews, diaries, journals, and newspapers can provide substantial amounts of information on past land uses, particularly when the interviewee has lived on the land for several decades or more. Where land has had frequent public use, such as for recreation, onsite interviews of the users may help you judge what level of support you can anticipate for the site's restoration, and obstacles that might occur if support cannot be garnered.

▶ 4.11 REGULATIONS AND PERMITS

As with many environmental activities, restorations and related activities must meet the regulatory rules of federal, state, and local agencies. During the site inventory process, it is not too early to begin exploring what regulations must be followed and permits acquired for the restoration to proceed. Figure 4.10 provides a small sampling of the kinds of permitting agencies and regulatory acts you may have to work with, depending on where the project is located. Nearly all countries have at least one government agency responsible for issuing permits for projects that have the potential to impact the environment, regardless of whether the impacts may be positive or negative. And most countries have several agencies or divisions within agencies that oversee different aspects of the environment, such as water, soils, animals, and plants. Permits required for restoration activities will vary widely with the type of restoration and the region of the world in which you are working.

The U.S. National Environmental Policy Act (NEPA) requires that all federal agencies consider the environmental impacts of their actions and projects. If the project has the potential to cause an environmental impact, the agency must conduct an environmental analysis. If an impact is shown to be likely, then a full environmental impact statement must be written. Federal actions are defined as "projects, activities, or programs funded in

United States Departments

U.S. Environmental Protection Agency

U.S. Natural Resources Conservation Service

U.S. Army Corp of Engineers

State Departments of Natural Resources/Environmental Conservation

    Wastewater management

    Air pollution

    Water quality

State Conservation Commissions

Departments of Fish and Wildlife

    Protection of endangered species (takings permits)

    Alteration of streams

Department of Forestry

    Chipping

    Harvesting

Department of Health Services

Agricultural Services

    Includes pesticide applications, nursery stock

Divisions of Water Quality, Fire Safety

    Alterations of streams, bank stabilization, filling, excavation

    Open burning

State Forestry and/or Fire Districts

State Departments of Transportation

United States Legislation

The Clean Water Act

    Addresses filling, inundation, or drainage of wetlands, dredging and discharge of dredged material, land clearing in all waters, shoreline stabilization of banks below ordinary high-water level

The Clean Air Act

The Endangered Species Act

The National Environmental Protection Act

The Coastal Zone Management Act

Australia

Australian Department of Environmental Climate Change and Water NSW Office of Water

Europe

The European Environmental Agency (information and policy advice)

Africa

South African Water Act (water reserved to maintain ecological functioning of rivers)

United Arab Emirates

EAD (Environmental Agency, Abu Dhabi)

New Zealand

Department of Conservation

Ministry for the Environment

Indonesia

The Directorate General of Forest Protection and Nature Conservation, Ministry of Forestry

China

The Ministry of Environmental Protection of the People's Republic of China (charged with the task of protecting China's air, water, and land from pollution and contamination)

FIGURE 4.10 **Permitting Agencies.**

whole or part under the direct or indirect jurisdiction of a federal agency, including those projects carried out with federal financial assistance; those which require a federal permit, license or approval, and those subject to state or local regulation administered pursuant to a delegation or approval by a federal agency" (*National Park Service DO-12 Handbook and Director's Order* 1982). The passage of NEPA has led to numerous legislative acts that require projects occurring on federal lands or using federal funds to obtain permits. Many states have their own related acts, such as the Wisconsin Environmental Policy Act, that require similar oversight.

The United States also established the Endangered Species Act (ESA) in 1973 to provide for the conservation of endangered and threatened species of fish, wildlife, and plants. The ESA is important to restoration because it provides for not only the conservation but also the recovery of populations that are close to extinction. This is accomplished largely through the protection and restoration of habitats. The U.S. Fish and Wildlife Service and National Marine Fisheries Service are mandated by the U.S. Congress to designate critical habitat and develop recovery plans for all listed threatened and endangered species. In some cases, this will require the development of restoration plans, but in other situations the ESA may restrict restoration activities, such as the use of fire, if it can be shown that the activity has even a small chance of killing individuals of a threatened or endangered species.

The Clean Water Act (CWA) and U.S. Army Corps of Engineers Section 404 permit have strict requirements and permitting for work in wetlands and aquatic systems on federal lands or those that use federal funds. The Clean Water Act led to additional programs, including the U.S. Estuaries and Clean Water Act of 2000, which seeks restoration of estuary habitats through a comprehensive planning approach to prioritizing projects, via the coordination of federal and nonfederal restoration activities, and by providing federal assistance. This Act builds upon the National Estuary Program, which was established in 1987 and has now been extended through 2016. States and the federal government also coadminister incentive programs, such as the Wetlands Reserve and Conservation Reserve programs. These programs provide incentives for landowners to take land out of crop production. In some instances these lands are planted to low-diversity native mixes that can lead to future restoration.

Along with encouraging restoration, the CWA provides legal protection for wetlands that can influence restoration activities, such as dredging and the relocation of dredge materials. The restoration of water bodies, such as streams, may require multiple permits from different government agencies. For example, projects to recontour streambanks or reestablish meanders that rely solely on the establishment of vegetation for stabilization may conflict with erosion and sediment control practices. Under these circumstances, the restoration team may need to work with regulators on methods and permits (Kabbes 1996).

The enforcement and monitoring of regulatory acts are performed by several government agencies, including the U.S. Department of Defense (DOD), the U.S. Army Corps of Engineers, the EPA, and the U.S. Departments of the Interior, Commerce, and Agriculture, as well as the National Oceanic and Atmospheric Administration and the Natural Resources Conservation Service. In addition, many state and local governments have their own set of regulations that affect wetland restoration, particularly when altering hydrology by filling ditches or redirecting the flow of water.

Another highly regulated activity is the use of fire. Federal, state, district or county, and local laws govern the use of fire, the majority of which are concerned with safety and liability. Smoke is regulated for the same reasons, as well as for its possible impact on air quality.

Throughout the U.S., as well as the rest of the world, regulatory rules are susceptible to frequent change. You and your restoration team will be responsible for knowing the current regulations and rules. Most countries throughout the world have some regulations that will need exploration.

During the site inventory process, you will develop a list of the rules and permitting requirements that can influence how you and your team go about planning and implementing the restoration. When working on federal or federally funded projects, you will often be required to conduct an Environmental Impact Assessment (see Sidebar; Figure 4.11).

Sidebar | Conducting an Environmental Impact Assessment

The restoration team should be cognizant of the potential environmental impacts of its projects. For many restoration projects that involve public lands, wetlands, and navigable waterways, an Environmental Impact Assessment (EIA) will be required. The EIA will reveal the laws and policies the team must consider.

Aside from identifying legal and policy requirements that the proposed project will trigger, there are other reasons for, and benefits that derive from, conducting an EIA. Conducting an EIA will help you to:

- Identify valuable environmental and cultural resources in the project area and adjoining areas that may be affected by the proposed project.

- Evaluate the restoration team's capacity to handle the project.

- Determine which resources will be affected by the project.

- Give the team an opportunity to consider alternative solutions.

- Ensure communication between stakeholders and the restoration team (stakeholders include local officials, interested citizens, and civic leaders) about potential impacts of the project on the community.

A convenient and reliable way to determine the need for an EIA is to assemble a simple checklist or environmental screening form. The checklist covers a range of environmental features upon which an impact is possible. The team completing the form answers yes, no, or "more information needed" to each potential impact category. The U.S. National Park Service has a formalized environmental screening form that asks managers to consider possible environmental impacts on a range of natural and cultural resources (see Figure 4.11).

Adapted from National Park Service DO (Director's Order) 12, January 28, 2002.

- Geological resources
- Air quality
- Water quality or quantity
- Stream flow characteristics
- Marine or estuarine resources
- Floodplains or wetlands
- Unique ecosystems
- Unique or important fish habitat
- Rare or unusual vegetation
- Wildlife species of special concern
- Visitor experience and aesthetic resources
- Cultural resources, including cultural landscapes and ethnographic resources
- Soundscapes
- Energy resources
- Human resource capacity for long-term management of resources

FIGURE 4.11 Cultural and Natural Resources That Can Require Environmental Impact Assessments.

▶ 4.12 AESTHETICS

Aesthetics influence many restoration decisions, although they are seldom recognized as doing so. Site location, species selection, trail design, implementation, and management methods—all of these are influenced by the desire to enhance aesthetics or to minimize the negative aesthetic impacts of activities. In the eyes of the public, aesthetics are significant attractors or detractors for accomplishing restoration goals.

Aesthetics are used to promote and protect resources, as well as enhance the public's interest in a conservation ethic. The glossy covers of newsletters and journals from numerous conservation and environmental agencies suggest that people are attracted to natural landscapes of great beauty. This is not to say that aesthetics equate to high-quality natural settings or restorations, but they can be used to entice people to learn about these areas and their characteristics. When conducting a site analysis, you can look for those aspects that enhance the natural setting.

When you are considering implementation and management possibilities, the aesthetic impacts from, as well as on, adjacent land uses can be important. Residential and institutional properties can be affected by smoke, noise, and smells that might occur during these activities.

Implications for restoration include:

- Site layout and location
- Trail placement
- Human interest and enjoyment
- Timing and establishment methods
- Species composition and proportions
- Site preparation techniques
- Management techniques and timing

▶ 4.13 THE ANALYSIS

As we discussed at the beginning of this chapter, you collect information about a restoration site and its context for several purposes. The first purpose is to determine the restoration goals and objectives that are most appropriate for the site. The second purpose is to use the information to assess how close the existing site conditions are to meeting the restoration goals and objectives, as well as what components and processes, the presence or absence of which, keep the site from meeting them. The third purpose is to determine whether, given the site conditions, the goals, and the available resources, the project can be done with a reasonable expectation of success. Addressing the questions raised earlier in the chapter will assist you in evaluating these purposes. The inventory by itself, however, will not provide the complete responses or answers. You need to analyze the data.

Several analysis approaches exist. The two that are most commonly used are (1) a suitability analysis and (2) an opportunity and constraints analysis. We will discuss each of these in turn. In both cases, the analysis begins by creating a base map that depicts property boundaries, topographic contours (landforms) and waterways, cultural features (e.g., buildings and roads), vegetation land cover, and regional context. Noting property and parcel ownerships is helpful, particularly when public participation in the planning process is desired. You will use this base map to get an initial impression of the restoration potential of a site and as the foundation for either type of formal analysis.

As discussed earlier, you can easily obtain much of this information from local and government sources, as well as air photos. Aerial and ortho-rectified photos are extremely useful in the initial evaluation of a landscape and its context. Since aerial photographs record a broad area of land that often has diverse topography, some distortion is inevitable, particularly as the distance and angle for points within the image increase from the photo point. Ortho-rectified aerial photographs are images that have been corrected for distortions caused by topographic relief, radial displacement with distance from the camera lens, and the tilt of the platform or airplane. The removal of these distortions allows the ortho-photo to be used as a map, by enhancing the accuracy of position, area, distance, and angle measurements.

In some instances, aerial photography of a site can be found for seven or more prior decades. Comparing land cover change for a particular land parcel over time can be quite useful. A time series of aerial photographs can be overlaid using computer software that allows for transparency so that several decades can be viewed. However, if these images are not ortho-rectified, the distortions will not allow the overlays to accurately line up.

4.13.1 Suitability Analysis

In a suitability analysis, you inventory and evaluate each resource completely and separately using a systematic, predetermined methodology. Suitability can be described as how well a site variable (i.e., soil pH, texture, or fertility), or a compilation of site variables, supports the various site alternatives. You look at whether the site's resources have few or many limitations for the proposed objectives.

Because land cover characteristics are located spatially, mapping methodologies work well for evaluating the collected data. Several steps are necessary in this process, including identifying the resources and their attributes to evaluate for each project objective, mapping the attributes, and identifying the site locations where these attributes are suitable to moderately suitable for your specific restoration targets. Prior to beginning this process, however, you will want to establish what information will be mapped and displayed in tables, photographs, and other formats, and how to use it in the analysis; determine the maps to be produced with what coordinate system and at what scale; consider all possible users of the data and develop a format or software solution that is easily accessed; and determine how the final products are to be displayed and used.

You start by identifying the resources of interest (i.e., soils, topography, vegetation) and the associated attributes (i.e., soil texture and pH; slope and aspect; plant species richness, individual plant species abundance) you wish to inventory. You then decide the levels of measurement precision at which to record and analyze each attribute. For example, you can measure slope aspect at levels of North, East, South, and West or at a somewhat finer detail of North, Northeast, East, Southeast, South, Southwest, West, and Northwest, or as a continuous variable using degrees (0°–360°). Plant density levels could be an exact count or they could be categorical, such as rare (<1/unit), somewhat common (2–10/unit), common (11–50/unit), and very common (>50/unit).

Simply because the ability exists to measure a resource at a highly detailed level does not necessarily mean there is an advantage in doing so. The decision on what to measure and at what level of precision must be carefully thought through; there are numerous

CASE STUDY   Using Air Photos to Guide a Restoration Plan

The Shea Prairie Grassland Preserve is one of several reserve units in southwest Wisconsin that will eventually be part of a much larger grassland complex. Former agricultural land that was grazed on its upland slopes, the site still retains an open character, with small patches of warm season grasses and native forbs that are mixed in with pasture grasses.

The right photograph in Figure 4.12 shows the lowland basin with an existing stream that no longer flows in its natural bed; its meanders were blocked in the past to create a straighter-flowing stream channel, thereby providing more land for growing crops. The stream has deeply cut steep banks that are elevated high above the original streambed, which suggests the need for sediment removal. Moreover, the image shows that the streambed is a corridor for reed canary grass (*Phalaris arundinacea*) (lighter tan areas) and that this introduced species is spreading onto former floodplain and agricultural land. At the lower right of the photo along a ridge are two small prairie remnants, although we cannot determine from the aerial photo any exact species they may contain. The image also shows a small highway along the west border and a road leading to a farmhouse, but the area is obviously rural and the housing density is low. The tree canopy along some of the slopes separates possible prairie remnants and restoration sites.

Based on the aerial photograph and a knowledge of this particular ecosystem, your restoration plan might involve eliminating the cool season pasture grasses and trees, removing the deposited sediments, modifying the stream channel, and controlling the reed canary grass—all which would occur prior to implementing any plantings. Air photos, in addition to satellite imagery of land cover, topography, and hydrology, provide considerable information about a site and its context, but these will not compensate for visiting the site for data collection.

The nonprofit conservation group that owns this property has secured funding and resources from several federal and state government agencies to restore this site. Plans have been made to remove several feet of sediment, which will involve removing the reed canary grass sod, removing drainage tile to help restore soil moisture levels, reconstructing several of the former stream meanders, and seeding with an appropriate native mix. ∎

FIGURE 4.12 **The Shea Prairie Restoration Project Area, Blue Mounds, Wisconsin, 1937 and 2010.** Left: The stream was straightened sometime after 1937 (black bars). The area frequently floods, and plans for taking the land out of crop production and restoring the former stream meanders are under way. Right: The red lines indicate the original meander courses to be restored. The area will eventually be restored to prairie, fen, and stream communities. Two prairie remnants occur on an upper slope in the lower right of both photographs. (Photos reprinted with permission from the Wisconsin Department of Natural Resources. Annotations courtesy of Robert Hansis.)

resources for which you will want to collect data and a large number of attributes per resource that could be evaluated. For every resource and its attribute, state why the data are needed and how they will inform the decisions that must be made in planning the restoration. Note when the data were collected, how they were collected, and why they were collected at that time and using that method. This information can be helpful in analyzing the data and also useful to those who did not take part in the inventory.

The next step is to determine the suitability of a resource for specific objectives (plant communities, trails, etc.). Begin by developing suitability maps for each resource attribute of interest. This step requires the assignment of suitability rankings to the levels of the attribute. These can be categorical, such as poorly suited, fairly suited, suited, or highly suited; or they can be numerical (ordinal), with 1 representing unsuitable and 10 highly suitable.

Suppose, for instance, you wish to find the areas within the project site having a high suitability for a community that is common to dry hillsides. You decide to begin with the criterion of slope and note that steep slopes are more likely to be drier than gentle slopes. You rank attribute levels for slope as >12% high suitability, 6–12% moderate suitability, and <6% poor suitability. You can then map those locations where this dry hillside community is highly and moderately suited, and eliminate those areas with poor suitability.

This process would be relatively simple if you only needed to rank and map one resource to determine where the highest suitability for a community type, trail, or road is on a given site. Unfortunately, there are many resources that can influence the placement of each of the above. You will want map layers that show suitability, at minimum, for topography, soils, existing vegetation, and adjacent land uses. As the number of resources requiring suitability rankings increases, the complexity of the analysis also increases. You may wish to locate several community and trail types on the site, which further increases the complexity.

Numerous map layers for each resource attribute can result for each community type. These can be overlaid manually using transparent sheets. As the layers build up, however, the ability to decipher and make sense of what is in front of you declines rapidly. Suitability mapping is more easily done with computers using software developed to analyze the information provided. Computerized geographic information system (GIS) software programs are widely used to determine or map the suitable locations for a community type (see Sidebar; Figure 4.13).

Sidebar | Geographic Information System Software

Ian McHarg introduced an overlay process for site planning that allowed data collected to be spatially mapped and weighted. This process, discussed in *Design with Nature*, was a basis for the development of geographic information system (GIS) software. GIS software enables computers to map spatial data (vegetation, topography, lakes and streams, soils, infrastructure, property lines, etc.) linked to a set of coordinates, weight the data in a variety of ways, look for compatibilities and incompatibilities, and assist with queries about the best fit of the land for a predetermined purpose. Users can query the GIS for vulnerability of sites to specific disturbances or events, the tolerance of the landscape to environmental and human activities, the adaptability of elements to the environment, and specific opportunities or constraints to restoration goals.

Figure 4.13 shows an example. Although mastering GIS software takes time, it is a tool that greatly enhances productivity in the restoration planning process.

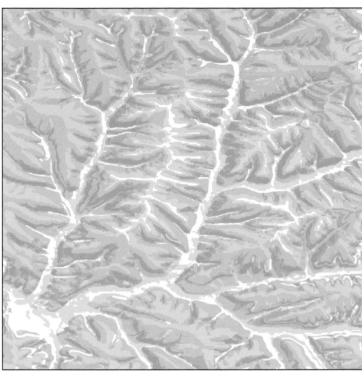

Combined Suitability

FIGURE 4.13 **A Suitability Map for Dry Prairie.** The darkest green areas have the highest suitability for dry prairie, the lightest have the poorest suitability. Suitability was based on a composite of soil attributes, sun exposure (aspect), and slope steepness greater than 12%. The location is southwest Wisconsin.

Soil Suitability

Aspect Southeast to West

Slopes Greater than 12%

A variety of methods can be used to compile the suitability factors once they are individually mapped—typically as polygons for those with area (vegetation cover, soil types, slope aspect), as lines for those that are measured by length or linear measures (fences, rivers), and by points for individual objects (trees, buildings). One analysis method ranks the suitability of several landscape features or criteria (soil texture, soil fertility, slope steepness, slope aspect, etc.) for a given purpose or land use. Suitability rankings are assigned to each polygon as numbers for each of the criteria, and these are then totaled.

Let's say we have a specific area of a site that appears to be suitable for two community types. For Community Type A, the soil texture is highly suitable; the slope, existing vegetation, and connectivity ratings are moderately suitable; and the access rating has low suitability. For Community Type B, the soil texture, the slope, and existing vegetation have high suitability; accessibility and connectivity have low suitability. We can convert these ratings to numerical rankings, such as 10 for high, 5 for moderate, and 0 for poor, and then add them to determine the suitability ranking of the area for a specific plant community restoration. The weighting process for this scenario is presented in Table 4.2.

Table 4.2 • The Weighting of Suitability Criteria

Resource	Suitability Ranking Community Type A	Suitability Ranking Community Type B	Weight
Soil Texture	High (10)	High (10)	5
Slope	Moderate (5)	High (10)	3
Existing Vegetation	Moderate (5)	High (10)	5
Accessibility	Low (1)	Low (1)	1
Connectivity	Moderate (5)	Low (1)	5

$$\left(\begin{array}{c}\text{Suitability Ranking of}\\\text{Resource "a"} \times \text{Weight}\end{array}\right) + \ldots + \left(\begin{array}{c}\text{Suitability Ranking of}\\\text{Resource "e"} \times \text{Weight}\end{array}\right) = \begin{array}{c}\text{Suitability Ranking}\\\text{for Community Type}\end{array}$$

$(10 \times 5) + (5 \times 3) + (5 \times 5) + (1 \times 1) + (5 \times 5) = 116$ Suitability Ranking for Type A

$(10 \times 5) + (10 \times 3) + (10 \times 5) + (1 \times 1) + (1 \times 5) = 136$ Suitability Ranking for Type B

However, without one additional step, the results can be misleading. Under the above scenario, each suitability criterion is weighted the same (i.e., soil texture and accessibility are each treated as being equally important, and they may not be). In such instances, we cannot simply assign a numerical ranking to the suitability levels. We must also weight the different suitability criteria as to their importance in our decision. The result is a map displaying areas that are most, moderately, and least suitable for a specific objective such as dry forest or sedge meadow. For this scenario to work, the assignment of suitability ratings and weights must be supported by science and your expertise, and then carefully documented. The key for any planner, however, is to document how the analysis was done and to monitor the success of that process during the restoration planning and implementation stages.

The site inventory and analysis process is goal- and site-driven; it is a diagnostic tool and does not simply describe a phenomenon or an observation. The analysis assesses suitability and constraints for a specific program through the evaluation of the inventory results. These results are then reviewed to determine whether the site and project goals are compatible, or if changes to the goals are necessary.

4.13.2 Opportunities and Constraints Analysis

Using this approach, you evaluate the site as a whole, based on your knowledge, your experience with similar situations in the past, and your vision for the site and restoration. You do not necessarily analyze the resources fully or in a systematic fashion. The results of the analysis uncover the positive features (opportunities) and the problematic features (constraints) of the site relative to the project goals.

Opportunities are attributes that can contribute to or aid the restoration process (such as how the site conditions help accomplish the restoration objectives). For example, planning for and implementing a forest restoration can often be prolonged if a canopy does not exist, particularly if it will be necessary to plant a canopy to provide shelter and shade for seedlings of the desired canopy trees and herbaceous species. Where a preexisting canopy of noncompetitive trees exists, these might serve as a cover crop. Seedlings of appropriate canopy trees can be planted beneath, and it may be possible to begin planting the herbaceous layer. Once the desired tree species begin to approach the canopy, the plantation trees can be removed and perhaps sold to support the restoration.

CASE STUDY The Cross Plains Ice Age National Scientific Reserve

In 2005, the U.S. National Park Service and Wisconsin Department of Natural Resources began collecting background information for a restoration and management master plan for the Cross Plains Ice Age National Scientific Reserve. The mission and purpose are to preserve the geologic, natural, cultural, and scenic resources of the CPIANSR and to provide interpretive, educational, and low-impact recreational opportunities. The major goal of the project is to interpret the site's geology and glacial history. A consulting team conducted a site inventory and analysis to assist with the site planning, and recommended a secondary goal of restoring and managing the landscape to represent the conditions that may have been found in this region prior to its settlement by Europeans in the 1800s. These conditions would have been savanna, prairie, and, to a lesser extent, maple woods in the deeper valleys.

The initial walk-through shows this landscape to be composed of ridges and rock outcrops sloping down to cropland in the surrounding river valley. Historically, the area was grazed and, since the 1920s, has been slowly transitioning from oak savanna to oak forest. Some pasture and mesic (maple) woods also occurred.

The inventory was recorded in a GIS database, and numerous inventory maps were produced, including those shown in Figures 4.14–4.21.

Upon completion of the inventory, the data were evaluated based on suitability and the opportunities to accomplish objectives. The attributes with highest suitability for a community type are listed in Figure 4.22. (For further discussion on attribute levels, see Sidebar.)

The result of the analysis was a landscape recommendation to restore much of the site to oak savanna and prairie, but with new components of oak forest and a limited zone of maple forest. These latter two components were, to some degree, in response to the amount of time and effort required to restore areas that are now entirely forested to their previous state of open prairie or savanna. Such a goal would have been prohibitive in terms of resources, and the project would probably not have moved forward.

Trail siting became an issue as the team discussed the project. Within this landscape exists a steeply sloped gorge with a diverse flora and excellent potential for further restoration. People walking into the gorge would certainly damage the quality of its landscape, and yet it is a unique geologic and natural feature that would be attractive to most visitors. Whether or not to construct trails in this area resulted in considerable discussion among agencies involved in the planning. The thought was that if formal trails were not permitted, ad hoc trails would spring up anyway. If trails were built to the edge of the gorge for overlooks, some individuals would likely continue into the gorge down some very steep and erodible slopes (see Sidebar). ∎

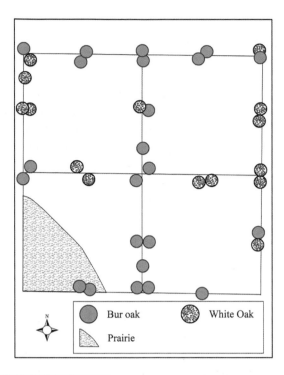

FIGURE 4.14 CPIANSR: Historic Land Cover of the 1830s. This map is based on the U.S. Federal General Land Office Surveys. The trees, along with their diameters, were recorded along the survey or mapped section lines. Surveyors also recorded descriptions of the surrounding land cover. (From Harrington, J.A. 2008.)

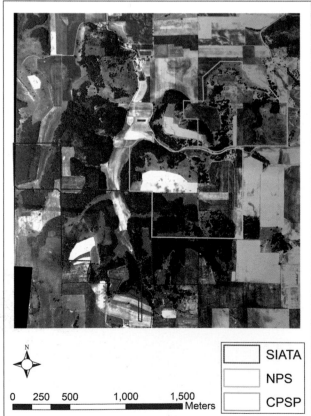

FIGURE 4.15 CPIANSR: 1937 Aerial Photograph. Can you detect the oak openings that are no longer present in the 2005 photo in Figure 4.16? (From Harrington, J.A. 2008.)

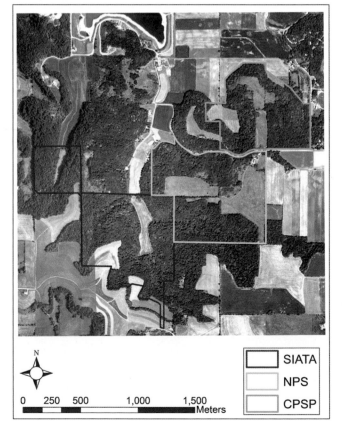

FIGURE 4.16 CPIANSR: 2005 Aerial Photograph. The canopy has closed in nearly all of the former savanna land cover. (From Harrington, J.A. 2008.)

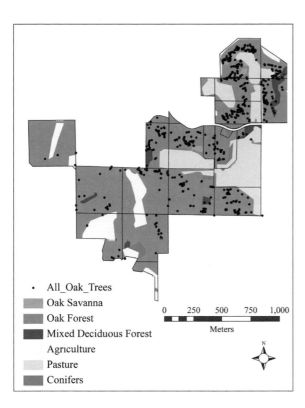

FIGURE 4.17 CPIANSR: Vegetation Cover and Mature Oaks. The oaks depicted on the 2005 land cover plan are growing in forested or closed canopy conditions. These oaks, however, developed under high sun exposure in oak savanna or open to semi-open settings. Much of the original flora have been shaded out. Tree thinning and restoration of the herbaceous ground layer are required for meeting the project goals of a savanna-grassland community. (From Harrington, J.A. 2008.)

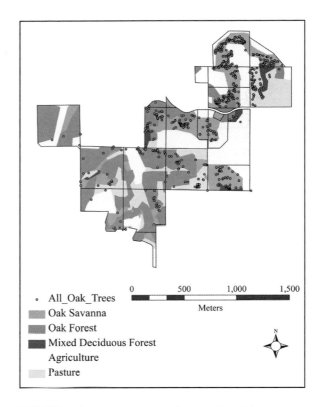

FIGURE 4.18 CPIANSR: 1937 Land Use and Land Cover. In 1937, the landscape in which the oaks depicted in Figure 4.17 occur was composed of semi-open to open canopy plant communities. (From Harrington, J.A. 2008.)

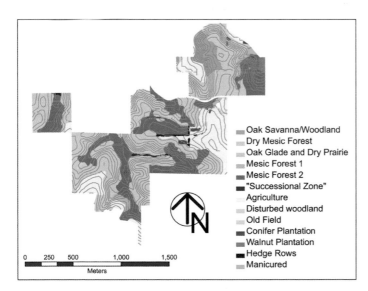

FIGURE 4.19 CPIANSR: Predicted Land Cover in 30 Years. Here is one scenario of future vegetation if no restoration or management were to occur. This map is based on ecosystem models and the site inventory and analysis. (From Harrington, J.A. 2008.)

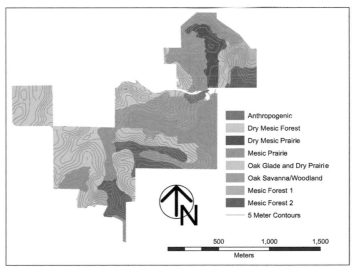

FIGURE 4.20. CPIANSR: Recommended Restored Communities. Here is one scenario of future vegetation with management. This map is based on ecosystem models and the site inventory and analysis. (From Harrington, J.A. 2008.)

Site Location

Property Boundaries

Property Ownership

Political Context (site in relation to municipalities, towns, watershed)

Ecological Context (location within ecoregions of the state)

Geology (Glacial)

Topography: Slope (see Figure 4.6)

Topography: Aspect

Hydrologic Conditions

Soils (see Figure 4.4)

Wildlife Habitat

Visual Resource Quality

Historic Land Cover: U.S. Federal Land Office Survey, 1830s (see Figure 4.14)

Historic Land Cover: Wisconsin Land Economic Survey (1938)

Aerial Photo of Project Area, 1937 (see Figure 4.15)

Aerial Photo of Project Area, 2005 (see Figure 4.16)

Existing Vegetation and Mature Oaks (see Figure 4.17)

1937 Estimated Land Cover (see Figure 4.18)

Predicted Land Cover in 30 Years with No Management (see Figure 4.19)

Predicted Land Cover in 30 Years with Management (see Figure 4.20)

FIGURE 4.21 **Inventory Maps for the Cross Plains Ice Age National Scientific Reserve.**

Community	Attributes of High Suitability
Savanna	Large-diameter open grown oaks (dbh >45 cm)
	Canopy and midstory remain partially open (20–50%)
	Traces of savanna forbs and grasses in the ground layer
	Well-drained soils (Basco, Dunbarton soil series)
	All aspects except north and east when slopes are greater than 8%
	Savanna was historically present
Prairie	Open areas (fields, croplands) on level and west and southern aspects
	Well-drained to moderately drained soils (Basco, Batavia, Edmund, Troxel soil series)
Oak Forest	Fully canopied oak woodland; more than 80% native
	Well-drained soils (Hixton, Kidder, McHenry, New Glarus, Seaton, Whalan soil series)
	Level or sloping hillsides with western and southern aspects
Maple Forest	Sugar maple present in understory
	Midstory shrubs and ground cover common to maple forests
	Silt loam soils, well-drained but moist (Whalan soil series)
	North- and east-facing aspects, no slope limit
	Upper and lower elevations

FIGURE 4.22 **Attributes of Highest Suitability for Select Community Types Within the CPIANSR.**

Sidebar	Deciding on the Attribute Level for Analysis

In this chapter, we have covered many of the resource categories (soils, topography, hydrology, etc.) and some of the attributes within each resource you'll need to understand about a site before setting restoration objectives. Yet there are also many we do not discuss in any detail. We cannot inventory resource categories per se; instead, we select resource attributes to inventory, such as soil texture, slope, or population size.

Most resources have several attributes that can be further divided, depending on the level of detail needed for the project, and the level of detail will be reflected in the scale of the attributes we choose. For examples under the resource soils, the composition of soil fauna at the restoration site could be one attribute. But that attribute could be subdivided into a focus on macrofauna, mesofauna, and microfauna, and these three attributes could be further divided into earthworms, beetles, ants; arthropods and nematodes; and protozoa; respectively. And each of these could be divided even further.

Similarly, for a particular restoration we may be interested in restoring habitat for migratory birds in general, or we may be interested in one specific migratory group of birds, such as cranes or even one specific species, such as the whooping crane.

The resources and level of attributes you'll have to explore depend on the restoration project goals and purpose. In the CPIANSR case study, for example, we provided a list of some attributes that were inventoried for high suitability (see Figure 4.22).

FOOD FOR THOUGHT

1. If you were the project manager for the CPIANSR restoration, what attributes under each resource category would you want to inventory, and why?

2. What sources would you use for data, and where or how might you acquire them?

Constraints are attributes that limit the success of the restoration or the ability to accomplish the stated objectives. If the preexisting tree species in the above scenario were non-native, aggressive, and difficult to eradicate, these factors could extend the length of time to complete a restoration, if one can be done at all. Constraints can also be logistical, such as when high costs would be incurred to implement plans on excessively steep slopes, or when policies prohibit the use of herbicides.

Opportunities and constraints are represented by what the site resources will support with ease, with some effort, or not at all. For example, soil type A provides opportunities for establishing Plant Community 1 but limits the establishment of Community 2 without substantial soil amendments. Typically, you base the opportunities and constraints for each target (e.g., mesic forest) on the resources investigated (soils, existing vegetation, topography, etc.) and present an annotated map composite of the site's features, together with a written report describing your decisions and how they were made. Preparing a table that lists opportunities (Table 4.3) for community restoration helps organize and communicate the information. For a table showing constraints, or limiting factors, see Table 8.1.

Regardless of the approach taken (suitability analysis, or opportunities and constraints analysis), what is most useful at this stage in the restoration process is to present a summary of opportunities and limitations to achieving the restoration goals and objectives. For example, if one of the general restoration goals is to maintain a minimum of two natural communities on site, the analysis would discuss what kinds of communities would be possible given the existing site features, and which of this set of possibilities would be the best fit. If one goal is to maintain a site as free as possible from pest species, the analysis would discuss the extent of the pest species currently on site, and the potential for such species entering the site from the surrounding land area. And if one of the goals is to accommodate

Sidebar | CPIANSR User Issues

The master planning team for the Cross Plains Ice Age National Scientific Reserve project consisted of individuals with the U.S. National Park Service, the Wisconsin Department of Natural Resources, the U.S. Fish and Wildlife Service, and the University of Wisconsin–Madison. Although the team agreed on the goals for the project, they were split between those who wanted to show the public the natural features of the gorge and those who wanted to completely steer the public away from it, fearing that visitation activities would severely impact this fragile landscape. The final consensus was to develop a limited set of trails that would skirt the top of the gorge and include one or two overlooks using existing outcroppings of rock as platforms. Monitoring would occur to see if visitor use resulted in damage to the gorge, and if so, alternative plans would be set in place for its protection, such as creating obstacles to entering the gorge.

Despite this final agreement, everyone on the team had questions.

1. If the trail were to lead down into the gorge, how would people be prevented from traveling along the fragile gorge floor?

2. If trails were taken to overlooks along the gorge, what would prevent people from attempting to climb down the slopes, which are very fragile and have a great potential for erosion?

3. Would fencing take away from the natural beauty of the gorge? Would it obstruct animal movement?

4. Some members proposed building a bridge over the gorge, but that led to several questions, including: Can the National Park Service or Wisconsin Department of Natural Resources afford the costs of a bridge spanning 20 m (65 ft) or more?

5. Would disturbance from the required bridge footings cause extensive damage?

6. Would shade cast from the bridge affect the plant composition and structure?

7. Would people throw items off the bridge?

8. And is the bridge an appropriate aesthetic for the site?

Others questioned whether trails should go to the gorge at all.

FOOD FOR THOUGHT

The above questions have no definitive answer, but are representative of the types of questions you may face when restoring a landscape in what is also part of a public facility. How would you answer each of these questions? Are there additional questions you can think of relating to how to handle trails in sensitive areas where visitors have the potential to cause damage? How would you go about finding consensus on this issue?

---

hiking, the analysis would discuss the presence and condition of existing trails, if any, and the suitability of the soils, topography, and other factors for adding trails if none were present. The summary should include a description of the sources and methods used in compiling the inventory (existing maps, fieldwork); this information helps decision makers who were not involved in conducting the inventory/analysis in interpreting the data.

The summary of existing conditions is also important for future site managers. This record of what was known, understood, or assumed about the site at the time the project was initiated served as the foundation of the restoration as originally conceived. Should problems develop in the future, knowledge of the original site conditions, and an understanding of how the information was interpreted, can help explain what might have occurred.

Once the site inventory and analysis process is completed, and prior to beginning the master and site plans, you should be able to address those questions that determine the feasibility of the project goals—what needs to be done onsite to achieve the goals, and what are the required means and resources for doing so. Examples of these questions are found in Figure 4.23.

Table 4.3 • Site Opportunities for Selected Terrestrial Community Types and Utilities

Community Type	Soil Moisture	Percent Slope	Aspect	Soil Type	Adjacent Land Cover
Dry prairie	Excessively drained	>12%	Southeast to west	Sand to sand loams	Crop, pasture, prairie remnant, savanna
Mesic prairie	Moist but well drained	3–12%	Level, southerly and westerly	Silt loam	Crop, pasture, prairie remnant, savanna
Wet prairie	Wet, dry in summer	0–3%	Level	Clay loams	Crop, pasture, prairie remnant, sedge meadow, savanna, stream
Oak savanna	Well to moderately drained	1–12%	All aspects but north when slope is >6%	Sand to clay loams	Crop, pasture, prairie remnant, savanna, wooded, stream
Oak forest	Moist but well drained	0–12%	All aspects but north when slope is > 12%	Silt loams	Wooded, savanna, stream

Utilities	Slope	Aspect	Soils	Views	Sensitive Features to Avoid (Constraints)
Primary trails	0–8%	Southeast to west	Sand to silt loams	High	Endangered flora Erosive soils
Secondary trails	3–12%	All aspects	All	High	Erosive soils
ADA* trails	0–3%	Southeast to west	Sand to silt loams	High	Erosive soils, steep slopes
Service roads	0–12%	All aspects	All	Not applicable	Erosive soils, steep slopes

*Trails complying with the Americans with Disabilities Act.

1. Do the physical conditions on the site support the restoration goals? If not, in what ways can they be modified, or will the goals need modification?

2. Are the physical factors of interest homogeneous or heterogeneous across the site?

3. If the latter, is the scale of these changes significant enough to influence the distribution of the community types or species that are feasible and desired for the site?

4. Is the size of the site sufficient to sustain the communities proposed for it?

5. How does the existing vegetation on the site support or hinder the restoration? Can it be used as an opportunity in establishing the desired vegetation? Would it support the wildlife species of interest?

6. What are the processes and functions necessary for the restoration's success? Are they present on the existing site? Can they be easily restored to the site? Can the restoration team assume that they will "restart" once site modifications and plantings are introduced, or do the processes need to be in place prior to planting?

7. Given the adjacent land use, what effect might the edge configuration have on plant and animal migration, and on the success of the restoration?

8. What support from the community or public is there for the restoration? Are there onsite human uses continuing into the future and that should be factored into the restoration? Are resources available to ensure that the restoration can be completed and managed long into the future?

FIGURE 4.23 Questions That Precede Master and Site Planning.

Perhaps more than any other step in the restoration process, the site inventory and analysis demonstrates the multidisciplinary nature of restoration ecology. The expertise to thoroughly understand the site parameters, and how each one influences restoration planning, requires a team effort. The planning team, as well as clients, can review and discuss the inventory and analysis documentation and relate it to the alternative site solutions with the goal of selecting one. Onsite data collection tools that are used in both site inventory and monitoring are similar and will be discussed in the next chapter.

KEY CONCEPTS

- The site inventory informs the restoration team about the physical, biological, and cultural conditions within and around the site.

- Site considerations extend beyond the physical site boundaries and include socioeconomic interests and potential external environmental changes that can influence restoration decisions and outcomes.

- The restoration team uses the site analysis to interpret the collected data and information in the context of the restoration purpose and goals, in order to match the restoration needs to the site.

- The site analysis considers the suitability of a site, or areas within a site, for a specific purpose or objective, such as plant community types, ecosystem functions, social experiences, or research.

- The site analysis determines the site conditions that limit the potential to achieve the restoration goals. The site and implementation plans will then explore possible strategies to work with the limitations.

FOOD FOR THOUGHT

1. What site data would you collect in the process of developing a plan for a stream restoration? Explain why you selected these data. How might they be used to determine opportunities and constraints for the restoration?

2. Explore the SSURGO files on the U.S. NRCS Soil Data Mart website. What data can they provide about a site? What is the level of accuracy of these data? What onsite data would you gather to supplement the SSURGO data?

3. Develop and then defend a list of data sources that can provide information on a potential site for restoration in your area, prior to your visiting it. What value might these sources have for you in learning about the site before actually visiting it? Where might they be inadequate without a site visit?

4. We presented a number of categories of potentially onsite resource attributes to be collected and analyzed, including climate, topography, soils, hydrology, and socioeconomic factors. Can you think of additional categories? Are their additional attributes within the categories provided that have not been discussed in the text?

5. This chapter also listed a number of offsite attributes that influence a proposed restoration site. What additional offsite attributes may influence the prospect of a successful restoration at a particular site? Think of a scenario in which each new attribute would be informative, and how it might influence restoration, implementation, or management plans and activities.

6. When beginning a restoration project, under what circumstances would you use an opportunities and constraints analysis? A suitability analysis? Would you want to use only one of these two approaches, or would both be useful in the analysis of site data?

7. The Internet has numerous data sources to help a restoration team learn about a site's characteristics. Are there attributes you can solely look to the Internet for? Are there attributes you would still need to visit the site to obtain and fully understand?

Resources for Further Study

1. Geographic resource information is available online from a wide variety of government agencies. The following list represents some of the websites we have found to be very informative:

Australian Bureau of Meteorology   http://www.bom.gov.au

Australian Department of Environmental Climate Change and Water NSW Office of Water http://www.water.nsw.gov.au

The European Environmental Agency (information and policy advice) http://www.eea.europa.eu

Geoscience Australia   http://www.ga.gov.au/

U.S. Army Corps of Engineers   http://www.usace.army.mil

U.S. Geological Survey (USGS)   http://usgs.gov/

USGS Water Resources   http://water.usgs.gov

USGS Geography   http://geography.usgs.gov

USGS Land Cover Institute   http://edcwww.cr.usgs.gov/programs/lccp/

USGS Mid-Continent Geographic Science Center   http://mcgsc.usgs.gov/

U.S. Fish and Wildlife Service   http://www.fws.gov

U. S. Natural Resources Conservation Service   http://www.nrcs.usda.gov/

Soil Data Mart   http://soildatamart.nrcs.usda.gov/

U. S. National Oceanic and Atmospheric Administration   http://www.noaa.gov

2. For more information about the role of site inventory and analysis in conservation and land use planning, check out the following publications:

Dramstad, W. E., J. D. Olson, and R. T. T. Forman. 1996. *Landscape Ecology Principles in Landscape Architecture and Land-Use Planning.* Washington, DC: Island Press.

LaGro, J. 2007. *Site Analysis: A Contextual Approach to Sustainable Land Planning and Site Design.* New York: Wiley.

McHarg, I. 1995. *Design with Nature.* New York: Wiley.

Steiner, F. 2008. *The Living Landscape: An Ecological Approach to Landscape Planning.* Washington, DC: Island Press.

3. These websites provide information on U.S. government land management programs:

NRCS (Natural Resources Conservation Service) Wetlands Reserve Program

http://www.nrcs.usda.gov/programs/wrp/

USDA Farm Service Agency Conservation Reserve Enhancement Program

http://www.nrcs.usda.gov/programs/wrp/

NRCS Conservation Reserve Program

http://www.nrcs.usda.gov/programs/crp/

Gathering Onsite Resource Information

Designing a restoration to match the physical, biological, and social environmental conditions of a site is one of the most important keys to the success of a project. In order to be able to create and maintain such a match, you need to know as much as possible about a restoration site and its surroundings. The need for site information starts at the very beginning of the restoration process, when you are deciding what kind of communities/ecosystems to restore, and continues even after your objectives have been achieved, and management begins (Figure 5.1). The site inventory and analysis process helps you collect and organize information at all of these stages.

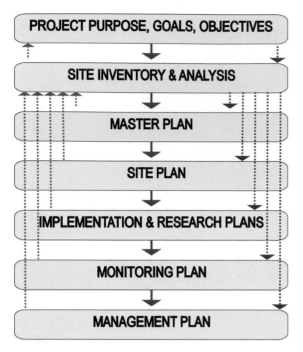

FIGURE 5.1 Using the Site Inventory and Analysis Process in Several Stages. As indicated by the arrows the results of a site inventory/analysis influence each step of a restoration project; the decisions reached at each step usually lead to the need to collect and analyze more (detailed) site information.

In Chapter 4, you learned about the kinds of resources that most influence restorations and how to locate and use some of the many sources of spatially referenced resource data that are readily available worldwide. Locating existing information (even if it was gathered for very different purposes) is always a good first step in creating an inventory and analysis of site resources, and it might be sufficient for completing a master plan. However, the scale and detail of available information is often not adequate for informing the decisions you will need to make during the site, implementation, monitoring, and management planning stages. This means that in most cases, you will need to collect your own onsite data.

The maps of soils and topography (discussed in Chapter 4) are very useful in describing overall site characteristics, but because of the spatial scale of the data used to create them, they often omit small variations. Topographic maps may not indicate local depressions or swales, and soil maps may not capture local variations in fertility and textures. Such variations are often very important for determining site plan objectives, concerning, for example, species composition and placement, as well as for making choices about implementation, monitoring, and management procedures. In addition, some information, such as the presence and abundance of plant and animal species, is best gathered onsite, close to the time the restoration project is to begin. And nothing can replace the experience of visiting a site to gain an overall impression of its potential, using the observation techniques discussed in Chapter 3. Such observations are fundamental for determining overall restoration goals.

In this chapter, we will explore the most common approaches to collecting resource information in the field. We begin with a reminder about the importance of establishing inventory protocols, and continue with a discussion of how to collect relevant data about the physical, biological, and social resources described in Chapter 4. Our aim is to acquaint you with some of the most commonly used inventory and analysis tools, and to suggest sources to turn to for more detailed information about the techniques and how to use them in different situations. Government agencies, professional societies, and private groups concerned with conservation and natural resource protection provide very helpful guides and "how to" manuals that cover both simple and sophisticated tools.

The technological revolution that has been changing our daily lives in recent years is also contributing greatly to advances in the tools available to restoration ecologists. In order to keep up with recent advancements in site inventory and analysis techniques, be sure to join the communication networks described in Chapter 3. By attending conferences, visiting the websites of public and private agencies engaged in resource conservation, restoration, or management, and reading journals, you will learn about new techniques, as well as the experiences of others who have tried them. The list of resources at the end of this chapter will get you started. In addition, equipment manufacturers provide specifications and suggested uses for the items they sell, and these can be another helpful set of resources.

▶ 5.1 GETTING STARTED

Before you begin to collect field data, it is necessary to establish protocols—sets of instructions that specify what is to be collected, which tools to use, when and where to apply them, and how to summarize and interpret the resulting information. The protocols are shaped by the nature of the resources they concern, the types of information you need, the size of the budget, and the number of available people-hours. Since each restoration project is unique, it is important for you to write the protocols to fit the situation so that you collect and analyze enough, but not too much, information. Because different levels of detail may be required as the restoration project moves from the design to the monitoring phase, you will usually have to write protocols to fit each stage.

Although it is logical to suppose that you can never know too much about a site, in practice "information overload" can confuse rather than clarify a situation. In addition, since most project budgets are limited, it is important that you focus the inventory and analysis procedures on capturing the information needed to address the issues at hand. In this way, you are less likely to neglect important information, and waste is avoided.

For most resources, there are several inventory and analysis techniques from which to choose. In addition to choosing tools that are the best match for the kinds of information you need that also fit the project's budget, you must match the tools to the skill levels of the people who will be using them. In many situations, your field crew may have little prior experience; in that case, you would choose tools that are relatively simple to use and/or do not require much training.

Most sites include several regions, each of which has different resource characteristics. A good way to capture this variation during an inventory is to divide the site into units and collect information within each. Depending on the stage of the restoration for which site information is being collected, these divisions can be implementation units, monitoring units, or management units, and the boundaries of each set may differ. (See Chapters 8, 9, and 10, respectively.) By cross-referencing site data and the locations from which they were collected, the information will be organized for later evaluation. You can always combine data collected from several units in order to describe the site as a whole—for example, to report the total plant diversity.

▶ 5.2 MICROCLIMATE

Whereas climate describes the regional patterns of daily weather, microclimate concerns local, site-specific weather phenomena. For many restorations, an understanding of climate using the data sources discussed in Chapter 4 is sufficient. In some cases, however—such as determining the annual water-holding capacity of a restoration for the purpose of managing local stormwater runoff, or determining sun and shade patterns in order to guide the precise placement of light-sensitive plants—you will need to collect microclimate data. This is particularly important for sites that are several kilometers away from a weather recording station, and in regions where weather events are often patchy or greatly influenced by elevation or topographic changes.

There are many instruments available for measuring and recording precipitation (amount and intensity); air and surface temperatures; solar radiation (total intensity or by wavelengths, i.e., photosynthetically active radiation, or PAR); air humidity; and wind direction and intensity in the field. Ranging from permanent "weather station" installations to portable units, the instruments vary in terms of the complexity and precision of the information collected (Figure 5.2). Most devices record the information in digital form, which you can download to a computer for analysis.

FIGURE 5.2 An Automatic Weather Station. A weather station at a remote location records and stores daily temperature, wind speed, and precipitation for researchers and managers who are unable to always be on site to take measurements. (Photo reprinted by permission from Steve Glass.)

Another microclimate factor that is important in planning restorations is the pattern of shadows cast by landscape features that block the sun. In Chapter 4, we discussed the roles that slope aspects of hills and other prominent topographic features play in creating patterns of sun and shade. Where available, you can use digital elevation models to predict the sunniest or most shaded portions of sites, based on their landforms. Other site features, such as buildings or trees and shrubs, also create shadows that can influence microclimate. It is certainly possible to use instruments to measure the light levels in specific locations, as discussed above, or to observe shadow patterns over the course of a year. Another approach is to measure the heights and extents of prominent site features; using these data, along with a basic understanding of the relationship between the Earth and the sun, you can model the shadows the features cast during the course of a year.

The length of a shadow is related to the height of the object that casts it, and to the sun's altitude angle—the angle created between the path of the sun's radiation as it approaches Earth's surface. The lower the sun is in the sky—at dawn or dusk—the longer the shadow. The direction of a shadow varies with the sun's azimuth angle—the angle between due south (or sometimes due north) and the projection of the line of site to the sun on the ground. Both the altitude and the azimuth angles vary with latitude, longitude, time of year, and time of day.

Shadow patterns are of particular interest when your restoration plan places a relatively low-growing, full-sun community, such as a grassland or marsh, next to a forest with a tall canopy. The shadows influence the microclimate of the edges, thus affecting species distributions (see Chapter 2). Depending on the location and height of the tree border, shadows may create a different microclimate at the edge of the open community than is true for the rest of its extent (see Sidebar; Figure 5.3). Unless you plan for this difference and choose species accordingly, such edges often create problems for a restoration team. If a forest is started "from scratch"—that is, in an area in which most trees are absent—it is particularly helpful to create models of what the shadow patterns at the border will be as the trees mature (Figure 5.4; Table 5.1). In this way, the restoration team can anticipate the need to modify the border area as the influence of shadows becomes more pronounced (see Chapter 7).

Sidebar | Modeling Shadow Patterns

To model the shadows for a particular place, date, and time, start by finding the sun altitude and azimuth angles. You can readily obtain this information for specific latitude and longitude locations using the Internet. Next, determine the heights of the shadow-casting features. You can estimate heights in the field by using a clinometer, a tool that measures angles relative to a baseline. Stand a known distance away from the feature, and sight the top through the lens of the clinometer (see Figure 5.3). Then, using the clinometer reading, you can estimate the height using trigonometry:

$$\frac{\text{Feature}}{\text{height}} = \left(\text{tangent of angle} \times \frac{}{\text{distance to feature}} \right) + \text{height of observer}$$

Then you can estimate the shadow lengths by using this formula:

$S_l = h/\tan SA$

where S_l = shadow length, h = height of feature, and SA = Sun angle.

The sun azimuth angle will describe the shadow directions, and the extent of the ground covered by the shadow-casting feature will determine the width. By modeling the shadows for early morning, noon, and late afternoon on the two equinoxes, and depending on the climate, on one or both solstices, you can then locate the sunniest and most shaded part of a site, as well as estimate the extent to which a particular location is in full sun or shade.

Figure 5.4 is a sun-shade analysis of the shadows cast by trees on June 21, 2011 (near the summer solstice) in a portion of the Wharton State Forest in the pine barrens region of New Jersey. In this simple example, the trees are all estimated to be 15.2 m (50 ft) in height and occur in one large patch, except for one isolated individual. The model shows the predicted length and direction of shadows cast in the morning (the sun is in the east, the shadows orient toward the west), afternoon (the sun is in the west, the shadows point to the east), and at noon. By looking at the overlap of the shadow patterns, you can find the areas that are covered in shade or in full sun for most of the day. Table 5.1 provides the data used in the model.

FOOD FOR THOUGHT

1. To test your understanding, try modeling the shadows for May 15 and September 15 to see how the shadows change with the seasons.

FIGURE 5.3 **Using a Clinometer.** The clinometer is an optical device for measuring angles of slope, elevation, or inclination above the horizontal plane. Common uses are to measure the height of trees or buildings, or degrees of slope of a hillside. (Photo reprinted by permission from Evelyn Howell.)

FIGURE 5.4 A Shadow Pattern Model. This model represents the shadow patterns for a stand of pitch pine (*Pinus rigida*) found in the Wharton State Forest, New Jersey. It shows the length and directions of the shadows cast by the trees at three different times of day on the day of the summer solstice in 2011 (June 21, 2011).

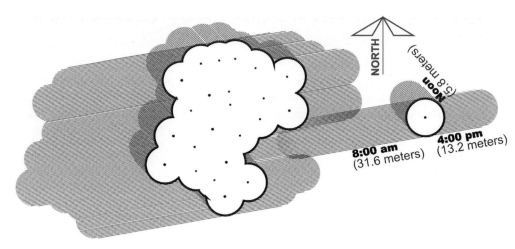

Table 5.1 • Data for Modeling Shadow Patterns

These patterns were cast on June 21, 2011, by trees in the Wharton State Forest, New Jersey, latitude 39.7°n, longitude 74.8°w.

Time of Day	Sun Angle (SA)	Tangent SA	Tree Height [m (ft)]	(S_1) Shadow Length [m (ft)]	Sun Azimuth Angle*
8:00 a.m.	25.73°	0.482	15.24 (50)	31.6 (104)	−100°
Noon	69.26°	2.641	15. 24 (50)	5.8 (19)	−42.91°
4:00 p.m.	49.05°	1.153	15.24 (50)	13.2 (43)	80.32°

*Azimuth angles are measured east of south if the measurement is a negative number, and west of south if the number is positive.

▶ 5.3 TOPOGRAPHY

For some restorations, you will need to obtain detailed information about patterns of small onsite elevation differences. Such information is essential, for example, in finding or designing small depressions to create ephemeral ponds in otherwise flat topography, or to match wetland plants to the appropriate water depth. The best way to obtain this level of detail is to conduct a topographic survey, which determines the elevations and spatial positions of points in the landscape with reference to a designated starting location.

Modern topographic surveys are conducted with sophisticated electronic measuring devices (total stations) that often connect to satellite global positioning systems (GPSs). In situations where precision is important, it is best to contract with a trained surveyor to obtain the necessary information. The surveyor can provide you with a digital elevation model (DEM) and contour map to analyze the topography (see Chapter 4).

Another technique for mapping relative elevations and slopes that is accurate and relatively easy to use is differential leveling, performed with a surveyor's level and rod. A surveyor's level is a small telescope with a bubble level, mounted on a tripod. The purpose of the level is to keep the line of sight perfectly horizontal. The rod is an extendable pole marked with distance units. The idea is to compare sightings of the rod positioned vertically above two points on the ground, and taking into account the height at which the readings are recorded, to determine the difference in elevation between the points. Here are the steps:

1. Place the rod at a point that will serve as the reference point for the survey (Point A). It is ideal if this point is an official benchmark of known elevation, but if not, for purposes of the survey, you can assign a relative elevation. (Official benchmarks are so designated by government agencies or private survey firms and marked with a plaque or some other symbol.)

2. Place the tripod and level at a point about halfway between the reference point and the point whose elevation you want to determine. Point B is the unknown point.

3. Use the level to sight the vertical rod at the reference point (Point A), and determine the reading. This is called the backsight. The elevation of the known point plus the backsight reading equals the height of the instrument.

4. Next, place the vertical rod on Point B and look through the scope to determine the reading. This is called the foresight. The height of the instrument minus the foresight reading gives the elevation of the unknown point (Point B).

You can use the differential leveling technique to create an elevation map of a single slope. Or by measuring the elevations at points in a grid pattern, you can create a contour map, if one is not otherwise available.

In some cases, you may be interested in knowing the small-scale slope changes in a particular portion of a site, rather than over the site as a whole. One of the best tools for measuring slopes on a site is with a handheld clinometer (see Figure 5.3). By facing up or down slope, you use one eye to sight on an object on the slope and the other to align the clinometer. The slope angle in degrees (or in slope percentage) is given by the scale on the instrument. (It is best if the object used for the sighting is about the same height as your eye level.)

When changes in elevation are occurring over very small distances—for example, in erosion gullies, in landscapes with ant or termite mounds, on dissected rock outcrops, or in wetlands with tussocks created by accumulations of roots, litter, and peat—elevation changes can be measured using a meter stick or surveyor's rod and level, using the relationship, percent slope = rise over run:

% Slope = Vertical distance/Horizontal distance × 100

To do this, place one end of a meterstick or rod at the top of the slope, using the level to be sure the stick is perfectly horizontal. The length of the suspended stick is called the run. Next, measure the distance from the bottom of the slope to the end of the level stick. This is the rise. Then calculate the slope using the above formula. Although this technique is prone to error, especially if the level is inaccurate, it can be useful because of its low cost.

▶ 5.4 HYDROLOGY: FRESHWATER SYSTEMS

When looked at from a global perspective, the hydrologic system is best represented as a closed cycle, but at the scale of an individual site, it is better represented as a flow (see Chapter 2). Water enters a site through precipitation, runoff from adjacent properties, and/or the upwelling of groundwater. It can be stored on the site (more or less temporarily) in topographic depressions, ponds, lakes, stream or river channels, or soil; and it leaves in the form of surface runoff, evaporation, or transpiration, or by entering the groundwater. Because of their importance in defining and shaping wetland ecosystems, as well as the importance of the availability of groundwater and surface water supplies for human development, freshwater hydrologic systems are often a focus of restoration. For freshwater systems, site-specific measurements are often designed to include information about (1) seasonal and daily cycles of groundwater and surface water levels; (2) sources and forms of onsite groundwater and surface water (precipitation in the form of rain or snow, surface runoff, groundwater flows, and infiltration rates); (3) water quality and how it varies annually and over a span of years; and (4) water flow rates in streams and rivers.

5.4.1 Water Levels

There are a number of approaches available for monitoring groundwater and surface water levels on a site, ranging from continuously recording digital devices to periodic measurements taken using a meterstick. You can track groundwater levels by placing a piezometer, a small cylinder with perforations, in a narrow hole bored into the ground. The level to which the water rises in the piezometer is a measure of the depth of the water table. By installing a number of piezometers on a site, you can also track the horizontal movement of the groundwater across a site.

Similarly, a relatively simple way to track surface water changes is to install a vertical pole perpendicular to the ground (using a level), and measuring the height above the ground surface reached by the water. You can easily install such poles in a restoration permanently, or put them in place each time you take measurements. This technique is usually accurate enough for many restoration projects.

Depending on the goals of the restoration, you can take water level measurements at daily, weekly, monthly, or even longer intervals. The frequency and timing of these measurements will also depend on the seasonal precipitation patterns of the climate of the region, as well as on such things as tidal cycles, the position of the restoration site relative to storm water movements, and the nature of the soils and geology on site.

5.4.2 Precipitation and Surface Runoff

The daily, seasonal, and yearly precipitation falling on a restoration site can be measured using rain and/or snow gauges. These are available from a variety of vendors, and you can easily install them individually, or as part of a weather station. On large sites with topographic extremes and variable canopy cover, it can be instructive to locate a series of gauges across the site to identify onsite differences in precipitation patterns. For example, precipitation amounts reaching the ground will likely be different underneath a dense tree canopy than in an open area. Even small topographical variations can produce rain shadows, and they can affect site microclimate, thereby influencing species survival.

In cold weather climates, it is also useful to document the depth and duration of snow cover. You can use time-lapse photography, or simply take pictures from the same spot at regular intervals.

Once you record precipitation data, it is important to summarize and interpret the information. The level of detail in the analysis, of course, depends on the type of gauge used. Precipitation data are often presented and graphed as seasonal, monthly, or weekly averages. Information in this form, especially when combined with temperature data, is very useful to understand; an example is, what happens to the course of a restoration during an unusually warm-dry or cool-wet year. Some types of equipment can record precipitation data from minute to minute during a storm. This level of detail is particularly useful when planning or assessing restorations designed to manage storm water.

Two of the factors used to determine the required capacity for the temporary onsite storage of storm water are precipitation amounts and intensity, in terms of seasonal average and extremes, but also in terms of the patterns of individual storms. If stormwater mitigation is one of the objectives of a restoration, precipitation data are an important consideration in the design. Once the restoration is implemented, precipitation data are among the essential inputs in the assessment tools used to evaluate performance with regard to stormwater management.

One of the best ways to locate and describe surface runoff patterns is to walk the site immediately after a precipitation event, and look for ponding or surface runoff channels. On large sites, to reduce the amount of effort involved, the walk-through usually focuses on areas of particular importance to the project goals—for example, species habitat, or areas that seem especially susceptible to precipitation patterns, such as flood-prone areas. In addition to mapping the presence of these features, their depth and width are recorded, and any evidence of erosion is noted.

5.4.3 Water Quality

A major goal of many restorations is to maintain or improve water quality—both on the restoration site itself, and also often in the surrounding region. The phrase "water quality" can refer to several attributes, and the optimum conditions vary with the goals of the restoration. Although restorations can and do contribute to the quality of groundwater systems, most restoration monitoring projects concern surface water quality. For example, many stream restorations done for the purpose of improving habitat for fish and other organisms focus on restoring water quality. The most commonly measured quality characteristics of surface water are temperature, turbidity, and pH, as well as levels of dissolved oxygen, nutrients, and heavy metals and other toxins. There are many different tools available, ranging from sophisticated and precise recording devices to those designed with low budgets in mind. Following are a few examples of the kinds of approaches that are in current use.

Temperature
Water temperature levels directly influence the growth and reproduction of aquatic organisms. Aquatic communities are often classified in terms of the ranges of tolerance of the dominant fish species. The two main groups are generally classified according to these criteria:

- *Warm water communities*: Temperatures reach 27°C (80°F) or more during the warm part of the year.
- *Cold water communities*: Temperatures remain below about 20°C (68°F) in warm weather.

It follows that each kind of aquatic system has different sets of organisms.

You can readily measure the temperature of surface water in the field using a simple thermometer, or a temperature probe that records digital data. Water temperature often varies with depth, across different seasons and storm cycles, and with landscape location (e.g., between sunlit or shaded parts of a pond or wetland). Your sampling protocols need to take these variations into account; readings in a number of locations are generally required. Continuously recording equipment may be desirable under some circumstances.

Turbidity

Turbidity is a measure of how transparent or cloudy the water appears, and therefore gives an indication of how many particles of sediment, or algae or other planktonic organisms, are suspended in the water. You can use this measure, for example, to evaluate whether or not a restoration is meeting expectations of reducing erosion or removing sediment from storm water. Often used as an indicator of drinking water quality, turbidity also influences how far into the water column light can penetrate, which, in turn, influences the species that can survive.

An easy way to measure turbidity in a lake or deep pond is to use a Secchi disc, which is a 20-cm-diameter circle with alternating black-and-white markings mounted at the end of a pole. You lower the disc into the water until it can no longer be seen. The depth at which it disappears is a measure of turbidity. It is important to take the measurements at the same time of day (between 10 a.m. and 2 p.m. is recommended), and to use the same technique of lowering the disc each time.

Turbidity can also be measured in the field by taking water samples and then analyzing them using a turbidity meter. This technique works well in shallow ponds or in areas of running water. The meter passes a beam of light through the water sample and measures the amount of light reflected by the dissolved particles. You then compare the reflection pattern of the sample to standard samples of known turbidity.

pH

The pH of water is a measure of its acidity, a factor that, among other things, influences the toxicity of heavy metals, and the survival and reproductive rates of fish and other organisms. For example, a small amount of dissolved iron may be tolerated by fish at one pH level, but be toxic at a slightly higher pH. You can get a good idea of water pH in the field by using a pH meter on a sample.

Dissolved Oxygen Levels

Dissolved oxygen levels in ponds, lakes, and streams influence the composition and diversity of the biological communities found there. Dissolved oxygen is a measure of the amount of oxygen available in the water column. In general, if water has saturation levels of under 30%, conditions are considered to be hypoxic, and many organisms are under stress. Although less frequently encountered than low oxygen levels, too much oxygen can also cause problems—gas bubble disease, for example. The amount of oxygen that can be dissolved in water relates to the water's temperature, atmospheric pressure (elevation), and levels of nutrients. Dissolved oxygen is most easily measured in the field using a meter created for this purpose.

Nutrient Levels

Nutrients are inorganic chemicals that are important for the growth and survival of organisms. Monitoring the nutrient content of water is of particular interest in restorations that refer to nutrient cycles in their project goals, or in those involved in protecting human health.

Nitrogen and phosphorus are important in evaluating water quality, in part because these are often limiting factors to the growth of algae and aquatic plants, but also because continuing human development has the potential to greatly alter the natural cycles of these elements. For example, stormwater runoff now enters surface water systems at an accelerated rate because of the increase in impermeable surfaces from development, such as roofs, paved roads, and sidewalks. These runoff waters often carry large quantities of nitrogen and phosphorous fertilizers from lawns and agricultural fields with them. In areas where timber has been removed from fields through which a stream or river flows, large amounts of nutrients may enter the water as the logging debris decays.

The addition of nitrogen and phosphorus to natural ponds, lakes, rivers, and streams often stimulates the growth of aquatic plants and algae, such that their population levels may exceed restoration goals. As these producers die, the process of decomposition uses up much of the oxygen in the system. As a result, fish populations may be greatly depleted and/or change in species composition, thus again moving a restored site away from its established goals. High levels of phosphorus, in particular, cause massive algal blooms. In some freshwater lakes, blooms of certain cyanobacteria (blue-green algae) release toxins into the water that can kill dogs or other animals.

Excess levels of nutrients can also have adverse human health effects. Not only are nitrates and nitrites from agricultural or lawn fertilizers carried with stormwater runoff into surface water ponds and lakes, but they also readily leach through the soil into underground aquifers. People relying on either groundwater or surface water in these situations may suffer from nitrate/nitrite poisoning. This occurs because nitrites combine with the hemoglobin that carries oxygen in the red blood cells to produce a new substance, methemoglobin, which cannot transport oxygen to the body tissues as hemoglobin does. This leads to internal suffocation, which can result in the death of fetuses and seriously harm the development of young children.

Levels of specific nutrients can be estimated in the field using a spectrophotometer. In addition, water samples can be collected and sent to a lab for a more detailed analysis.

Heavy Metals and Other Toxins

Heavy metals are so-named because they have a high atomic weight. Some heavy metals (e.g., zinc, iron, and manganese) are essential in small amounts to maintain the health of organisms. Others are nonessential, and when present in large amounts in water can be toxic (e.g., mercury, lead, and arsenic). These elements can come from natural sources or be introduced by human activities, such as the dumping of waste. They can be found in the substrate of lakes, streams, and rivers, and are sometimes present in dissolved sediment.

The action and availability of heavy metals and most other waterborne toxins is also influenced by water chemistry. Thus, heavy metals and other toxins are most accurately measured by water chemists. The best way to test for these materials is to collect onsite samples and send them to a lab for analysis.

5.4.4 Flow Rates in Streams and Rivers

The amount of water and the rate at which it flows in a stream or river can be an important factor in aquatic habitat restoration, as well as in managing water quality downstream. Some organisms are adapted to fast-flowing currents; others do best in calm waters. Streams and rivers with fast currents generally have more dissolved oxygen than slow-moving currents; sediment will settle more quickly to the bottom in slow streams. Flow rates vary with the seasons in climates with annual precipitation patterns, often reaching their peaks after snowmelt and major storm events.

Stream flow is described by the rate at which a volume of water moves through the channel and is measured in cubic feet per second. The simplest way to estimate the velocity of a current is to stand on the channel bank and use a stopwatch to measure the time it takes an object to move downstream from one point to another. By averaging several such

FIGURE 5.5 **A Stream Flow Spinner.** A stream flow probe (left photo) has a small propeller that is placed in the stream current to be measured. In the right photo, stream velocity in cubic feet per second is registered on the small computer atop the flow probe. (Photo reprinted by permission from Steve Glass.)

downstream passes, you can get a reasonable estimate of speed (feet per second). You can estimate the water volume by measuring the average depth and width of the channel and then multiplying these together. An easy way to estimate the dimensions of shallow, narrow streams is to measure a line stretched across the channel from bank to bank (width) and then measure the depth of the channel using a meterstick at several points along the string. You can then obtain a rough estimate of volume by multiplying the average of the depth measurements by the width of the channel.

Several types of instruments are also available for measuring stream velocity. One mechanical device that is commonly used is a stream flow spinner, which involves a spinner on the end of a probe connected to earphones or to a small computer (Figure 5.5). The spinner is inserted into the stream at a series of points across the channel, and each revolution of the wheel is counted as heard in the earphones, or recorded by the computer. The number of clicks per minute indicates the velocity of water at that point. Flow rates are found by multiplying the velocity of the water by the area measured, and then by summing the flow rates of the various points across the spring channel.

▶ 5.5 HYDROLOGY: COASTAL MARINE SYSTEMS

Site-based understanding of hydrology is essential for the success of the restoration of the streams and tidal marshes of estuaries, as well as other coastal marshes. As is the case for terrestrial and freshwater systems, the major categories of information include seasonal and daily cycles and water quality. Two additional factors of particular importance in coastal marine systems are tidal fluctuations and water salinity.

5.5.1 Tides

Oceanic tides are cyclical changes in seawater levels caused by gravitational forces between the sun and the moon and the effects of the rotation of the Earth. Most coastal areas have two high tides and two low tides each day. Tidal patterns are also influenced by the shape of the ocean floor, and by less predictable events, such as storms and earthquakes. Information about maximum and minimum water levels, and the duration of the cycles, is of particular importance to the restoration of intertidal zones—areas that are exposed to the air at low tides—and of estuaries—coastal areas exposed to ocean tides and freshwater outflow from streams or rivers.

Because tidal information is very important to human commerce, navigation, and fishing, it is available on the Internet for many major continental coastlines. For example, the U.S. National Oceanic and Atmospheric Administration (NOAA) has posted online interactive maps of its tidal recording stations, displaying information about the hourly predicted heights of tides during a particular day. You can use information from the nearest recording station for a general idea of what to expect on site. When resources permit, however, it is helpful to collect onsite tidal information. You can measure tides with tide gauges, consisting of pipes open to the water that record the rise and fall of water levels either mechanically, or by using sound waves. Either type of gauge can be connected to a recording device.

5.5.2 Salinity

Oceans and estuaries differ from freshwater systems in the salinity or amount of salts—especially salts of chlorine and sodium—that are dissolved in the water. Several kinds of salinity measurement systems are available, each of which involves collecting and testing water samples: conductivity meters, hydrometers, refractometers, and chlorinity measures.

Using a probe, a conductivity meter measures how well a sample of water is able to conduct electricity, a characteristic that is influenced by the salt concentration. Hydrometers measure specific gravity—the ratio of the density of a sample (mass per unit volume) to the density of pure water. Density increases as water becomes saltier. To measure salinity, you place the hydrometer in a jar containing a water sample, and record the level at which it floats. A refractometer measures the deflection of light as its velocity slows down while passing from air through a solution. The salt concentration of water affects the deflection or "refraction" in predictable ways. You place a few drops of the water to be sampled in the refractometer, and look through an eyepiece to read the measurement scale. Chlorinity measures use chemical techniques to determine the amount of chlorine in the sample. Chloride ions usually are a large portion of the salts in saltwater systems, and hence provide a good approximation of total salinity.

► 5.6 SOILS

Soils are of interest in restoration for their importance in supporting the growth of plants, in determining the nature of surface water runoff and/or onsite groundwater infiltration, and in supporting trails and other facilities. Eight soil characteristics that are important to measure onsite are texture, nutrient levels, organic matter, pH, plant litter, compaction, drainage, and erosion. Chapter 4 discussed the importance of each of these to restoration in some detail. The purpose here is to review some common onsite and laboratory inventory and analysis techniques.

5.6.1 Soil Texture

One way to obtain site-based information on soil texture is to collect soil samples and send them to a soil-testing laboratory for analysis (as described below for nutrients). Another way, and one that can yield immediate results, is to pick up a handful of soil and feel its consistency. A sandy soil will feel gritty; a soil that is composed of fine clay or silt particles will feel smooth and silky. An organic soil is often dark with humus, and, in the case of peat, will contain undecomposed plant litter. To make further distinctions, you can perform a soil hand test (see Sidebar; Figure 5.6).

5.6.2 Level of Soil Nutrients

The status of the nutrients in the soils of a site is best evaluated by collecting samples in the field and performing tests in a laboratory. Nitrogen, phosphorus, and potassium are usually relevant, because of their importance in plant growth. Information about micronutrients, those minerals needed by plants in lesser amounts (e.g., boron, zinc, and manganese), is sometimes also of interest, as is the presence of heavy metals linked to soil toxicity.

In collecting soil samples, it is, of course, important to obtain material that is representative of the site. The first step is to divide the site into relatively homogenous subunits, based on initial soil conditions, current land cover, and use history. Depending on the size of the area, you can use the soil mapping units of the U.S. Natural Resources Conservation Service (NRCS) to help create the subunits (see Chapter 4). Then, select a series of sampling points within each subunit. Soil scientists recommend distributing the samples at random locations within each area, or, on relatively uniform sites, in a W pattern.

It would be ideal to record geographical coordinates of each sampling point with a GPS device. At each point, use a soil probe or auger to collect ten soil cores within a few feet of each sampling point. The depth(s) at which the samples should be taken depends on the rooting depths of the plants being considered for the restoration, which is usually in the range of 10–23 cm. Rooting depths, the depths to which water and oxygen penetrate, are influenced by soil texture and the composition of the soil profile. Combine the cores and place them in a soil-sampling bag for later analysis.

There are many public and private laboratories that will evaluate soil samples and provide detailed information on the nutrient status. Because nutrients such as carbon or nitrogen exist in several forms (e.g., nitrogen is commonly found in the form of nitrates, nitrites, and ammonia), it is important to specify in the protocol which formulations you want to analyze. Many of the standard analysis procedures were developed to evaluate agricultural soil fertility, and most of them can be adapted to inform general restoration goals. For restorations that are being proposed to provide ecosystem services related to nutrient cycles, however, more specialized tests, such as rates of mineralization, may be required.

In some cases, the soil samples need to be either analyzed immediately or frozen in order to slow down or stop the continuing actions of microbes. It is important to take this kind of timing into account when writing the protocol.

5.6.3 Soil Organic Matter

The organic matter content of soil is usually estimated in a laboratory by measuring the organic carbon present in a sample. This can be done by igniting a sample in an oven and measuring the carbon dioxide emitted, or by weighing the sample before and after ignition.

Sidebar | The Soil Hand Test

The soil hand test is a quick way to classify the soils of an upland site into general texture classes, such as sand or sandy loam, loam, and clay. The only equipment you need is a trowel or soil probe, a small bottle of water, and a rag to use to wipe your hands. The idea is to try to mold soil into three shapes. How well you can succeed in doing so depends on the soil texture. Here are the three basic shapes:

1. Soil cast: formed by holding the soil in the palm of your hand and closing your fingers around it, and then letting go. The cast includes an impression of your fingers (see Figure 5.6).

2. Soil thread: a long, thin cylinder formed by rolling the soil between your palms.

3. Soil ribbon: a flat shape like a long stick of gum, formed by pushing the soil out of your hand between your thumb and forefinger.

To perform the test, collect a soil sample, using the trowel or soil probe. In most cases, you will likely be interested in the topsoil, so a depth of 10–20 cm is appropriate. Next, moisten the soil with a small amount of water, and try to form the soil into the three shapes. If the soil is very sandy, it will fail to hold any of the shapes. If the soil is a sandy loam, it will form a cast and a thick, short thread, but these shapes crumble easily; it will not form a ribbon. If the soil is a loam, it will form a cast that holds together well; a thread the diameter of a pencil lead, but which breaks easily; and a short, thick ribbon that breaks when it extends too far from your hand. At the other extreme, if the soil is composed largely of clay, the cast will easily hold a number of shapes, the thread can be rolled into thin, long strands, and the ribbon will be long and flexible.

This hand test does not provide the detailed quantitative analysis of a lab analysis, but it can be performed quickly and inexpensively. You can use it, together with data provided by a general soil survey (see Chapter 4), to do a rapid assessment of many portions of a site, saving the lab tests for areas of particular concern.

FIGURE 5.6 The Soil Hand Test. This is a soil cast, one of three shapes used in the hand test to determine soil texture. This cast easily keeps its shape, indicating that the soil sample contains a high percentage of clay-sized particles. (Photo reprinted by permission from Steve Glass.)

5.6.4 Soil pH

Soil pH is a measure of how acid (pH <7) or alkaline (pH >7) the chemical environment of the soil is. This feature is important for restoration because the pH influences the degree to which plants can access the nutrients in the soil. For this reason and others, some species survive better in acidic soils, and some thrive in more neutral conditions. Certain plants also tolerate alkaline conditions. The pH level also influences the survival of both beneficial and harmful microorganisms.

Soil pH is a product of the mineral substrate on which it is formed, as well as of the kinds of plants and microorganisms associated with it. Precipitation causes chemicals to leach through the soil, thereby changing the acidity levels, and the decay products of plant litter also contribute to the overall pH.

You can measure pH with relatively inexpensive field test kits. One type uses a chemical, powdered barium sulfate, which, when mixed with a soil sample dissolved in water, changes color in a predictable way, according to the acidity. Another type, also based on observing a color change, uses litmus paper that you insert into a vial containing a soil sample mixed with water. A third type of test kit uses an electronic meter that measures the concentration of hydrogen ions in a moistened soil sample. Soil-testing labs will also measure pH in the soil samples sent for texture or nutrient analysis.

5.6.5 Plant Litter

The amount and depth of the soil litter layer is often correlated with the germination and survival of plants, and with fire intensity and frequency; therefore, objectives related to litter depth are often included in restoration plans. One way to measure litter depth is to use a measuring stick pushed vertically into the litter at several predetermined sampling locations. This technique is quick and easy to use, but litter depth can be influenced by a number of factors, including moisture content, compaction, and variations caused by wind movement, which piles up some litter while leaving other spots bare. Of course, the very existence of such variation may be typical of particular communities.

A more precise way to estimate the amount of plant litter in soil is to measure the dry weight of samples. Dry weight is the weight of the litter after water is removed by heating the material overnight in an oven set to a low temperature—38°C (100°F). Dry weight is usually a better indication than wet weight of the amount of litter present, as wet weight is influenced by rainfall events and/or soil and air humidity.

As in measuring soil nutrients, litter samples should be collected separately for each soil sampling subunit, and distributed to represent the conditions in the area. At each sampling point, collect the litter over a specified area, such as 0.25 m², and place the sample in a labeled bag. If it is inconvenient to have the samples processed immediately, you will need to specify the storage conditions to prevent change, such as loss of material due to ongoing decay.

5.6.6 Compaction

Soil consists of minerals, living and dead organic matter, water, and air. The minerals and organics form aggregates, which cluster together to create a three-dimensional structure punctuated by pores. The pores are filled with water or air. One consequence of human disturbance is soil compaction, a process by which the structure of the soil collapses, essentially eliminating the pores. Without the pores, plant roots die, and water floods the land surface, rather than entering the groundwater.

There are two relatively simple ways to measure soil compaction onsite. Using an instrument called a penetrometer, you measure the amount of pressure needed to cause the tip to drive into the ground. The more pressure needed, the more compact the soil.

Another technique for measuring compaction is by calculating the bulk density of the soil. Bulk density is the weight of the soil (usually the dry weight) divided by the volume, expressed, for example, as grams per cubic centimeter. Bulk density is influenced by the kinds of particles that make up the soil, and also by compaction. As the pore spaces are reduced in size and number by compaction, the volume of soil of a given weight and texture is reduced, and the bulk density is increased.

To calculate bulk density, collect soil samples as you would for determining nutrient status, only this time, use a large core or small tins of known volume. Dry the soil at low temperature to remove the moisture, and determine the weight. Calculate bulk density by dividing the weight by the volume of the sample.

5.6.7 Soil Drainage

The rate at which water drains from the soil is easy to measure using a percolation test. A simple, qualitative percolation test that works in most restoration situations consists of these two steps:

1. Dig a test hole 15–20 cm in diameter and 20–30 cm deep, fill with water, let drain.
2. Fill the hole again, and measure the time it takes to drain. If it doesn't drain within 45 minutes the second time, the site has poor drainage.

Drainage rates will vary across large sites, so percolation tests should be done in several locations.

For restorations that require more exact soil drainage measurements, percolation tests can be done with more precision. In this case, you have to be sure all the test holes are of the same dimensions, measure the water added, and time the rate of drainage.

5.6.8 Soil Erosion

Monitoring the extent and rate of soil erosion is of particular importance on sites with evident erosion problems. In situations where erosion gullies are present, measure the length, width, and depth of the channels. Such measurements can be taken with relative ease using survey instruments or, if the gully is small, more simply with a tape measure. In other situations where sheet erosion is the issue, you can measure the loss or gain of soil by examining the soil profile using a soil core.

▶ 5.7 FAUNA

As discussed in Chapter 4, the word "fauna" (or "wildlife") can refer to a variety of taxa, including mammals, birds, reptiles, amphibians, invertebrates, and soil organisms. Wildlife species and communities are important considerations in restoration projects for many reasons, ranging from being the focus of the project (e.g., a restoration of endangered species habitat) to being a stressor that interferes with restoration goals (e.g., an herbivore that

CASE STUDY | Analyzing Mine Tailings Prior to Restoration

The process of soil inventory and analysis is of particular importance prior to initiating a restoration on a brownfield site (a location such as a landfill or industrial site that may contain chemicals or other pollutants hazardous to human health), or on waste materials generated by mining activities. Although many states had laws regulating mining prior to its passage, the U.S. Surface Mining Control and Reclamation Act of 1977 (SMCRA) spurred a growing interest in using native species to reclaim mine sites. The goal of the legislation is to protect surface water and groundwater and to return the land to productive use. In 1978, the Jackson County (Wisconsin) Iron Mine awarded a grant to the University of Wisconsin–Madison to study the potential of using native prairie species to revegetate the tailings piles generated by its open pit mining process. Tailings consist of what remains of ore-bearing rocks after the desired mineral has been extracted. Mine operators spread the tailings, containing fine particles of rocks and minerals, onsite at the edge of the mine pit. At the time the study was initiated, no plants were establishing on their own on the Jackson County tailings piles.

The initial goal was to establish a series of field and greenhouse studies to test whether native prairie plants could grow on the tailings. The investigators, Julie Hardell and Gretel Hengst, began by comparing the physical characteristics of the tailings with those of native prairie soils. They established a grid on the study site and collected two soil cores within each grid cell. They sent the samples to the Wisconsin Soil and Plant Analysis Lab for analysis. The particle size analysis indicated that the tailings were sandy loams, which matches the textures of native dry and dry-mesic prairie soils. However, compared with native prairies, the pH of the tailings was considerably more basic than that of native prairie soils (e.g., 8.5 compared with 6.2 for dry and mesic prairies), and available phosphorus and nitrogen were considerably lower. Based in the lack of natural revegetation and the results of the soils analysis, Julie and Gretel hypothesized that the success of using prairie species to restore the site would be significantly enhanced by adding phosphorous and nitrogen fertilizer.

To test this hypothesis, they set up an experiment using nine prairie species, including grasses and forbs (non-grasslike flowering plants) and treatments consisting of no fertilizer, a low and a high rate of ammonium nitrate only, a low and a high rate of monocalcium phosphate only, combinations of both phosphorus and nitrogen at low and high rates, and sewage sludge from a nearby waste treatment plant at both a low and high. At the end of three years, they found that the prairie grasses could establish under any of the fertilizer treatments, but that the treatment using low (88 kg/ha) nitrogen and high (112 kg/ha) phosphorus produced the highest plant cover. The forbs were slow to establish under any of the treatments. Based on these results, they recommended the addition of nitrogen and phosphorus to the tailings before proceeding with the full restoration (Hardell 1980; Hengst 1982).

Building on these and other studies, the restoration firm Applied Ecological Services established an ecological services prairie restoration after mining operations ceased in the 1980s (Ludwig et al. 2000). ∎

destroys desired plants). A site inventory concerning wildlife has two components: describing the presence of fauna on a site (the focus of this discussion), and collecting the information needed to run habitat suitability models. Habitat suitability models are often used to determine the probability that particular species could exist on a site, during the implementation of a restoration or after it is completed (see Chapter 4). The model describes the minimum resources (food, shelter, breeding territories/facilities, etc.), site size, and climatic conditions required for sustaining a certain species. The information you will need to run specific models is an important consideration in establishing your inventory protocols.

The fauna on a site can be described in a number of ways. Depending on the goals of the project, information about overall species diversity, the status of a single species, or the dispersal of a population of species may be important.

Birds, mammals, reptiles, amphibians, and butterflies are the most common kinds of organisms, other than plants, considered in terrestrial restoration plans. Aquatic restorations might also include fish and a number of invertebrates, such as mollusks, corals, and echinoderms. As we learn more about the importance of soil organisms to the functioning of ecosystems, inventories of these organisms will probably also become routine.

Many animal species are a challenge to monitor because they are difficult to find. For example, some insects are very small and remain hidden in tall vegetation. Some, such as the larvae of the regal fritillary butterfly and many moths, are active primarily at night.

Many faunal species are a challenge to inventory because they are highly mobile. Not only do they move around onsite; they also may move offsite for extended periods. Some of these movements are predictable, such as seasonal or daily migrations, but others are less so—the necessity to move during a drought, for example. This mobility is one of the reasons the nature of the physical context of a site is often a key to the success of a restoration (see Chapter 4). Habitat conditions may be ideal on the site, but if organisms are unable to return if they leave, or are unable to thrive in a portion of their habitat located offsite, the restoration of the wildlife community will not succeed.

5.7.1 Walk-Through Transects

Most restoration projects require, at minimum, a record of the wildlife species found on the site. Such presence data provide information on overall site diversity (species richness), and can verify the status of a particular species of interest. For example, if one of your restoration goals is to provide habitat for ground-nesting birds, it would be important to verify that the birds occupy the site.

Inventories of fauna usually start with a qualitative survey. Qualitative approaches are techniques that involve careful observations, impressions, and interpretations rather than numerical measurements. We described site immersion qualitative techniques in Chapter 3. These techniques are particularly useful in gaining an overall impression of the wildlife found in an area.

Another common inventory technique is developing a species presence list by walking the site, and recording any birds, mammals, or insects you observe directly—by sight or by sound. You can also use tracks, scat, nests, browse marks, and so on to infer the presence of additional species. You can collect qualitative information about wildlife, or quantitative data, by keeping track of the numbers of individuals you encounter.

It is important to standardize the walk-through, in terms of any factors that might influence the results. Daily weather conditions, seasons of the year, and time of day all influence the activities of organisms, such that existing species might not be recorded as present. The protocol should specify the times of year the surveys are to be done, thus accounting for seasonal activity, as well as weather conditions when wildlife species are most likely to be active. The area covered and time spent also influence the observations, particularly if the site has a very diverse array of habitats. It is also a good idea to standardize the time spent in the walk-through to help standardize observational detail among subunits.

Using transects, or specified routes, to guide the paths of the surveys is an excellent way to begin to systematize the observations. If the site contains subunits because of the presence of diverse potential habitats, large size, or restoration activities, for example, you should place transects in each subunit. It is helpful to use a system—such as compass directions, chosen at random in several portions of the site—to locate the transects to help avoid bias. However, you choose the route, mark the starting points and any points where the direction changes on a map of the site, and/or record the positions of each observation using a GPS device. In this way, you can tie your observations to specific portions of the site; spatial information is important for site inventory and analysis. You can also repeat the process in the same location year after year—an activity that is important to monitoring.

The walk-through takes little time, and requires very little in the way of equipment. All you need are a means of recording information and binoculars.

5.7.2 Photography

Onsite photography is an excellent means of documenting the presence of animals on a site. You can take photographs of all kinds of species, as well as their traces—for example, scat, tracks, or chew marks. Low-level aerial photography is also useful for some applications, such as documenting the size of herds of mammals in open grassland or in leaf-off conditions in winter.

Three photographic technologies that are especially useful in monitoring fauna have become more widely available in recent years: (1) miniature cameras you can place within underground burrows and nests, (2) monition-sensitive cameras for capturing images automatically when an animal passes by, and (3) infrared techniques for taking images in the dark to record nocturnal activities and behavior. Most of these let you collect images remotely, without disturbing the animals, and often in locations that would be otherwise inaccessible. They can usually be be set to collect both still images and videos. Generally, they can be connected to the Internet and streamed from a remote location.

5.7.3 Systematic Collection of Specimens

Walk-through transect surveys can work quite well to determine the presence of birds, mammals, reptiles, amphibians, and large flying insects; in other words, any organism that is easily visible, leaves large traces, such as tracks, or makes distinctive sounds, like bird or frog calls. However, many species—soil organisms and small insects, for example—cannot be easily detected in this way. In such cases, the inventory protocol will include strategies to collect the organisms to hold for further identification and study.

Techniques range form collecting small samples of the habitat containing the organisms—soil samples or volumes of water—to using various traps, nets, and lures. For example, you can use sweep nets to collect organisms in the air or water, and on land, you can employ fences or other devices that allow organisms to enter, but not leave.

As is the case with determining how to locate transects, it is important to cover the range of possible habitats (and all the site subunits, if present). In addition, you must limit the effects of the collections on the populations. It is best to use traps, for example, that do not harm the organisms, so that they can be released back into the population. If the technique does remove the sampled individuals from the site, the protocol should take into account the potential for impact on the populations, and designate the number of samples and frequency of collection accordingly.

5.7.4 Quantitative Samples

Quantitative techniques involve taking measurements that result in numerical data. Although it generally takes longer to collect quantitative field data systematically than it does to take photographs or compile a qualitative species list, the contributions of such data sets to restoration ecology are invaluable. This is particularly true in attempting to document and understand detailed long-term trends. Quantitative techniques generally have more interresearcher reliability than qualitative ones. This means that the accuracy of the information collected will be less affected by changes in personnel than would be true with qualitative data.

In addition, numerical data generally require less interpretation than qualitative measures. For example, consider the difference between the interpretation of two measures of abundance: "a density of 24 robin nests per hectare" versus a statement that "nests are very dense." A particular restoration ecologist may very well know what she means by "very dense," but subsequent generations of mangers probably will not.

In many cases, it is important to collect information about the size, distribution, age structure, and reproductive potential of one or more species. For instance, a coastal wetland restoration designed to support sustainable populations of shorebirds may have a goal of supporting minimum population sizes. In such cases, quantitative data are essential.

A complete survey of all individuals within the population can be very difficult to accomplish, especially because most animal species move and many are hidden from view, but also because populations can be too large to count with accuracy. Therefore, most quantitative wildlife population-level surveys rely on data taken from a subset of the area—a sample.

The collection techniques described above are often a first step in collecting quantitative information. Once captured, you can hold and examine the organisms—measure, weigh, and in some cases determine age and sex. As mentioned above, you can also count individuals encountered along a walk-through transect, creating a sample of sorts. Additional commonly used sampling techniques are quadrats (for relatively stationary organisms), mark-recapture techniques, and tracking marked individuals (e.g., using telemetry). Each technique can be used to count and measure the organisms themselves, and most also can be used to survey the physical traces left by their activities. Counts of tracks, nests, and scat are often used to estimate species presence and even population sizes.

Quadrats

Quadrats are commonly used to monitor vegetation (see Chapter 3 and the vegetation section below). However, this technique also has a more limited use in monitoring wildlife. A quadrat is an area of defined shape and size that is placed at each sampling point. Usually, a survey collects data from several sampling points distributed across an area. Data are collected on the organisms found within each quadrat. Quadrats are most useful for monitoring "rooted" animals, such as some marine invertebrates (anemones, barnacles), or animals that have a relatively small range or are very slow moving. For example, quadrats work well to count and measure butterfly larvae feeding on plant tissue, or pupae, or egg masses attached to the plants. You simply look at the plants within a quadrat to find and count the immature butterflies. Quadrats are also often used to sample nests, tracks, food caches, and the like.

Quadrats can yield information on species presence and on abundances and demographics (for those organisms for which sex and age are readily determined).

Mark-Recapture

The collection methods discussed above are useful for identifying the species found on the site, and some can also be used to estimate population sizes, such as by using a mark-recapture technique. Mark-recapture is a technique for estimating the population size of a species, the individuals of which move around. The idea is to (1) use live traps, sweep nets, and so on to capture a sample of the members of a population; (2) place a mark, such as a speck of paint, a tag, or a band, on each individual collected; (3) release the marked individuals back into the population; and (4) set up the traps for a second time, and note how many of the new captures had been caught the previous time. Assuming the released individuals distributed themselves randomly within the population, the population size can be estimated using the following relationship:

$$\text{Population size} = \frac{\text{Number marked} \times \text{Total catch second time}}{\text{Number marked recaptures}}$$

Several assumptions underlie the use of the mark-recapture technique: for example the idea that individuals do not become either trap-shy or attracted to the traps, no marks are lost, and that the population does not change in size between first and second capture events. Some risk is involved for the captured animals, from either injuries or shock in handling, or from the kind of mark used. In addition, a number of modifications to the basic equation have been suggested in order to increase the accuracy. It is important to check out the details before using the mark-recapture technique to estimate population size for a particular species.

CASE STUDY | Measuring a Butterfly Population

The grassland-dependent butterfly known as the regal fritillary (*Speyeria idalia*) is considered to be endangered in Wisconsin and is therefore the focus of several grassland restoration projects. Graduate student Katie Beilfuss developed two monitoring protocols for tracking the size of butterfly populations through time, as well as for studying the distribution of the adults across specific sites (Beilfuss 2001).

COUNTS AND DISTRIBUTION

The first method involved walking ten randomly located linear transects, counting all individuals observed within 20 m on either side of the transect line, and recording the location, behavior (flight, resting, nectaring), and, when possible, sex of each individual. The protocol specified that the surveys be conducted between 11:00 a.m. and 6:00 p.m. CST on clear or partly cloudy days when wind speeds are less than 24 kph and temperatures are between 24°C and 34°C (75°F and 93°F). The protocol further specified that two-person teams conduct the survey, one person making the observations and the other recording the data. By matching the locations of each of the transects with a vegetation map of the site, it was possible to correlate the density of butterflies with vegetation measures such as shrub cover or the abundance of specific plant species.

MARK AND RECAPTURE

The second method used a mark-and-recapture technique, as a second estimate of population size and, by observing the marked individuals, to determine the dispersion patterns across a site and between locally discrete sites. This protocol called for capturing butterflies using a sweep net between 8:30 and 10:30 a.m. CST on cool, cloudy days when the butterflies are expected to be less active. The captured individuals were marked with a unique identification number on the underside of both hind wings, using an ultra-fine permanent marker and an extra-fine Marvy/Uchida DecoColor Liquid silver opaque paint marker (Figure 5.7). In addition, the sex and location of capture were recorded. Once marked, individuals were inserted into individual glassine envelopes and placed in a cooler with an average temperature of 19.6°C (67°F) until the end of that day's capturing event. The purpose is to reduce "capture-release trauma" and to prevent same-day recapture.

The recapture event then was to take place within the next 5 days. In addition, the protocol called for observations to be made across the prairie, to see how far marked individuals have spread from the original capture site.

RESULTS

Using the first method, Katie discovered that the density of adult regal fritillary butterflies was significantly higher in areas of lower shrub cover, abundant native grasses, and large numbers of violets, an important food source. This finding is important for creating a habitat suitability model for this endangered species.

Using the second method to study the daily population sizes of *Speyeria idalia* in two different prairies, Katie found that the populations of males were greatest in July, reaching an estimated peak of 372 individuals at one site (density of about 22 individuals per hectare) and 171 (8 individuals per hectare) at the other site. (The number of females captured was too small to use for these estimates.) The greatest distance recorded for an individual between the original capture location and the site of its recapture was 1,030 m. ∎

FIGURE 5.7 **Marking Butterflies.** *Speyeria idalia* were marked with a unique identification number on the underside of both hind wings. (Photos reprinted by permission of Megan Foss.)

Tracking Marked Individuals

One way to understand how organisms are moving about a restoration site, as well as where they go when they leave the site, is to fit a transmitter to an animal, using a collar or other device, and then using a radio or GPS device to record the animal's movements through space and time. This technology is becoming more and more affordable, and the transmitters are becoming smaller and more lightweight. As a result, these devices will likely become widely used in surveying restorations, especially those established as habitat for specific wildlife species.

If the animals have been tagged, using the same kinds of marks discussed above for the mark-recapture techniques, and are subsequently found in another location, this information helps in estimating travel ranges. Bird banding and butterfly marking have contributed much information about the migratory habits of species in this way.

▶ 5.8 VEGETATION

The vegetation of a site can be described in a number of ways, but restorations are generally concerned with plant communities and/or individual species populations. Given that most of the vegetation measures concern species, it is important to establish, at the very beginning, the taxonomic classification system to be used during the inventory. Usually this is done according to the nomenclature adopted by state, regional, or national herbaria, or by referring to local university experts or to the taxonomic publications used by the professional botanists in the region.

As with animals, you can use both qualitative and quantitative techniques to collect onsite vegetation information. The choice of technique depends on desired accuracy and reliability, and on the time and resources available.

5.8.1 Qualitative Techniques

The qualitative techniques that are used most often in onsite vegetation assessment are the site immersion technique and the walk-through survey. It is helpful to have a checklist of items to look for during the walk-through in order to make the best use of the visit. This approach is particularly helpful in compiling species presence lists, noting signs of disturbance (recent blow-downs in wooded areas, patches of pest species invasions, etc.), and locating populations of particular species (e.g., rare species or signature species).

5.8.2 Quantitative Techniques

Data on species composition and abundance are the most common quantitative factors used to evaluate the vegetation community as a whole. When population information is important, you may want to include measures of size, pattern, reproductive status, and age class for specific species or life forms (such as trees). In some circumstances, ecosystem functions mediated by vegetation may also be of interest.

With a few notable exceptions, such as tracking the fate of all individuals within a population of a rare species, most quantitative descriptions of vegetation rely on data taken from a subset of the area—a sample. The two most common techniques are the quadrat method and the line-intercept method.

Theoretically, quadrats can take any shape, but they are usually circles, squares, or rectangles, chosen for ease of construction, among other things. Sizes range from 0.25 m^2 in dense vegetation cover, such as grasslands, to 100 m² or larger for use in sampling trees in a forest. Information is collected about the plants within the quadrat, always including the species present, sometimes including counts of individuals (density), estimates of cover, and, for trees, diameter. You can also use quadrats to map the locations of individual plants in order to understand the patterns of their distributions, calculate blooming frequencies and seed set, and determine survivorship (Figure 5.8). (You can see examples of vegetation maps in Chapter 3.)

The line-intercept method involves stretching a line of known length in a specified direction from the sampling point, and recording the length of line that is touched by different plants, or sometimes by their shadows. Vegetation cover can be estimated by dividing this vegetation intercept by the total length of the line.

Here are two more vegetation sampling techniques that may be useful. In the plotless point-quarter method, developed for sampling trees in a forest, each sampling point is divided into four equal sections or quarters. In each of the four quarters, the distance from the sampling point to the nearest tree is measured and recorded, as is the species of the tree and its diameter. The point-intercept method is especially useful for sampling plant cover in shrubland, grassland, and lichen communities. A thin pin or wire is pushed vertically from above toward the ground, and the investigator records the number of points that touch vegetation compared with the number of points that do not. (Information about each species is often recorded separately.)

The choice of method depends on the kind of information needed, as well as the kind of communities that are present. For example, the line-intercept and point-intercept methods were developed to measure cover directly. If cover information is of primary importance, one of these techniques might be a better choice than the quadrat method, in which cover measurements are often less precise.

FIGURE 5.8 **Sampling a Forest Using Quadrats.** These students are inventorying the vegetation by recording and counting the stems of all of the species found within a 1-m² quadrat. (Photo reprinted by permission of Evelyn Howell.)

Distributing Samples

After specifying the particular sampling technique to use and what information to collect, the next decision is to determine how to distribute the samples. You need to decide how many sampling points to include, and where to place them. These decisions depend on the variability of the vegetation, as well as on the desired level of detail. For example, many more samples will likely be needed if it is important to take measurements on all species present on the site than if information only about the common species is needed (see Chapter 9).

Most restoration projects are expected to last for many years. Depending on the situation, it may take decades for plants and populations to mature; and once the restoration goals have been achieved, restorations often need periodic management interventions (see Chapters 8 and 10). In addition to conducting an inventory at the beginning of a restoration project, it is necessary to collect site vegetation information as the restoration enters the site design, implementation, and management phases of the process. Therefore, it is helpful to design sampling protocols so that the same system can be used for informing each phase. (For a more detailed discussion of sampling design, see Chapter 9.)

5.8.3 Identifying Unknowns

Sometimes it is hard to identify all the plants encountered in the field during an inventory. Perhaps all that is showing at the time of the visit are a couple of leaves, or the taxonomic key you are using requires the presence of flowers, and the inventory is being conducted before the flowers will appear. In these cases, there are several options available.

One approach is to mark the plants in some way, such a placing a flag in the ground next to each plant. Then, return later when the flowers or more parts of the plant are showing. Another possibility is to photograph the plant; take the image to a local expert for identification, or to a local herbarium in order to compare the image with preserved specimens. Many herbarium collections have searchable online image collections—both photographs of living plants and scans of preserved specimens. In the next few years, it will become increasingly possible to access such collections onsite using wireless technology.

As a last resort, it is possible to simply acknowledge that in making the inventory, you have encountered several unknowns. Depending on the goals and objectives of the restoration project, it may be a problem if a large number of the plants are unknown—for example, if an objective sets expectations for a proportion of all the species on the site to be natives. In other cases, such as when the objectives specify a minimum plant cover or the presence of certain life forms without regard to species, having a number of unknowns may not be an issue. Regardless, you should establish protocols for handling unknowns.

5.8.4 Collecting Specimens

In some situations, it is helpful to collect plants in whole or in part, either to provide reference collections for future site stewards, or to take to experts for help with identification. Reference collections usually use dried materials that need to be carefully housed and cataloged. Such an undertaking requires both time and money for preparing and preserving the specimens, and providing proper cabinets, climate control, pest control, and so on. As an alternative to housing the materials onsite, it is sometimes possible to persuade an established herbarium to accept and manage the collection.

Whatever the reason for collecting specimens, it is important to consider the impact on the site. There is the potential of damaging the individuals themselves (especially if the whole plant, including the roots, is removed). But there is also the potential loss of growth and reproduction (even if only a portion is removed), as well as the possibility of disturbing the soil or neighboring plants. In most cases, the benefits of collecting will outweigh the potential problems. Nevertheless, the monitoring plan must be clear about whether and under what conditions collecting is allowed.

▶ 5.9 CULTURAL FEATURES

The phrase "cultural features" refers in the broadest sense to anything that shows evidence of human interaction. These can be physical structures, such as buildings, fences, or roads; landforms shaped by human activities; gardens or other vegetation features planted by people; or even a location considered to be important because it is the site of archaeological remains or a historic event. As we discussed in Chapter 4, information about many of these features can be found from aerial photographs, books, and other archives.

Onsite inspections can yield information about condition and provide more precise measurements (necessary when considering a structure for preservation because of historic value); inspections can also can locate cultural features not readily apparent from existing maps. In the case of suspected archaeological significance, it is often necessary to call in teams of cultural resource professionals to do a detailed investigation.

▶ 5.10 PEOPLE

It is important to know how people are using and experiencing a site, during all phases of restoration—at the beginning and throughout the initial planning stages, while the project is being implemented, and once the goals have initially been met. The interactions of people with a site are especially important to understand for those projects that welcome people to participate in the site, but this information is also important for projects that hope to limit human use (see Chapter 1). The kinds of information that can be most useful include:

- The extent to which people are currently visiting and using the site, the extent that they have done so in the past, and their reasons for doing so. For example, community members might rely on the site for food, fuel, recreation, or other functions, such as transportation routes.
- The extent to which these onsite human activities conflict with the biological and physical goals of the restoration.
- The extent to which offsite human activities influence the attainment and maintenance of the biological and physical goals of the restoration.
- The extent to which a community supports a project.
- Any specific concerns or questions about the restoration that exist within the community.

Similarly, it is important for projects that rely on volunteers for fund-raising; assisting with onsite implementation, monitoring, and management activities; or leading educational tours to understand participant satisfaction.

Four approaches that can provide information on how different groups of people (visitors, volunteers, and community members) interact with a site are direct observations, interviews and questionnaires, public meetings, and participatory photo mapping.

5.10.1 Direct Observations

Techniques for observing people using and interacting with a site have a long history in the fields of landscape architecture and environment-behavior studies. A variety of such techniques exist, ranging from simply watching people onsite and taking notes about their numbers and behavior, to employing a standardized checklist that focuses attention on particular kinds of activities. Observations can be quantitative, as in head counts, or qualitative—for example, sketches, photographs, and notes.

One of the benefits of direct observations is that this approach gives you firsthand information about how the site is being used. One of the drawbacks is that the presence of an observer can influence user behavior. One strategy for minimizing this drawback is to be as unobtrusive as possible. For example, dress or behave like a typical user, and take mental notes and write them down later, or sit quietly for a while before starting to record information. Another option is to use photography or videography from a remote location to record information for later analysis.

In addition to deciding how to conduct the onsite observations, it is important to think about where and when the observations will take place. To a large extent, the nature of the site guides these decisions. If the site has specific seasonal patterns of use, for instance, it would be important to conduct the observations at least once a season. If use patterns are expected to be different on weekdays from those on weekends, or under different weather conditions, these ideas should be reflected in the observation schedule.

It is also helpful to conduct two rounds of observations. In the first round, take note of anything you see, regardless of how relevant you think it might be to the project at hand. Try to capture as much use variation as you can. Then, back at the office, analyze your notes to see if any patterns emerge, and organize the data to create a checklist of things to look for onsite. Then return for a second round of observations, using the checklist to focus your attention on the human–environment interactions you have determined to be most relevant to your goals. You can also use the checklist to record the numbers of vistors who are using the site in similar ways.

5.10.2 Interviews and Questionnaires

Firsthand observations can be very valuable in determining numbers of visitors and patterns of site use, and in identifying conflicts among uses or with site integrity. However, they generally are not effective in addressing restoration goals concerning visitor, volunteer, or community satisfaction; nor can they provide much information on what the site means to people, or how participating in a restoration changes or adds to the understanding or appreciation of biodiversity, a common restoration goal. Interviews and questionnaires are effective for getting at such issues focusing on values, opinions, or impressions.

Interviews
Interviews, of course, involve direct contact with people. Usually, you ask the subject a series of questions, and record the responses by taking notes and/or using a recording device. The interview can be open-ended, letting the conversation flow in the direction

the subject takes it. For example, you might begin an interview by asking a visitor how and why she learned about the site, and then ask follow-up questions pertaining to her initial responses. Open-ended interviews are particularly valuable in uncovering meanings and sources of satisfaction.

Alternatively, the interview can be tightly focused, asking for short answers to specific questions. You might want to know, for instance, what kinds of activities people enjoy at the site. By asking visitors to name three things that they engage in, you could generate a short list of the most common things people do at the site.

Interviews are particularly valuable if you are interested in the rich texture of associations and meanings attached to people's interactions with a site. Because you are working directly with someone, you can interpret body language and tone of voice in addition to what is spoken. And, should the situation warrant it, you can modify the questions as the interview proceeds. However, interviews are time-consuming to administer, require well-thought-out and perhaps pretested questions, and with the open-ended format, it can be difficult to interpret narrative responses.

Questionnaires

Questionnaires differ from interviews in that they are administered remotely. They are similar in that the questions can call for either open-ended or brief responses. Questionnaires are particularly useful for generating quantitative data, such as by asking respondents to rate the degree to which they agree or disagree with a statement, or to list items in order of preference.

You can distribute questionnaires onsite or to randomly selected members of the community. You can design them to be filled out and collected onsite, or to be returned via regular mail or e-mail, or to be completed online. It is advisable to do a practice run first to be sure the questions are worded in an understandable way and are generating the kinds of responses you expect.

It is also a good idea to solicit feedback in less formal ways—for example, by posing questions in neighborhood newsletters or on a website, if one is available. Though less structured than answers to a questionnaire, such responses can draw attention to emerging issues.

Respecting Privacy and Cultural Traditions

No matter which technique you use, it is important to respect privacy rights, either by avoiding the use of names or other identifying characteristics in summarizing the information and/or by obtaining the consent of the participants. Restorations that are conducted by public groups or agencies may have to follow explicit procedures, mandated by law. It is also important to respect cultural traditions, especially if you are from a different part of the country or region.

Timing

Because of the logistics involved, interviews and questionnaires are usually not used as frequently as direct observations. The timing and coverage (how many people are included) depend on the goals of the project, and may vary among different groups. For example, if the site has a volunteer coordinator, he or she might want to interview a subset of the volunteers once a season, and all of the volunteers at least once every two years. Managers of long-established restorations might do visitor surveys less frequently, perhaps every three years, unless something has changed.

5.10.3 Public Meetings

Meetings in which members of the public are asked for their concerns and opinions are a vital component of the restoration planning process. During these meetings, you can gauge the extent of community support and uncover concerns, which, if left unaddressed, could undermine the long-term survival of a project. (We discuss several techniques for conducting public meetings in Chapter 13.)

It is particularly important to hold public meetings in advance of any anticipated change—either a change in restoration or management activities, or a change in programs or policies—or immediately after any unanticipated events (e.g., wildfires or floods). Also, because people invariably move in and out of neighborhoods, it can be a good idea to meet with members of the public, at least briefly, once every year or two. The updates could be part of the agenda of a regular neighborhood meeting, or they could be part of a field day, during which the public is invited to explore the restoration site.

5.10.4 Participatory Photo Mapping

Participatory photo mapping (PPM) is a relatively new approach to understanding how people experience a site. The technique has been made possible by the advent of digital photography and GPS technology. The idea is to equip people with digital cameras linked to a GPS system, and ask them to take pictures of any feature of the site that has either a positive or a negative meaning for them. You can then download the images into a GIS database and map the locations of the observations.

You then invite the participants to describe each of their pictures in order to understand what site features are important to them and why. People of all ages—from very young to very old—can participate with ease. Because of its participatory nature, PPM has great promise for helping establish patterns of use, as well as for designing restorations that appeal to people.

5.10.5 Monitoring Human Impact

Regardless of a site's use-policy, there is always the potential for human impact, from both on and offsite activities. Many of these impacts involve changes to vegetation, wildlife, soils, and hydrologic systems, and will be noticed in the inventory protocols for these resources.

Two additional things to be aware of are vandalism to signs, benches, or other kinds of infrastructures, and physical traces—items such as trash, footprints, broken branches, fire rings, or initials carved in trees that are left behind in the wake of human occupancy. The most direct way to look for these kinds of impacts is to do the same sort of qualitative walk-through inventory as described earlier in this chapter for animals and vegetation. In addition to taking notes about the location and types of impacts detected, it is also very helpful to take photographs to document the extent of the damages.

▶ 5.11 INVENTORY PROTOCOLS REVISITED

One of the major themes in both this chapter and Chapter 4 is that, in order to produce the most useful results, your inventory and analysis protocols should match your questions. Inventory and analysis occurs at several stages of a restoration, starting with the design of the master and site plans, and continuing with implementation, monitoring, and management. Moreover, the adaptive approach to restoration often requires inventory and analysis in the course of evaluating built-in experiments. The protocols describe what is to be done, where, when, and by whom.

The information you need for conducting a restoration at each of these stages will often differ in scale and detail. For example, you may be able to determine which communities you can restore on a site by knowing the general climate of a region, but you have to know more about specific soil, microclimate, and hydrological variations in order to establish particular species within these communities (see Sidebar). Therefore, in creating your protocols, planning ahead is essential.

As you have seen, there are often many different data collection and analysis tools from which to choose, each of which provides particular kinds of information. Thinking about the pros and cons of several alternatives before making a decision will give you confidence in your choice, and your reasoning will be clear. For example, if you need to know that a forest has a density of 50 trees per ha, your inventory protocol would not call for a qualitative inventory, as such a method is not designed to collect numerical information. On the other hand, if you need to know if a particular rare plant species is present on a site, a thorough qualitative inventory, rather than a quantitative sampling technique, will likely be your best choice. Other things to consider include the past experience, training, and expertise of your restoration team (including volunteers) in using the tools; the time and money that are available; the availability of storage facilities; and the durability, ease of use, safety, and longevity of the equipment. Always remember that it can be a good idea to hire outside experts to conduct portions of your inventory and analysis, especially when special expertise is involved. If you do so, it is especially important to work with this team to be sure they follow your protocols.

Sidebar | Designing an Onsite Inventory Protocol

One of the restorations at the University of Wisconsin–Madison Arboretum is the Leopold Pines. This 8.5-ha community was created as an example of a community that is native to northern Wisconsin, but not to the region where the Arboretum is located. The idea was to provide the opportunity for students to study this community, without having to travel far from campus. The restoration began in the 1930s with a planting of red and white pines (*Pinus resinosa* and *P. strobus*), followed by underplantings of northern Wisconsin species of shrubs and herbs. As of 2011, although the trees have established well, the understory plantings have met with limited success.

The Arboretum managers have several ideas about why the understory is not doing well. These include concerns that, with regard to the restoration site:

- The soils do not match the requirements of the northern species.
- The snow cover in winter is too shallow and too uncertain to protect the northern species.
- The litter layer is too deep.
- The light levels under the trees are too low.

FOOD FOR THOUGHT

1. Assume you have been hired by the Arboretum to conduct an inventory of the restoration site so that the managers can evaluate whether these concerns are plausible. Describe what information you would like to collect, what techniques you would use, and why.

KEY CONCEPTS

- Every restoration project relies on information that is collected onsite using protocols designed specifically for the situation at hand.

- The protocols are sets of instructions that specify what is to be collected, which tools to use, when and where to apply them, and how to summarize and interpret the resulting information.

- Onsite inventory techniques include systematic observations, photography, simple instruments such as metersticks and handheld thermometers, and sophisticated battery-powered digital recording meters. In some cases, samples are collected in the field and taken to laboratories for chemical and other types of analysis.

- Features that are commonly measured onsite include microclimate; topography; water quality, quantity, and movement; soil composition and structure; plant and animal community composition and structure; species population sizes, structure, and dynamics; evidence of contemporary and historic human presence; and human experience and perceptions.

FOOD FOR THOUGHT

1. You work for a conservation organization that has acquired a property in your region for the purpose of creating a nature study preserve. List the steps you would follow to design onsite data collection protocols for the initial site inventory and analysis.

2. For each of the following, name two ways by which you could measure each of the following site features. Which one would you choose, and why?

   A. The number of different bird species present

   B. Soil texture

   C. The areas with the most/least sun exposure

   D. The rate of spread of an invasive species

   E. The ways in which people interact with a restoration

3. Would you use the same protocols for an initial site inventory and for checking on the progress of a restoration as it is being implemented? Why or why not?

4. Under what circumstances would the soil maps described in Chapter 4 provide sufficient information for a restoration plan? Under what circumstances would you need to collect additional onsite soils information?

5. Do you think it is possible to design a restoration without visiting the site yourself? Why or why not?

Resources for Further Study

1. You can find information about a variety of physical site features and their measurements in the following handbook, written for land use planners of all kinds:

   Marsh, W. M. 2010. *Landscape Planning: Environmental Applications*, 5th ed. San Francisco: Wiley.

2. The following website contains helpful information about measuring microclimate, including instructions on how to use and site equipment:

   United States National Weather Service Cooperative Observer Program (COOP) http://www.nws. noaa.gov/om/coop/

3. For information on measuring the hydrology of streams and rivers, check out the following:

   Hauer, R. F., and G. A. Lamberti. 2007. *Methods in Stream Ecology*, 2nd ed. San Diego, CA: Elsevier.

   U.S. Environmental Protection Agency. *Voluntary Stream Monitoring: A Methods Manual*. 1997. http://water.epa.gov/type/rsl/monitoring/stream_index.cfm

4. For an overview of plant and animal field ecology inventory and evaluation methods, you might like to try:

   Henderson, P. A. 2003. *Practical Methods in Ecology*. San Francisco: Wiley.

   Hill, D., M. Fasham, G. Tucker, M. Shewry, and P. Shaw, eds. 2005. *Handbook of Biodiversity Methods: Survey, Evaluation and Monitoring*. New York: Cambridge University Press.

   Southwood, T. R. E., and P. A. Henderson. 2000. *Ecological Methods*, 3rd ed. San Francisco: Wiley.

   Young, L. J., and J. H. Young. 2007. *Statistical Ecology*. New York: Springer-Verlag.

5. If you are interested in learning about participatory photo mapping, check out a website developed by Sam Dennis at the University of Wisconsin–Madison. Professor Dennis focuses on studying links between children and the environment from a health perspective.

   http://www.la.wisc.edu/ppm/

The Master Plan

Project Purpose, Solutions, and Goals

> **LEARNING OBJECTIVES**
>
> *After reading this chapter, you will be able to:*
>
> - Recognize the differences between master, site, implementation, monitoring, and management plans and where they occur in the restoration process.
>
> - Explain what a master plan is and the attributes shared by all master plans.
>
> - Identify the structural components of a master plan and their importance to the restoration planning process.
>
> - Assess when during the restoration process collaboration between interested and affected parties needs to occur.

Restoration plans can and, in practice, do take several forms. Some reside only in the minds of the restoration team, and others consist of sets of informal notes and sketches (organized or not). The most useful plans, however, are more formal; that is, they are developed in collaboration with neighbors and advisors and use text, maps, photographs, and graphics to describe the situation in enough detail that someone other than the original authors can understand how to proceed. Because restorations often take decades to reach their initial goals and project directors often move on before such benchmarks are reached, documents that describe the vision and assumptions of a project are invaluable to future site managers. Also, most restorations involve public lands and/or rely on public support in the form of monetary gifts and grants; donations of time, materials, and equipment; and

FIGURE 6.1 **The Restoration Process: The Master Plan.**

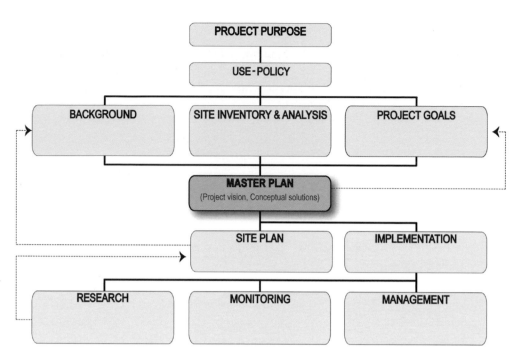

political support. An easy-to-understand planning process greatly facilitates communicating and working with people. The resulting document also provides a benchmark against which to determine whether the investment of time and effort has been well spent.

As discussed in Chapter 1, a restoration project involves the creation of five interconnected plans:

1. The master plan, which presents the project vision in general terms, suggests several possible solutions, and then identifies the communities to be restored and their relative locations according to the preferred alternative (Chapter 6; Figure 6.1).

2. The site plan, which adds detail to the project vision, and designs and gives form to the communities described in the master plan (see Chapter 7).

3. The implementation plan, which provides detailed instructions on how to realize the site plan (see Chapter 8).

4. The monitoring plan, which explains how to determine whether the goals and objectives described in the site plan have been met (see Chapter 9).

5. The management plan, which describes how to ensure that the restoration continues to meet the goals and objectives for many years to come (see Chapter 10).

The thinking behind the steps of these various plans works, even if circumstances do not allow all the ideas to be preserved in written or graphic form. And what is a plan? Simply put, a plan is a guide telling participants what is to be done—where, when, and why.

▶ 6.1 PLAN FEATURES

Three features are applicable to all plans in the restoration process (i.e., master, site, implementation, management, and monitoring plans): collaboration, format, and flexibility.

6.1.1 Collaboration

Restorations are most successful in the long run if they are planned using a collaborative process involving neighbors, members of the community interested in conservation issues, and invited area professionals. Participation, when done well, builds enthusiasm, interest, understanding, and support. In the United States, a procedure that allows for public commentary is usually required for projects on public lands. Such a process is highly desirable, perhaps with a more limited number of participants, on privately held properties. We will discuss various procedures for obtaining public input during the planning process in more detail in Chapter 12. Some of the more common approaches are initial brainstorming sessions, in which problems are articulated and various solutions proposed; public hearings to discuss potential alternative solutions; and opportunities for submitting written commentaries on plan drafts.

Regardless of the techniques or approaches used, it is important to include all the stakeholders or interested parties during several stages of the project (Table 6.1). At minimum, the collaborative stages include:

1. The very beginning, when information is first being gathered and general goals are being agreed upon.
2. The stage when alternative solutions are being discussed.
3. The point when the final plan is ready for adoption.

In our experience, the earlier in a project you involve people, and the more chances you give them to comment on the process, the more likely it is that they will become supporters—even enthusiastic ones. If a planned action ultimately goes against the wishes of a

Table 6.1 · Potential Stakeholders of Restoration Projects	
General public	Landowners, client (if not landowner), adjacent landowners and neighborhoods, potential site users, civic groups, private sector partners (landowners, special interests, industry, and agricultural groups), resource user groups (cyclists, runners, skiers, hunters)
Project team	Restoration ecologists, engineers, landscape architects, geologists, soil scientists, hydrologists, botanists, wildlife ecologists, entomologists
Project volunteers	Team leaders, volunteer coordinators, big event volunteers, repeat volunteers, on-the-ground volunteers
Site manager and staff	Site managers, office staff, research staff, ground maintenance staff, restoration specialists, public relations staff, seed collectors, pest plant removers, botanists, wildlife ecologists, plant propagators
Government units and authorities	Federal (Forestry Service, National Park Service, Bureau of Land Management, U.S. Army Corps). Similar agencies exist in many countries; state (Departments of Natural Resources and similar governmental units, oversight and regulatory units); regional and tribal (Regional Development Commissions, Conservation Districts); taxing bodies, elected officials, government partners (city, township, parish, county, regional, state, commonwealth, federal, and tribal governments)
Nonprofit environmental organizations	Tourism, business and commerce groups, friends groups, area colleges and universities, contractors involved with implementation

portion of the participants, the fact that the restoration planners carefully considered their points of view can help minimize their concerns. Of course, the best situations are those in which all parties are satisfied.

It is important to note, however, that in some circumstances the need to protect the resources being restored outweighs the desirability of public collaboration in all steps of the planning process. For example, you may want to keep the locations of certain endangered species out of the public arena. In such cases, collaboration with a more select group is still helpful.

6.1.2 Plan Format

One very important format of all five types of restoration plans is that they include plan graphics—representations of the physical location, size, and shape of the communities to be restored and/or the actions to be taken. Maps and pictures help convey at a glance the vision of the restoration designers. They both supplement and complement the text. Depending on the situation, plan graphics can be informal sketches or accurate scale drawings. You can hand-draw or create them using any of a number of computer applications, especially those using computer-aided design (CAD) or geographic information system (GIS) software. You will find examples of each of these formats throughout this text.

6.1.3 Plan Flexibility: Midcourse Corrections

Another essential characteristic of formal restoration plans is flexibility: They have to be flexible, which means including built-in strategies for review and for making modifications—midcourse corrections—if necessary. Restoration plans and the thinking behind them need to be able to develop through time, as will the communities they support. As discussed in Chapter 3, this idea is sometimes referred to as adaptive restoration. The importance of this approach cannot be overstated.

Flexibility is essential because, for one thing, as we discussed in Chapter 2, our knowledge about the natural world and about restoration practice continues to increase. Just as doctors switch treatment regimes as new medications come on the market, you and your restoration team need to be able to change the specifics of a plan in light of new information. Similarly, an unanticipated drought or windstorm, or a land use change on the borders of a site, would call for a reconsideration of the plan and, potentially, a change in procedures. Global climate change will make flexibility even more important in the coming decades as the microclimates of restoration sites shift, with as yet unpredictable effects on community composition and structure.

Flexibility does not, however, imply modifying a plan without careful consideration. A formal review process helps ensure that any proposed changes in a plan have been carefully discussed before actions are taken. The review process you use should do the following:

1. Identify the review participants (site managers, members of the public, etc.).
2. Specify who is authorized to approve and/or adopt any recommendations made as a result of the review.
3. Include a schedule whereby a restoration plan, in whole or in part, is updated on a regular basis. For example, the whole plan might be evaluated every ten years, whereas some of the objectives might be reviewed more frequently.
4. Specify situations, such as a flash flood, that would initiate an immediate review.

The social programming that is part of many restorations should follow a similar review process.

Restoration plans can be prepared in several formats, including bound paper documents, loose-leaf notebooks, and digital files. In most cases, all five plans (master, site, implementation, management, and monitoring) are included in a single document; sometimes each one is kept separately. Plans must be able to last over the life of the project—in other words, over many decades. Therefore, the chosen medium or media should be designed to last. Because technology is changing rapidly, if a digital format is chosen, all the pertinent files (contemporary as well as historic) will probably have to be transferred to new formats every few years in order to remain accessible. These changes involve both software and hardware, and program updates, as well as decisions to switch programs. For example, plans written using one kind of publishing software may not be readable with another. Documents stored on floppy discs—the secure storage media of choice in the 1980s and 1990s—can no longer be read by most computers without obtaining an external floppy disc drive, a device that is now hard to come by.

One way to maintain plan flexibility and track the updates through time is to produce new plan documents after every review, and archive all the previous versions. In the case of digital plans, it is helpful to convert the materials to be archived to paper documents, thus avoiding the obsolescence issue described above. A second approach would be to continue to attach amendments to the original plan.

▶ 6.2 THE MASTER PLAN STRUCTURE

The main goals of a master plan are to (1) describe the layout, vision, and need for the proposed restoration; (2) explain why it is the optimal solution to the situation at hand; and (3) provide enough information so that future managers will have a theoretical and practical context within which to proceed with the plan, or to make changes, if necessary. A formal master plan generally includes seven structural components (Figure 6.2).

Different restoration projects may vary with respect to the components they include. In some cases, components may be combined, and in other cases you may wish to further divide the components in order to place emphasis on a particular aspect of the project. And, in some instances, the master plan will include general descriptions of components that are more detailed in the site, implementation, monitoring, and management plans (see Sidebar).

A. Project Overview
  1. Project purpose
  2. Use-policy
  3. Restoration goals
B. Description of the Site Inventory and Analysis, Including the Methods and Sources Used
C. A Set of Alternative Solutions
D. Choosing and Justifying the Desired Solution

E. Goals for Each Community or Restoration Unit (in the Solution of Choice)
F. The Estimated Budget
  1. The planning budget
  2. Personnel
  3. Materials
  4. Contingencies
G. Procedures for Plan Adoption, Review, and Modifications

FIGURE 6.2 Structural Components of the Master Plan.

Sidebar | **Components of the Hog Island and Newton Creek Ecological Restoration Master Plan**

The following master plan components were used in the 2007 Hog Island and Newton Creek Ecological Restoration Master Plan by Biohabitats, Inc. (These components are discussed conceptually in Chapters 4, 6, 7, and 8.) Hog Island and Newton Creek contain important habitats for local and migratory fish and wildlife populations located in Lake Superior. Stakeholders for this project include federal and state agencies, the City of Superior, the St. Louis River Citizen Action Committee, and local citizens and interest groups. This plan also serves as a possible template for ecological restoration for the Great Lakes watershed.

Here are the master plan components:

- Refined vision, goals, and measurable objectives (Project Overview)
- Ecological and cultural descriptions of the sites (Site Inventory and Analysis)
- Description and location of potential reference ecosystems (Site Inventory/Background)
- Description and conceptual site plan (Master Plan Solution)
- Description of how the proposed restoration initiatives will integrate with regional landscape processes (Alternative Solutions and Desired Solution)

- Implementation strategy, phasing plan, and a timeline of the restoration process (Goals for Each Community or Restoration Unit in Solution of Choice/Implementation Plan)
- Framework for performance standards, each with suitable monitoring protocols (Monitoring Plan)
- Description of the ecological institutional strategies that will be required for the long-term protection and management of the restored ecosystem (Procedures for Plan Adoption, Review, Making Modifications)
- Regulatory permitting requirements (Procedures for Plan Adoption, Review, Making Modifications)
- Framework for ensuring that stakeholders are involved in the continued development and management of the restoration initiatives (Procedures for Plan Adoption, Review, Making Modifications)
- Planning level cost estimates for final design, implementation, and management (Estimated Budget)

Adapted from Biohabitats 2007.

▶ 6.3 PROJECT OVERVIEW AND PURPOSE

The project overview introduces the intent of the project by discussing why it exists, who is it for, how it is to be used, and what the broad-based targets are that the restoration proposes to achieve. The project purpose is a simple statement of the reason for a project. It summarizes the problem that the restoration is attempting to solve. The statement is action-oriented and refers to the situation as it is at the time the project begins. Here are some examples:

- Establish habitat for whooping cranes.
- Develop a vegetation complex representative of natural conditions.
- Increase the native biodiversity of the Leopold Reserve.
- Reverse the eutrophication of Kettle Pond.
- Return the river to a previous course.
- Establish a system that increases the carbon storage capacity of a site.
- Create a nature study laboratory.
- Provide a nature-based recreation facility.
- Establish native communities for research and education.

In addition to presenting the purpose per se, it is most helpful if your plan describes the circumstances that have led to the project. For example: "The site has just been purchased by a private conservation group in order to be able to further its goals of rare species preservation." "A college has received a donation of land on which the faculty plan to create a restoration field research laboratory." "A community has decided to remove a millpond dam that has been in place for more than a century." "Heavy grazing by domestic livestock has denuded the landscape to such an extent that it promotes flooding in wet seasons and can no longer support the needs of native wildlife species, some of which are threatened or endangered." By including as much information about the issue as possible, you make the context of the restoration clearer (see Chapter 4).

The purpose defines which type of restoration (complete, process-based, experiential, etc.) to create, and it helps to define the use-policies (see Chapter 1).

6.3.1 The Use-Policy

The use-policy is a description of the ways in which people will interact with the site. In some cases, the ways that people will or will not be allowed to use the restoration are known at the start of the restoration project. In other cases, the use-policies are set following the initial site inventory and analysis, and are determined to a large extent by the results. Generally two sets of policies are included: those pertaining to visitors and those governing the choice and implementation of restoration and management activities. These two sets of policies both influence and arise from the restoration purpose and from planning decisions and considerations, such as:

- The kind, size, form, structure, function, and composition of the communities to be restored.
- The sensitivities (as discussed in the community/ecosystem models) of the proposed communities to different kinds and degrees of human use (sites containing rare species might be closed to the public).
- Whether or not to include visitor amenities, such as trails or benches (such design features can help confine visitors to small portions of a site, thereby minimizing impacts, but also introducing human elements to the setting).
- The physical context of the restoration (an urban context might preclude the use of herbicides or prescription burns, and be exposed to many potentially invasive species).
- The sociopolitical context of the restoration (sites owned by the public or projects supported with public funds often require public access to site resources).

Discussions about use-policies start at the beginning of the master planning process and continue as the site and implementation plans proceed. It is often helpful to begin a restoration plan with the idea that all visitors will be allowed, and that only hands-off implementation or management strategies will be employed. Next, the consequences are listed, both good and bad, of such a position. Finally, only those uses are proposed that are needed for the project to succeed. By considering what the situation would be like without them, this approach ensures that core uses are identified.

The Visitor Use-Policy
The visitor use-policy describes who will (or will not) be able use the site, the kinds of activities that will be encouraged or will not be allowed, the nature of the desired

experiences, any limits as to when and how often the proposed uses can (or cannot) occur, and any facilities that should be provided to accommodate the desired uses. Specifying the use-policy is crucial and directly related to the success of a restoration for both biological and sociopolitical reasons. The visitor use-policy represents a delicate balance between welcoming people and protecting the resource.

People can be major disturbance agents within natural communities (see Chapter 4). While on site, people may trample, harvest, or vandalize vegetation; kill or harass animals; import exotic pest species; or leave behind trash. The more people and kinds of activities allowed onsite, and the less attention paid to the creation of accommodations through design (e.g., by creating trails) or to supervision or education, the more likely it is that harmful impacts will occur. On the other hand, without the political, monetary, and sweat-equity support of neighbors and other interested citizens, most restoration projects cannot succeed. An important way to gain such support is by welcoming people to visit and enjoy the restored communities.

In most cases, visitor use-policies follow from the project purpose and range from situations in which no visitors are allowed, such as a site dedicated to protecting an endangered species, to restorations created explicitly to encourage people to interact closely with natural systems—a publicly owned conservation park, for example (see Chapter 1). In some situations, people have inhabited a potential restoration site for centuries, living within its borders, and/or living off its physical and biological resources. Use-policies in these cases are carefully crafted to accommodate continued traditional uses and restoration goals. The policies are also influenced by the site conditions (a wet site may suffer more trampling impact than one with moderately wet conditions) and by the relative fragility of the communities to be restored.

To determine who can use a site, it is helpful to address whether:

- Anyone except the site managers should use the site.
- Use should be restricted to specific portions of the site, or to designated facilities (e.g., a requirement to stay on the trails).
- Use should be by invitation or permit only, based, for example, on demonstrating an understanding of how to avoid impacts.
- Limits on total user numbers should be imposed; annual totals or limits on numbers allowed at any one time should be defined.
- Restrictions should be placed on who can use the site.
- Some or all portions of the site should be accessible to people of all ability levels.

In the context of restoration planning, an "activity" has a fairly straightforward meaning: an undertaking or pursuit often, but not always, involving physical exertion. For instance, hiking, jogging, biking, skiing, birdwatching, taking photographs, collecting research data, and hunting are activities that might be enjoyed by visitors to restorations. As indicated above, a visitor use-policy generally describes which activities will be encouraged, which ones allowed, and which ones prohibited. In addition, it specifies where the designated activities may take place. As an example, the University of Wisconsin–Madison Arboretum actively encourages nondestructive research and nature study activities; it allows hiking, jogging, and cross-country skiing, and prohibits dog walking. Appropriate research and educational activities may take place anywhere on the Arboretum with a permit, but such activities as hiking may take place only on designated trails.

An "experience" is a little more difficult to describe and plan for, since to a large extent experiences are very personal and individual. To experience a site is to be engaged in events; to create mental images, perceptions, and memories; and to feel emotions. It is helpful to think about the adjectives we use to describe experiences, like these: memorable, unique, relaxing, energizing, exciting, social, happy, unpleasant, stimulating, and educational. Planners then have to consider what it is about the attributes of a site that makes an experience memorable, educational, and so on.

It is also advisable for planners to establish policies that enable the managers to place restrictions on when and how often visitors can use all or portions of the site when particular situations arise. Such actions would be for the safety of the visitors, as well as to protect the natural communities. Things to consider include nesting seasons, wet soil conditions, and the application of potentially dangerous management tools (such as prescription fires). The policies could describe the circumstances under which visitors either can or cannot visit a site, whichever is clearer. Here are some examples:

- Trails are for hiking only.
- Access is by permit for nonintrusive research only.
- The site provides solitary nature study experience.
- Collecting plant and animal materials is not permitted.
- Trails will be closed as necessary to protect the resource or the safety of the visitors.
- Trails and gathering places are provided for use by organized educational classes.

Once the visitor policies have been adopted, they need to be communicated to potential users. The policies can be in the form of opportunities and/or constraints (rules). Figure 6.3 shows the visitor use-policy of the University of Wisconsin–Madison Arboretum.

Implementation and Management Activities Use-Policies

As we will see in Chapters 8 and 10, the tools available for implementing and managing restorations include several techniques (fire, herbicides, tree removal, selective hunting, biological control agents, the introduction of plants or animals from distant locations) that are controversial in some situations. For example, many people are protective of trees and do not like the idea of "destroying" one part of nature to favor another. Although widely used, herbicides raise issues of health and safety among many members of the public. It may not be possible to use fire in the form of controlled burns in some situations where smoke is hazardous, or the surrounding land uses would be extremely vulnerable to an accidental escape. Similar to the visitor use-policies, the implementation and management use-policies should be explicit about what can be done where, when, and by whom. As an example, Figure 6.4 presents the Wisconsin State Natural Areas pesticide policy.

Discussing and resolving potential issues associated with these tools at the beginning of the restoration planning process helps planners determine which communities to restore and anticipate public concerns that may arise later. If the proposed project is in an area with strict fire regulations, other things being equal, it might make sense to decide to restore communities that are not fire-dependent.

6.3.2 Restoration Goals

Goals and objectives generally refer to something toward which we are directing our efforts, in other words, something we are aiming to reach or achieve. Both words refer to achieving

The University of Wisconsin–Madison Arboretum, which has more than 1,200 acres (486 ha) of restored or protected plant communities among several properties throughout Wisconsin, allows the following uses.

RECREATION OPPORTUNITIES

Here are a few ideas for enjoying and experiencing the Arboretum:

Observing Nature: Be it bird-watching, looking at spring wildflowers, or gazing at the fall colors, there is always something to marvel about within the Arboretum's many ecological niches. For recent field observations of interest, peruse the *Naturalists' Notes.*

Hiking and Walking: When using our paved roads, travel in single file, facing traffic, and along the edge of the road.

Biking and Jogging: Bikers and joggers can travel over four miles of paved road through woodlands, wetlands and next to Curtis Prairie by following McCaffrey Drive. Although you can run on the unpaved fire lanes and footpaths within the Arboretum, biking on unpaved areas is not permitted.

Skiing and Snowshoeing: Although skiing is prohibited on some trails, most of the 20+ miles of trails are available during winter. Enjoy yourselves, but please refrain from going off trail.

Art and Photography: Plants and animals throughout the Arboretum make excellent subjects for drawing or photographing. Our *Naturalists' Notes* may help you find a unique subject.

RULES

We hope you enjoy the Arboretum's many attractions. While you are here, please obey the following regulations. Your cooperation will help protect the Arboretum's plants and wildlife as well as ongoing research projects.

- For your safety, remain on trails, firelanes, lawn areas or the paved road at all times.
- Use bicycles and cars on the paved drives only.
- Run and ski only on designated routes.
- Leave pets at home — they are prohibited on Arboretum grounds and the road through it.
- In-line skates, roller skates and roller skis are prohibited throughout the Arboretum.
- Do not picnic, build fires, or hunt, trap, or disturb animals.
- Recreational games, radios, and portable stereos are not permitted.
- Unauthorized removal of natural materials or Arboretum equipment is strictly prohibited.

Contact the Visitor Center at 608.263.7888 for more information about the Arboretum's activities or to participate in restoration efforts.

Permission granted by the University of Wisconsin-Madison; credit to the University of Wisconsin-Madison Arboretum

http://uwarboretum.org/

FIGURE 6.3 University of Wisconsin–Madison Arboretum Visitor Use-Policy.

an end, but when applied to restorations, they also have separate and more explicit meanings. Restoration goals are usually general statements; objectives are measurable statements, qualitative or quantitative, that clarify the goals. Every major step of the restoration process includes the creation of a plan that contains goals and/or objectives that summarize the intentions of the restoration team.

The goals of the master plan are descriptions of what the site will be like in physical, biological, and cultural terms when it has been restored. Here are some typical restoration goals:

- Wetland complex with both recharge and discharge areas.
- The presence of at least two different community types.
- Wet-mesic prairie, mesic prairie, and savanna present.
- Coastal salt marsh vegetation to support breeding clapper rails.
- Accommodation of site users with minimum impact.
- Stormwater runoff leaving the site that contains minimal silt.
- Trails ranging from easy to difficult access ratings.

The primary objective of natural areas is to preserve whole communities and their inherent natural processes. Any decision to control unwanted species, either plant or animal, needs careful consideration based on clear, specific management objectives. The use of pesticides, in particular, should be considered only as a last resort.

The Natural Areas Preservation Council recommends that the Department of Natural Resources implement the following guidelines on State Natural Areas (SNA).

Guidelines

1. Use of chemical pesticides on SNA has been avoided almost completely in the past. The Council reaffirms that general approach.

2. In all cases where the decision has been made that the invading species must be controlled, non-chemical controls should be attempted first, e.g., hand pulling or picking, mowing, local cultivation, fire, and, perhaps, temporary regulation of the moisture regime.

3. Before a pesticide will be considered for use on a SNA, the following conditions must be met:

    a. Other control methods have been tried and found ineffective.

    b. Persistence of the unwanted species is demonstratively inimical to management objectives. Cases of succession in buffer zones, on unclassified sites, and in certain forest and other communities may fail this test.

    c. It must be possible to treat the target species and only that species by direct contact, e.g., stump painting or application to individual leaves or bodies.

    d. The pesticide to be used must be of proven efficacy and one that is quickly degraded into benign, naturally occurring substances.

    e. The pesticide must be applied by individuals trained in the application and handling of pesticides and instructed in the use of such substances in State Natural Areas.

4. Recommendations on the use of pesticides on SNA will be considered, on a case-by-case basis, by the Council upon written application, and after other control measures have been exhausted or justification has been provided as to why other control measures will not be effective. Ordinarily, application must be made in advance of each proposed treatment program. However, the Council will consider requests to grant site managers limited discretionary use of pesticides to control certain species that are capable of rapid expansion.

5. Policies and practices that reduce the incidence of invasion by exotic species should be developed and implemented. Prevention is preferable to attempts to cure problems. Research on invasion and establishment of exotic species is critically needed and should be promoted.

Adopted by the Natural Areas Preservation Council: April 1988, amended April 1998 Wisconsin Natural Areas Preservation Council Guidelines.

Permission granted by the Wisconsin Natural Areas Preservation Council.

FIGURE 6.4 Guidelines for the Use of Pesticides on Wisconsin State Natural Areas.

The site plan contains objectives that work with the goals of the master plan. The site plan objectives are expressed as measurable outcomes that describe what the site will be like once it has been restored. For example:

- Chaparral containing a minimum of 30 native plant species as found on the attached list.

- Aerial plant cover >90%.

- Less than 5% cover from exotic species.

- An average of 2–8 research projects per year hosted by the site.

The master plan goals and site plan objectives (outcomes) explain the vision you have in mind for the restoration. They describe the solution to the problem identified in the purpose statement. Being time-neutral, they identify the desired results, not the actions to be taken. Unlike the purpose statement, goals and objectives do not refer to current site conditions. The idea is that these statements will serve as benchmarks and remain current for the restoration long after the initial implementation, and through transitions from one site manager to another. Time frames and deadlines, however, are applied to activities and desired achievements during the implementation phase, in the form of performance standards (see Chapter 8).

The goals guide the direction of the restoration and are general enough to allow several different restoration designs to meet them. For example, the goal of having "at least two different community types present" means that a proposal to include four communities meets this goal, as would a proposal that includes two.

Site plan objectives are more detailed and explicit than goals and can be used as performance standards by which to judge the success of a project. Objectives are most useful if the outcomes they describe have built-in flexibility—that is, if they specify a range of conditions ("a minimum of 30 native plant species") rather than an exact target (35 species). Flexible statements account for natural variation and for the dynamic nature of communities, and they provide some leeway for our imperfect understanding of communities and ecosystems.

Goals and objectives can be stated in terms of what is or is not desired for the restoration:

- Woods having 25–35% canopy cover of maple. (Desirable.)
- Woods canopy composed of any species *except* Norway maple (*Acer platanoides*), green ash (*Fraxinus pensylvanica*), or box elder (*Acer negundo*). (Undesirable.)

The master plan consists of three levels of goals, but no objectives. This is because the specificity of the project increases as the planning process proceeds. The three master plan goal levels are (1) overall project goals, (2) goals for each master plan alternative, and (3) goals for each community included in the adopted master plan.

We will explain the role of each of these kinds of goals as our discussion of the master plan proceeds. But first, how does one arrive at restoration goals? It is helpful to consider this question in terms of both the natural communities to be restored and the ways in which people will interact with the site.

Natural System Goals

Biological community/ecosystem restoration goals arise from a consideration of (1) the project purpose and use-policy, (2) the results of the site inventory and analysis, (3) the study of restoration precedents and existing remnants, (4) ecological theory, (5) community/ecosystem models, and (6) resource availability (including sources of materials, time and money), with additional input from the stakeholders (potential users, neighbors, clients, etc.).

More specifically, in drafting these ecological goals, you will usually consider these questions:

- Which communities are possible, given the current nature of the site, the project purpose and use-policy, and the available resources, especially human capital?
- Which attributes of composition, structure, and/or function must be provided by the potential communities in order to achieve the purpose and use-policy?

An additional question that has arisen over the past decade is: How will global climate change affect the site?

Depending on the type of restoration (complete, ecological services, or experiential; see Chapter 1), the natural system goals will focus more or less on biodiversity, the relative absence of signs of human use, the presence and extent of particular ecosystem services, habitat structure, the presence of several successional stages, uniqueness or rarity, an abundance of showy flowers or spectacular wildlife, and so on (see Chapter 7). The idea is to capture the vision of the restoration planners—to explain the essence of what they are hoping to achieve.

User/Social Goals

Many restoration projects include social programming—designed visitor facilities (trails, buildings, benches), classes, informal nature-based appreciation events, and volunteer or friends groups. In such cases, the master plan includes goals describing how the restoration will accommodate and/or facilitate the activities and experiences called for in the use-policies. In some cases, the desired outcomes of people interacting with a restoration site are relatively simple. For example, a goal might be simply that people visit the site, or that when people do visit the site, they will not harm its natural features while they are there.

In many cases, however, the restoration purpose calls for more extensive site–people interactions, and the restoration goals describe relatively complicated experiences, both tangible and intangible. It is relatively easy to set goals for the kinds of activities mandated by a use-policy. For example, goals can specify that a site will support hiking, skiing, or an educational nature trail. It is more difficult to describe hoped-for knowledge gain, increased appreciation for nature, or a fulfilling experience.

Professionals in the social sciences and humanities have amassed a wealth of information about how to develop the kinds of social programming supported by restorations (Adult and K–12 Environmental Education, Nature-Based Recreation, Service-Learning and Volunteer Programs, in particular). One type of goal-setting approach that has been used in many settings is called the logic model, and it is very similar to the restoration planning process used in this text.

Several kinds of logic models are available. Researchers at the University of Wisconsin–Madison Extension developed one that has worked well in many situations (Taylor-Powell and Henert 2008). This model, shown in Figure 6.5, takes the form of a flow chart featuring Inputs, Outputs, and Outcomes/Impacts. Inputs are resources and investments that go into the program. Outputs are the tools used to deliver a program: activities, services, events such as hiking and skiing, and creating a nature trail. Outcomes/Impacts are the "results"—knowledge gained, happy experience, and so on. The outputs and outcomes together compose the restoration goals. The idea is to use the model to clearly identify the hoped-for impact of policies and activities designed to engage the public with a restoration project, as well as to understand what circumstances might influence the likelihood of successful programs.

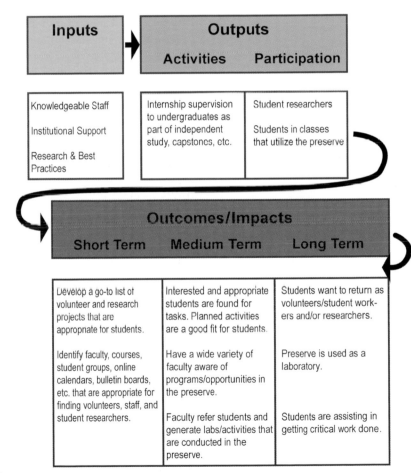

FIGURE 6.5 A Logic Model for Student Participation in the Lakeshore Preserve, University of Wisconsin–Madison. The logic model organizes restoration social programming. It lists inputs, desired outputs, and planned outcomes/impacts for involving students in active learning in restoring a portion of the campus. (Adapted from Kagle 2007; based on the University of Wisconsin Extension logic model.)

The overall project goals are very general and usually few in number. They describe the focus of the restoration—for example, supplying an ecosystem service such as water quality protection, providing habitat for endangered butterfly species, or maximizing the biodiversity of an area. It is best if the overall goals can be met in several different ways.

▶ 6.4 DESCRIPTION OF THE SITE INVENTORY AND ANALYSIS

The site inventory and analysis step happens several times during the restoration process (see Chapter 4), each time with a slightly different purpose and level of detail. As we have noted before, information about the natural and cultural resources of a site, and its physical and cultural context, is crucial in order for restoration planners to be able to create successful microclimate–species matches and biological interactions, minimize cross-boundary issues, and build a network of social support. At the beginning of the master plan, the purpose is to provide a descriptive overview of the site and its context (inventory) and to highlight the site resources that help determine the direction of the restoration. These are the resources that could either enhance or detract from the project purpose, use-policy, and general goals (analysis).

The inventory topics that are most helpful in developing the master plan, and therefore should be addressed in the summary, include:

- Description of site boundaries and current ownership and use patterns.
- Description of surrounding land use patterns.
- Brief biological and human site history.
- Overview of the landforms onsite and in the surrounding area.
- Overview of hydrological systems.
- Overview of soils, particularly with regard to texture and moisture.
- Overview of existing vegetation and animals.
- Description of any onsite cultural resources.
- Description of pertinent policy issues (zoning, deed restrictions, etc.).

The inventory information is presented in the form of both maps and text (see Chapter 4). The analysis that follows assesses the degree to which the current conditions of the site already match the general restoration goals, and/or how they might stand in the way of achieving them.

▶ 6.5 A SET OF ALTERNATIVE SOLUTIONS

The next step is the beginning of what is the most satisfying and most creative part of the master planning process: (1) selecting communities and cultural facilities (trails, etc.) that both match the site conditions and advance the restoration purpose, use-policy, and overall goals; and (2) designing the general spatial layout of the site. This is where your vision begins to be translated into three-dimensional form, and the solution becomes part of the landscape.

During this step, the planners propose several different alternative plans and discuss the merits of, and issues raised by, each one. Every alternative should be a feasible solution—that is, be able to be implemented within the limits of current knowledge, expertise, and resources. Each alternate plan should meet the project goals.

The alternative plans can be created informally on tracing paper and reviewed only by the original restoration designer, or by one or two additional people. They can be presented formally in a public meeting, with a full discussion of the costs and benefits of each plan. Or the review can take place in other ways. In any case, the idea is to find more than one way to design a site before settling on a solution.

The purpose for creating, assessing, and comparing the relative merits of master plan alternatives is fivefold:

1. In order to develop alternatives, you have to think carefully about the site and project at hand, and minimize the chance of overlooking important factors and avoid the temptation to "recycle" restoration design solutions.

2. Just as the act of creating several drafts helps a writer perfect a story, the act of designing alternatives helps you perfect the restoration plan.

3. An external review early in the design process can uncover misunderstandings, erroneous assumptions, or incorrect information at a stage when the restoration can be easily modified.

4. Assessing the relative merits of alternatives clarifies the reasons behind the choice of the adopted plan. This information can be very useful in the future when considering midcourse changes.

5. In many situations, giving members of the public an opportunity to provide meaningful input (comments that result in modifications or influence the choice of one alternative over another) is a good way to gain support.

The alternative plans cover variations in the number, extent, and kinds of communities proposed for the restoration, and/or the way in which users will be accommodated. It is helpful to summarize the intent of each alternative by writing goals, which will highlight the differences among the alternatives. The case study below includes three alternative scenarios for the Kishwauketoe Nature Conservancy.

CASE STUDY | The Kishwauketoe Nature Conservancy

The Kishwauketoe Nature Conservancy consists of 230 acres near the shore of Lake Geneva in the village of Williams Bay, Wisconsin. It contains several natural communities common to southeastern Wisconsin prior to European settlement. The site is owned by the village and managed by a nonprofit land trust. The Kishwauketoe Nature Conservancy is open to the public for recreation and education. A condensed summary of its master plan and three alternatives are provided in Figures 6.6, 6.7, 6.8, and 6.9. ■

Project Purpose:

Increase native plant and bird species biodiversity; reduce dominance of pest species.

Use-Policy:

Solitary hiking, birdwatching, small-group informal and class-based nature study.

Site Inventory and Analysis:

The rolling topography and mesic to wet soils can support several native forest and grassland communities. Existing land cover is disturbed woods and pasture.

Restoration Goals (applicable to all alternatives):

The land cover of the site is predominantly that of native communities.

Native bird species use the site for nesting and/or other habitat needs.

Access exists for solitary hiking and birdwatching, and for small-group informal and class-based nature study.

Alternative Solutions:

Alternative 1: Land cover of the site consists of two community types (oak savanna and shallow marsh) and a trail system. This alternative restores the entire 230-acre site to its presettlement vegetation, as determined by using the federal public land survey records of the 1830s. Visitor amenities are limited to trails, but this network allows for access to most areas of the site and places learning emphasis on a single community, the oak savanna. (See Figure 6.7.)

Alternative 2: The site is restored to three communities (oak forest, oak savanna, and marsh) with the highest cover in oak forest. Emphasis is placed on visitor use and experience. The trail system includes educational signage and small-group gathering places. The project includes onsite naturalist-led classes. (See Figure 6.8.)

Alternative 3: Land cover of the site consists of five community types (oak savanna, oak forest, mesic prairie, wet-mesic forest, and marsh) that blend from one to another across the site and are representative of the major communities that once occurred throughout southern Wisconsin. This alternative closely matches the existing physical site conditions and requires the least disturbance to existing vegetation of the three alternatives. Visitors have access to the site via a well-developed trail system that includes boardwalks, interpretive signs, benches, and an entrance kiosk. (See Figure 6.9.)

Adapted from Carlson et al. 2004.

FIGURE 6.6 The Kishwauketoe Nature Conservancy Master Plan.

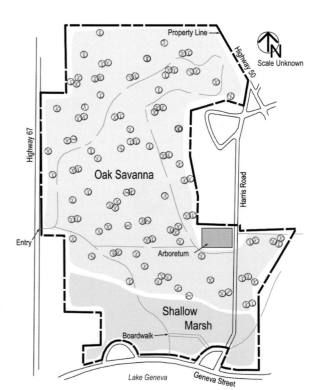

FIGURE 6.7 The Kishwauketoe Nature Conservancy Master Plan, Alternative 1. Two community types typical of the U.S. Upper Midwest, savanna and shallow marsh, are established across the project site, and they transition from one to another. (Adapted from Carlson et al. 2004.)

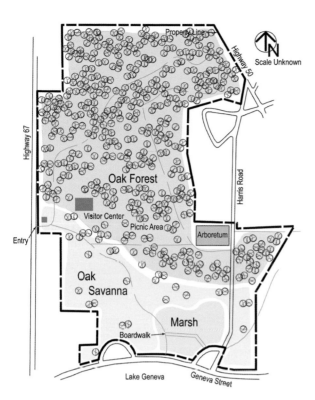

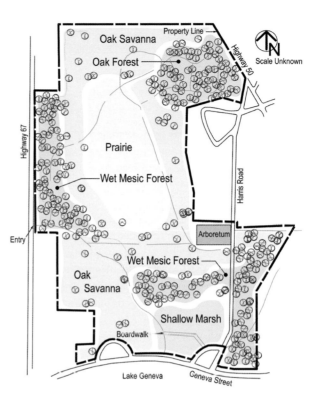

FIGURE 6.8 The Kishwauketoe Nature Conservancy Master Plan, Alternative 2. In this alternative, the oak savanna and marsh areas are reduced, and oak forest is reintroduced into the site. Note the manner in which canopy density is illustrated between community types. Areas for visitor use are delineated. (Adapted from Carlson et al. 2004.)

FIGURE 6.9 The Kishwauketoe Nature Conservancy Master Plan, Alternative 3. In this alternative, several community types are restored on the site. The density of trees represents whether the land cover is open, semi-open, or closed canopy and relates to communities of grasslands, savanna, forest, and marsh. Trail locations are approximate. How the community types transition from one to another will require further development in the site plan. (Adapted from Carlson et al. 2004.)

6.5.1 Choosing and Justifying the Desired Solution

The result of analyzing and comparing alternative solutions for a master plan is often a new plan consisting of pieces from different alternatives—for example, a solution using the plant communities from Alternative 3 and the educational programming from Alternative 2 from the Kishwauketoe Nature Conservancy master plan.

Even when the restoration purpose is more narrowly defined than the case study above, it is still helpful to think about alternatives where possible. If a site is small or has a unique setting, rather than consider different land cover types, you might consider different ways of accommodating use. Different options may not always make sense at the master plan level. For example, the restoration purpose may be to restore a specific community type. You and your restoration team need to decide whether that community can be restored to the site. If it can, you might then consider variations of species choices or arrangements when doing the site plan (see Chapter 7). The important thing to remember is that discussing the merits of different alternatives leads to a better restoration (see Sidebar).

Once you have agreed on the master plan, you provide a written summary in the plan document that explains how and why one alternative was selected over other solutions. This justification will serve as a benchmark reference for future managers. You would also

---

Sidebar | Choosing the Desired Solution

The three alternatives presented in the Kishwauketoe Nature Conservancy restoration plan are different in terms of both the land cover and, to some extent, the degree of development of people accommodations; yet all three proposals meet the purpose, use-policy, and goals (assuming all potential native community types provide good bird habitat). In this case, there would be several issues to discuss before deciding on the final plan:

- Which alternative plan would be easiest to implement, given the existing vegetation?
- Would Alternative 3 result in community patches that are too small to support desirable species?
- Which alternative would provide the best biodiversity for educational or birdwatching purposes?
- Is the presence of grassland bird habitat more desirable in this region than the presence of forest bird habitat?
- Would the more developed amenities of Alternative 2 detract from a solitary experience?

- How difficult will it be to manage each alternative once established? For example, is the difference in the disturbance regimes necessary to maintain the communities more feasible for one alternative than the others?

FOOD FOR THOUGHT

1. How would you respond to the above questions?
2. What information would you need to make a decision?
3. What additional questions might be addressed in the selection of one of several alternative master plan solutions?
4. You will want to consider the similarities and differences between each scenario, and then evaluate how well each meets the project purpose, use-policy, and objectives.
5. Consider what other criteria you might use to evaluate and select the desired solution.

---

include a set of elevation graphics, showing the relative locations of communities and cultural features in relation to one another (Figures 6.10). The layout is still general at this point in the process; site-level detail is added in the site plan to come.

▶ 6.6 GOALS FOR EACH COMMUNITY OR RESTORATION UNIT

The next step is to create goals for each community type or cultural resource included in the plan. Depending on the purpose of the restoration, these goals address such things as community/ecosystem size, dynamics, composition, habit quality for specific animal species, structure, and function, as well as the nature of such cultural features as trails and volunteer or educational programming. The site plan to follow includes measurable objectives that describe the desired outcome for each community/ecosystem goal.

Here is one goal/objective pair for the Kootenai River Habitat Restoration in Idaho (2009):

- Goal: The Kootenai River provides habitat for Kootenai sturgeon migration and spawning.
- Objective: Intermittent depths of 16.5–23 ft (5–7 m) or greater occur in 60% of the area of rocky substrate from RM (river meander) 152 to RM 157 during peak augmentation flows.
- Objective: Velocities of 3.3 ft/s (1 m/s) and greater occur in approximately 60% of the area of rocky substrate from RM 162 to RM 157 during post-peak augmentation flows.

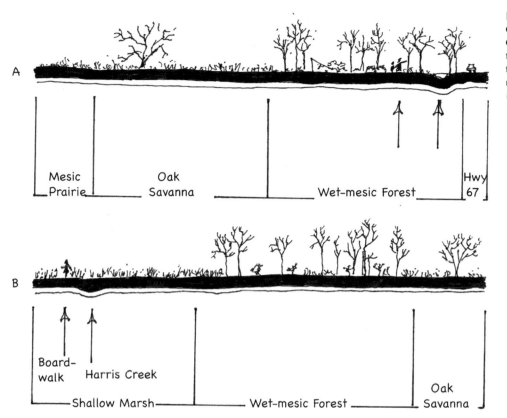

FIGURE 6.10 The Kishwauketoe Nature Conservancy Master Plan: Cross-Section Corresponding to Alternative 3. This shows the topography of the site and illustrates tree densities of three of the four community types and their landscape position. (Adapted from Carlson et al. 2004.)

As discussed above, the idea is to express a vision for the restoration in concrete terms, and provide a means of assessing whether or not the project is succeeding.

▶ 6.7 THE ESTIMATED BUDGET

During the master planning process, after the general restoration goals are established, you should begin to develop general budget categories for the project's expenses. You will not yet have enough information to develop detailed line items or to attach specific dollar amounts to them. Detailed expense estimates can only be developed after all of the restoration plans have been completed. But at this point, you should have sufficient information to set up general expense categories with estimated costs that will cover what needs to be done, who will do it, how to do it, and how long it will take. Although these estimates will be quite broad, they will be sufficient to determine whether the master plan is realistic, given the budget available for the project.

The three main budget categories are time, money, and resources (people, materials) required to get the job done. These budget categories cover several phases of the restoration: planning, implementation, monitoring, management, and review/evaluation. Developing a budget ahead of time will tell you if you have the resources you need to begin. A shortfall in any of the three main budget categories will require a compensating adjustment elsewhere in the project. For example, if the project is short on time, either more money or more people will have to be allocated; if staffing is in short supply, the project timeline may have to be extended.

6.7.1 The Planning Budget

The expenses involved in creating the master plan support a number of activities, including:

- Conducting the site inventory and analysis.
- Conducting "project scoping" meetings with interested members of the public, neighbors, clients, and project staff—the stakeholders.
- Finding information and assembling potential ecosystem models.
- Developing alternative plans and conducting meetings with the stakeholders to obtain feedback.
- Producing the text and plan graphics for the report and distributing copies as needed.

Perhaps the biggest commitment at this and the other planning stages of a restoration is time. But depending on the location of the restoration site, travel expenses can also be considerable. Supplies and equipment for conducting the site inventory and analysis, and producing the report, need to be considered at this stage as well.

6.7.2 Personnel

Budgeting begins by estimating what kinds of restoration tasks are necessary given the general direction of the master plan and current site conditions. This leads to an estimate of the kinds of skills needed and the number of people-hours required to do the work. At this stage, the budget consists primarily of general estimates; the details will be added in the implementation plan.

If the project site includes a permanent restoration field staff, as is the case for some government-owned properties, the budget allocates all or part of their time and talents to the project at hand. In some cases, the staff time available is set before the details of the implementation are known, and thus becomes a factor in choosing the details of both the site and the implementation plans. Roles that will have to be filled by the restoration team include restoration planner, overall project manager, onsite supervisor, field biologist, plant propagator, wildlife ecologist, GIS specialist, database administrator, and, depending on the restoration requirements, engineers, geologists, soil scientists, hydrologists, and marketing or public relations personnel.

Many projects rely on unpaid volunteers, student interns, or university classes to help with various parts of the project. University students, or in some instances neighboring residents, can develop management plans and conduct biological inventories, and volunteers can collect, clean, and plant seeds. For reasons of coordination, communication, and inclusiveness, it is a good idea to include these potential helpers right at the start of the restoration planning.

These "free" labor sources are not without cost, because you need to have people on staff who are skilled in project management and experienced in recruiting, training, and managing volunteers. As discussed in Chapter 12, restorations are most successful in the long run if they are planned, implemented, and managed using a collaborative process—in other words, a team effort. The budget should include the cost of regular team meetings, updates, travel, and administrative costs.

6.7.3 Materials

If necessary tools, equipment, construction materials, and seeds and plants are not on hand, the budget has to include the cost of purchasing, borrowing, or trading for them. Natural resource supply company catalogs are good sources of restoration tools, equipment, and pricing information. Costly equipment that might be used only once (e.g., a tractor) can generally be borrowed or rented. Sometimes project managers arrange an exchange of in-kind goods and services—for example, trading a prescribed fire or site preparation for GPS services and a plant community inventory and monitoring.

Plants and seeds may be purchased from commercial sources or collected, with permission, from the wild. Restoration projects are also increasingly creating plant nurseries onsite, not to reduce costs but to ensure species availability and origin. In any event, it is important to factor in the need for such materials from the beginning, even though, again, the details cannot be known until the site, implementation, and management plans have been developed (see Chapters 7, 8, and 10).

6.7.4 Contingencies

Monitoring and managing the restoration will reveal places where the project goals should be modified, or the implementation and management plans adapted. Because these changes can result in unanticipated expenses and because restorations are moving targets, you should expect surprises and budget for contingency plans. Even if the project runs smoothly, rising prices and cost overruns are common. It is important to build a contingency fund into the budget. This is generally done once the rest of the budget estimates have been made, and it is usually a percentage of that total—about 7–10% of the overall budget. At the master plan level, the point is to include this line item as the budget-building process begins.

▶ 6.8 PROCEDURES FOR PLAN ADOPTION, REVIEW, AND MODIFICATIONS

The last section of the master plan describes the procedures for adoption, review, and modifications of the restoration. This section addresses who reviews and okays the plan, as well as what forms of documentation must be provided for the review (i.e., maps, tables, written proposals, budgets, timelines). Included here would be recognition of any time constraints as far as when the documents must be submitted.

Often plans will be changed as the restoration proceeds, and the process by which modifications are made and accepted should also be noted here. These procedures will be similar in each of the plan types that follow: site, implementation, management, and monitoring. Each will need review by someone or some group, and that process must be understood prior to the initiation of planning.

KEY CONCEPTS

- The restoration master plan presents the project vision in general terms, using both text and plan graphics; it suggests several possible solutions, and then identifies the kinds of communities to be restored and their relative locations according to a preferred alternative.

- The restoration master plan is created using a collaborative process; it is flexible and adaptive, with built-in strategies for review and modifications if necessary.

- The restoration master plan includes a statement of project purpose and goals, a use-policy, a description of the site inventory and analysis, a set of alternative solutions with a description of and justification for the chosen solution, and a preliminary budget.

- Developing alternative solutions to a problem at hand provides opportunities to consider their advantages and disadvantages relative to the restoration goals; the process ensures that several possible solutions for the restoration have been given due consideration.

FOOD FOR THOUGHT

1. Why is it helpful to include both text and plan graphics in a master plan?

2. Assume you are in charge of developing a conservation park, a publicly owned property established to conserve native communities while at the same time providing opportunities for hiking and nature study. Create a use-policy and a set of natural system and user goals for the park, based on the forest ecosystem model presented in Chapter 3 or a native community of your choice.

3. Design a public participation strategy to encourage understanding and support for the conservation park. Explain who you would invite to participate, how you would pique their interest, and what activities you would plan.

4. Would the master plan components we listed change depending on where you are conducting your restoration? Or depending on different community types?

5. How would you differentiate the project purpose, use-policy, and goals? How does the master plan graphically exemplify each?

Resources for Further Study

1. The following website presents the process and standards the U.S. National Park Service uses to develop master plans and use-policies for its properties:

    U.S. National Park Service. 2008. *General Management Plan Dynamic Sourcebook*, version 2.1. http://planning.nps.gov/GMPSourcebook/GMPhome.htm

2. The following book presents several real-world examples of a restoration master planning process at a watershed scale; the examples illustrate the importance of working with diverse sets of stakeholders, using a flexible, adaptive planning approach:

    Doyle, M., and C. Drew, eds. 2008. *Large-Scale Ecosystem Restoration*. Washington, DC: Island Press.

The Site Plan

Design and Plan Documents

The master plan, discussed in Chapter 6, is a guide to the types of communities and infrastructure you are proposing for a project, along with their relative locations. The site plan (Figure 7.1) adds the details to the components and layout of the master plan by describing the desired community composition and structure, physical site conditions, and landscape infrastructure; using a graphic plan layout, it illustrates where you want the communities and infrastructure components (trails, signs) to occur. The site plan includes a number of supporting documents, such as species lists, engineering drawings, and design specifications for the infrastructure, as well as recommended physical landscape alterations. Last, but not least, the site plan provides a rationale for why this solution or restoration vision was selected.

FIGURE 7.1 **The Restoration Process: The Site Plan.**

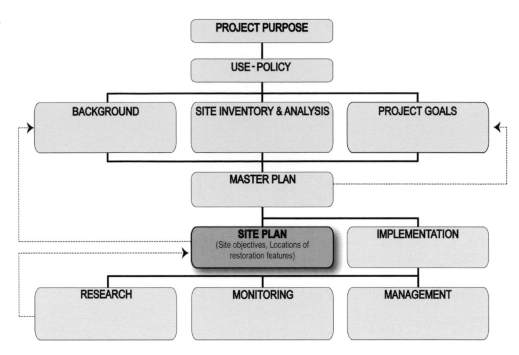

Although the site plan describes a vision for the restoration when completed, it is likely that you will alter the original restoration design as the project proceeds, because of changes in the environment and in response to human activities. During the development of the site plan, you need to consider how your plans may affect implementation activities (i.e., site preparation, planting methods, reintroduction methods), but these will not be specified until later.

▶ 7.1 SITE PLAN OUTCOMES

All the plans, which together guide a restoration, include goals and objectives. Like all parts of the restoration planning process, the site plan sets specific and measurable targets to be achieved. But, unlike the objectives in the implementation and management plans, the site plan objectives are more like outcomes. The site plan includes measurable descriptions of what the restoration will be like once it is completed.

Our preference is to write these descriptions without an action clause or a time frame. For example, a site plan might specify that the canopy cover of a site is "between 30% and 50% cover." If the site at the beginning of the restoration has a 60% canopy cover, the relevant implementation plan objective might be: "Thin the trees so that the canopy will have between 30% and 50% cover." Once the restoration has achieved this objective, there may come a time when the canopy cover decreases to 10% cover. The site plan description remains the same, but the now management objective might be: "Plant 5 to 6 trees so that the canopy cover will have between 30% and 50% cover."

The site plan outcomes are not linked to the status of a restoration site at a particular point in time, but to the intended vision of the restoration goals. The time-neutral status ensures that the descriptions are current in year 1 of the project, as well as in year 20 or year 50.

Site plan outcomes are based on the restoration purpose and goals. You can write them to specify what you do want, based on, for example, an ecosystem/community model, ecological theory, the results of your site inventory and analysis, and other background information discussed in the preceding chapters. Outcomes usually include information about species numbers and composition; community structure, dynamics, and size/shape; human use; and infrastructure. Or you can write them to specify what it is that you do not want—exotic species, dense canopy cover, eroded gullies, for instance.

There should be no ambiguity or need to interpret whether or not an outcome has been achieved. In certain situations, however, it will be necessary to question whether it is realistic to achieve your original vision. In some cases your solution will be to try a different implementation or management approach, or to allocate more time. In others, it will be necessary for you to revise your plan.

CASE STUDY   The Longleaf Pine Ecosystem Restoration

At the time of European settlement, the longleaf pine (*Pinus palustris*) savanna community in the southeastern coastal plain of the United States was believed to occupy 33–38 million hectares, or 40% of the Atlantic and Gulf coastal plain from eastern Texas to Florida and Virginia. By 2000, after a long history of timber harvests, conversion to agriculture, and suppression of natural fire regimes, only 0.07% of the original expanse of longleaf pine community remained. Longleaf pine occupied several habitats from upland to lowland, and the community composition changed relative to each. In addition, several other communities interacted with longleaf pine across this landscape. The suppression of fires led to the colonization and proliferation of woody species from nearby communities, as well as the development of a thick pine needle litter mat. These structural changes have resulted in increased fuel loads, so the risk of a catastrophic wildfire is much greater.

As with savanna systems throughout the world, small and scattered remnants of longleaf pine communities still exist. In addition, areas remain where large mature specimens of longleaf pine still stand, but the community structure around them has dramatically changed, and without intervention, these stands will not be replaced. The loss of the longleaf pine ecosystem has led to the loss of habitat for a number of plant and animal species that are now endangered. Over the past few decades, government agencies and environmental groups have been working toward the restoration and recovery of the longleaf pine ecosystem.

Van Lear et al. (2005) describe five restoration factors in the recovery of the longleaf pine ecosystem: developing a general understanding of the historical condition of the ecosystem, initiating and expanding a fire regime, enhancing herbaceous species diversity, continued research on habitat requirements and distribution of rare species, and encouraging a multiowner partnership approach to promote conservation.

Each of the target factors is intended to restore specific aspects of the ecosystem, including adding species components that facilitate natural disturbance regimes. Outcomes can be used to specify these targets, and in Figure 7.2, we provide four main goals for restoration of the dry-mesic savanna longleaf pine community. A set of outcomes accompanies each goal (see Sidebar). Research by public and private agencies continues on the longleaf pine ecosystem. Topics include its composition, structure, and environmental processes prior to European settlement, as well as methods for protecting remaining stands, methods to restore former stands, and criteria that can be used to evaluate restoration success. ∎

1. Maintain a complex of upland forest, savanna (longleaf pine), and flatwood communities common to the southeastern United States.

2. Keep dry-mesic savanna to be a minimum of 16 ha (40 acres).

   a. Longleaf pine represent at least 70% of the basal area and 60% of the tree species present. Basal area average of 7.0 m²/ha (30 ft²/acre); at least 95% of species native to the longleaf pine savanna, as listed in Boyer 1990.

   b. Southern red oak (*Quercus falcata*), blackjack oak (*Q. marilandica*), and sweetgum (*Liquidambar styaciflua*) represent less than 10% of the canopy; loblolly pine (*Pinus taeda*) and shortleaf pine (*P. echniata*) represent less than 5% of the canopy.

   c. At least 9 species common to the longleaf pine community, as listed in Boyer 1990, in shrub layer with less than 5% cover over the entire area.

   d. Sedges and grasses represent approximately one-fifth of the ground layer species. Wiregrass (*Aristida stricta*) with at least 40% cover; big bluestem (*Andropogon gerardii*), little bluestem (*Schizachyrium scoparium*), piney woods dropseed (*Sporobolus juncea*), and sand sedge (*Carex muhlenbergii*) together less than 20% of the cover.

   e. Minimum species density of 60; composites at least 25% of the species; legumes next largest representation.

   f. Non-native species less than 5% of the ground layer cover and less than 5% of the shrub and tree stem density.

   g. Stable populations of a minimum of 10 nesting bird species native to the longleaf pine community, including the red-cockaded woodpecker (*Colinus virginianus*).

3. Manage with "natural" disturbance events.

   a. Sufficient fuels to support prescribed surface fires every 2–8 years.

   b. Use of silvicultural techniques (e.g., irregular shelterwood) to mimic wind throw and lightning strikes in order to provide moderate-sized canopy gaps for the release of longleaf pine seedlings for canopy regeneration.

4. Open the site to the public, and provide learning experiences about its ecology and the pre-European settlement landscape.

   a. Grass surface parking for 30 cars located on disturbed land without directing surface runoff into the restored areas.

   b. Presence of at least one trail meeting the requirements of the American Disabilities Act; 40% of all trails adhere to principles of universal design.

   c. Trails avoid areas that contain species listed as sensitive or threatened and endangered.

FIGURE 7.2 **Site Goals and Outcomes for a Longleaf Pine Ecosystem Restoration.**

7.1.1 Creating Site Plan Outcomes Related to Community Composition and Structure

Most restorations include plant communities, so your site plans will probably include measurable outcomes related to species composition and structure. For each type of community/ecosystem being restored, you will have to review its composition and structure, and the related processes and functions. The kinds of features that are most often described in restoration site plans include species presence and relative abundances; taxonomic composition; life form composition; species' functions or roles, such as keystone species, colonists, phenology; and species–species interactions.

Species Presence and Relative Abundances

Most restoration site plans include descriptions of desired species composition. Chapter 2 listed and discussed several questions you should consider. For example:

- Do communities need to have a minimum number of species or organisms in order to be viable or sustainable? Or, what is the average number of species found in remnants and similar restorations?

Sidebar	Writing Restoration Outcomes

Site outcomes can be written to be quite specific or broad, depending on the targeted vision. For example, many outcomes allow for flexibility in the choice of species that will compose a community, thereby recognizing the dynamics that occur in ecosystems and the unpredictability inherent in restoration. If you look at the outcomes listed in Figure 7.2, you'll notice they do not prescribe a specific set of species to compose the restoration, but instead refer to a list from which selections are made (i.e., Outcomes 2a and c). Nor do these outcomes specify an exact number of species; instead, they provide the minimum number expected (i.e., Outcomes 2e and 2g).

In certain situations, however, outcomes must be specific, such as when a restoration calls for the recovery of an endangered species or the reintroduction of specific ecosystem processes. The set of outcomes for the longleaf pine ecosystem restoration specifically calls for longleaf pine to dominate in the canopy (Outcome 2a) and wiregrass in the herbaceous layer (Outcome 2d). There is only limited flexibility in pine density, and none as to which species are dominant. The Kootenai River Habitat Restoration Plan is an attempt to restore river habitat, particularly the habitat for the endangered Kootenai River white sturgeon (see Chapter 14). That plan included specific outcomes to be achieved for restoring the dynamic hydrologic regime of the river, and for restoring sturgeon populations.

Goals and outcomes for some projects aim to restore ecosystem services, such as stormwater or erosion control, that prevent land degradation or jeopardize human populations. In some instances, you will need to integrate human activities, such as agriculture, grazing, timber production, or recreation, with restoration (Goal 4, Figure 7.2), crafting outcomes that allow for both, while minimizing the potential for conflicts.

Lastly, you must monitor the outcomes periodically to see if they remain applicable to the restoration and its context. As environments change due to internal or external forces, outcomes may change. Climate change, extinction of predators, increased human dwelling densities, and the invasion of new pest species can all require modification of outcomes (see Chapter 11). Outcomes should be only as specific as you can comfortably predict what your implementation methods can achieve, and no outcome is necessarily permanent.

- Are some species, life forms, or roles pivotal to the success of a particular community restoration, and if so, what are they?

- How variable are communities in composition, structure, or function across space and through time? To what extent can the resulting spatial and temporal patterns be mimicked in the restoration?

Ecosystem/community models are one way to provide a reasonable approximation of the required species numbers for a given restoration. A second way is to visit remnants of the desired community type that occur in locations similar to the restoration project site. For example, to determine how many plant species you might expect to find in a functioning community or ecosystem, visiting a single remnant and inventorying the number of species found will provide a good start. However, as noted in Chapter 2, stands of the same community type will vary in the number of species they contain. In part, this is due to their recent history (grazing, logging, fire, etc.), as well as the heterogeneity of the landscape in which they are found. Therefore, if you can sample more remnant stands, all of which are similar in physical attributes to the project area, calculating the average (and minimum) number of species found will give you a better estimate. You can calculate, for instance, the average number of species found (also called the species density) in 20 remnant stands as follows:

$$\text{Average Species number} = \frac{\left(\begin{array}{c}\text{\# of species}\\\text{in Remnant 1}\end{array}\right) + \left(\begin{array}{c}\text{\# of species}\\\text{in Remnant 2}\end{array}\right) + \cdots + \left(\begin{array}{c}\text{\# of species}\\\text{in Remnant 20}\end{array}\right)}{\text{Total remnants sampled (20)}}$$

To determine the average species number this way, you have to use the same inventory scheme in each remnant—using the same quadrat size and number of samples, or doing a walk-through inventory using the same amount of time. In most situations, your result would be a conservative estimate of the minimum number of species to provide. Unless you visit each site and sample several times throughout a year, realistically you would not be able to inventory all or even the majority of plant species in each remnant. The species first observed are more likely to be common or abundant and less likely to be rare. In other words, sampling will probably not record every species that occurs in a stand.

Because communities are variable in space and through time, rather than specifying a specific set of species that must be in a restoration, a better approach is referencing a list of a broad array of species that are appropriate to the physical and biological landscape setting. The expectation would be that the vast majority of species growing in the restoration would be from this list, not that any one species in particular would occur unless required to meet a specific goal.

The list of species from inventories of similar remnants is an excellent starting point from which you can build your own list of desired species. You can also refine your list by determining which species are most widespread or common. One way to do this is to calculate the average occurrence or "presence" of each species over all the remnants you visit. You calculate the percentage of remnants each species is found in. For example, a species you find in 10 of 20 sampled bogs has a 50% presence, making it a common, but not consistent, member of the bog. In contrast, a species with 98% presence is even more "representative" of a bog, but a species with a presence value below 10% would be considered rare. Using this information, in addition to describing a desired minimum number of species for a restored community, you can specify a desired minimum number of representative species, or of rare species.

You can also use supplemental texts for background information on species composition. One example is *The Vegetation of Wisconsin,* by John T. Curtis (1959). This book provides extensive summary tables for each of that state's major communities (Tables 7.1–7.3). It also illustrates the level of detail that can be useful in developing species lists for a restoration project. Plant atlases exist for many geographic regions, describing the species occurring there. University and college herbaria, as well as local experts, are additional sources of information about species found in a given locality. As with species density and presence, you can sample remnants to determine the stem densities, biomass, or cover that is typical of a community and use these to describe the desired restoration outcomes.

In many cases, it is important to also include, in your desired restoration outcomes, information about species that fill specific niches and functions, as determined through ecosystem model studies. This would be, for example, species that are tolerant to one or more environmental events (e.g., fire, drought, flooding, wind throw); species that serve ecosystem functions (e.g., nitrogen fixation, carbon uptake, wildlife support, soil building); species that are dominants after which a community is named (e.g., maple forest); and species that have broad environmental ranges of tolerance, to respond to long-term climate change.

Table 7.1 • Average Tree Composition of Wisconsin Southern Mesic Forest					
Species	**Av. I.V.****	**Constancy****	**Species**	**Av. I.V.****	**Constancy**
Acer saccharum*	126.0	100.0%	Carya ovata	1.8	19.1%
Tilia americana*	34.1	95.7	Celtis occidentalis	1.6	12.8
Fagus grandifolia	30.3	34.0	Betula papyrifera	1.2	6.4
Ulmus rubra*	25.5	72.3	Populus grandidentata	0.8	12.7
Quercus borealis	21.2	55.4	Juglans nigra	0.7	8.5
Ostrya virginiana*	15.8	78.8	Acer rubrum	0.6	4.2
Fraxinus americana	7.0	49.0	Fraxinus nigra	0.5	6.4
Quercus alba	5.7	38.3	Quercus bicolor	0.2	2.1
Carya cordiformis*	5.3	46.8	Fraxinus quadrangulata*	0.2	2.1
Juglans cinerea*	5.2	25.6	F. pennsylvanica	0.1	4.2
Ulmus americana	5.0	40.4	Quercus macrocarpa	0.1	2.1
Prunus serotina	1.9	31.9	Q. velutina	0.1	2.1
Gymnocladus dioca*	1.8	6.4	Ulmus thomasi	0.1	2.1

*Species that reach their optimum importance in Wisconsin in this forest.

**Quantitative figures for tree composition are given as average importance values (sum of relative density, relative frequency, and relative dominance) and as constancy percentages, based on the measured sample of 80 trees per stand rather than on a single sample of a fixed area.

Adapted from Curtis, John T. *The Vegetation of Wisconsin.* © 1959 by the Board of Regents of the University of Wisconsin System. Reprinted by permission of The University of Wisconsin Press.

Composition and Structure: Family

One criterion that can be useful when considering the species composition for a community is to match the relative proportions of their taxonomic families. Some of these families represent species with well-known ecosystem functions, such as the Fabaceae and Rhamnaceae, both of which fix nitrogen. Other families, such as the Poaceae, provide the energy to maintain expansive grazing herds and are one factor in the development of the rich agricultural soils in temperate zones around the world. Species composition within a community will vary between individual sites; however, family presence for a specific community type often remains relatively consistent. For example, the dry forest ground layer in southern Wisconsin is highly represented by the Asteraceae, Liliaceae, Fabaceae, and Poaceae, while in the southern mesic forest, the Poaceae and Fabaceae are replaced by Ranunculaceae and Cyperaceae.

Composition and Structure: Life Form

Plant life forms are based on size, shape, presence or absence of woodiness, leaf structure and orientation, and morphological characteristics. They include, but are not limited to, lichens, ferns, grasses, sedges, herbs, shrubs, and trees. Life forms help define a community type; the grasses and monocots that compose the various grasslands around the world are examples.

In forested communities, life forms include grasses, sedges, herbaceous dicots and monocots, vines, shrubs saplings, conifer trees, and hardwood trees. These are distributed within vertical layers: the ground layer, midstory, and one or more layers of canopy. The proportions of these categories vary between different forest community types, but they are relatively consistent within a single community type. For example, the shrub layer in a

Table 7.2 • Prevalent Ground Layer Species of Wisconsin Southern Mesic Forest

Species	Presence*	Average Frequency	Species	Presence	Average Frequency
Actaea pachypoda	38%	1.4%	Geum canadense	43%	1.9 %
Adiantum pedatum	43	4.0	Hydrophyllum virginianum*	60	12.3
Allium tricoccum*	78	9.3	Laportea canadensis	43	7.7
Amphicarpa bracteata	35	4.9	Maianthemum racemosa	76	15.7
Anemone acutiloba*	62	15.3	Osmorhiza claytonii	89	19.8
A. quinquefolia	54	5.6	Parthenocissus vitacea	65	12.8
Arisaema triphyllum	57	12.9	Phryma leptostachya	46	3.8
Athyrium filix-femina	49	3.2	Podophyllum peltatum*	84	14.8
Botrychium virginianum	65	8.6	Polygonatum pubescens	70	19.2
Brachyelytrum erectum	41	2.0	Prenanthes alba	41	3.0
Carex pensylvanica	60	8.3	Ribes cynosbati	46	3.1
Caulophyllum thalictroides*	68	6.7	Sanguinaria canadensis*	65	19.1
Celastrus scandens	38	1.9	Sanicula gregaria	65	12.3
Circaea quadrisulcata	65	12.5	Smilax ecirrhata	49	2.5
Claytonia virginica*	38	20.8	Thalictrum dioicum	57	7.6
Cryptotaenia canadensis	35	6.3	Trillium grandiflorum*	65	14.5
Galium aparine*	70	26.0	Uvularia grandiflora	54	10.4
G. concinnum	54	7.6	Viola cucullata	65	15.4
G. triflorum	41	2.9	V. pubescens*	65	23.4
Geranium maculatum	78	15.7			

*Species are also modal, since their presence values are higher here than in any other Wisconsin community. Number of modal species: 35. Number of prevalent modal species as percentage of total prevalents: 25.6%. Additional model species occur infrequently or are rare and do not appear in the above list.

Presence is the number of stands sampled in which a species is found, as a percentage of all stands sampled. Prevalence represents the species with highest presence. Modal species are those species that have their maximum presence value in a given community.

Adapted from Curtis, John T. *The Vegetation of Wisconsin.* © 1959 by the Board of Regents of the University of Wisconsin System. Reprinted by permission of The University of Wisconsin Press.

North American conifer or maple forest is relatively undeveloped, compared with that found in an oak forest, a pine savanna, or the shrub-dominated Southeast U.S. pocosin community.

Within tropical forests, life forms, in addition to trees, shrubs, and herbs, include epiphytes, hemiepiphytes, climbers, lianas, and stranglers. The vertical stratification of these forests is also subdivided into layers of different species groups: ground, shrub, lower canopy (tree height 18 m), upper canopy (tree height 24.5 m), and emergent layer (tree height 30.5 m or more).

Species Functions: Keystone Species
A keystone species is one that has a disproportionate influence on its environment, including the survival of other plant and animal species and the ecosystem processes within the

Table 7.3 • Community Summary for Wisconsin Southern Mesic Forest

Major dominants (I.V.): *Acer saccharum* (126), *Tilia americana* (34), *Fagus grandifolia* (30), *Ulmus rubra* (26), *Quercus borealis* (21).

Most prevalent ground layer species (P%): *Osmorhiza claytoni* (89), *Podophyllum peltatum* (84), *Allium tricoccum* (78), *Geranium maculatum* (78), *Maianthemum racemosa* (76).

Leading families (% of total species): Liliaceae (9.3), Compositae (8.7), Ranunculaceae (7.8), Cyperaceae (6.8), Caprifoliaceae (5.4).

Related communities (index of similarity): Southern dry-mesic forest (66), Southern dry forest (56), Northern mesic forest (49), Southern wet-mesic forest (49), Northern dry-mesic forest (44).

Species density: 39.

Index of homogeneity: 54.4%.

Number of stands studied: 47.

Number of species: Trees 25, Shrubs 33, Herbs 172, Total 230.

Stability: Very stable—a terminal forest.

Climate: Total ppt. 30.6″ (78 cm), Snowfall 45.8″ (116 cm), January temperature 16.9°F (–8.4°C), Growing season 154 days. Typical weather stations: Milwaukee, Richland Center, West Bend.

Catena position: Middle. No gley layer.

Soil group: Brown forest and gray-brown podzolic.

Major soil series: Parr, Waupun, Warsaw, Elba, Fayette, and Downs.

Soil analyses: w.r.c. 75%, pH 6.9, Ca 3655 ppm, K 95 ppm, P 33 ppm.

Approximate original area: 3,432,500 acres (1,389,000 ha).

Typical examples: Wyalusing Scientific Area, Wychwood Scientific Area, Mauthe Lake Scientific Area, Grant Park, Milwaukee, Petrifying Springs Park, Kenosha.

Major publications: Curtis and McIntosh (1951), Gilbert and Curtis (1953), Randall (1953), Ward (1956).

Geographical distribution: Southern Minnesota and eastern Iowa, south and east to optimum development in southern Appalachians.

Adapted from Curtis, John T. *The Vegetation of Wisconsin.* © 1959 by the Board of Regents of the University of Wisconsin System. Reprinted by permission of The University of Wisconsin Press.

landscape (see Chapter 2). The reintroduction or loss of a keystone species will result in significant changes in the ecosystem. For example, the demise of the wolf (*Canis lupus*) in the Greater Yellowstone ecosystem resulted in large-scale changes to both flora and fauna that are only now beginning to be understood with the return of the wolf to this location. The reestablished wolf population is thought to have changed the behavior of elk (*Cervus canadensis*) and, in turn, that of the grizzly bear (*Ursus arctos horribilis*), golden eagle (*Aquila chrysaetos*), and coyote (*Canis latrans*) in the ecosystem. In response to changes in animal populations and behavior, vegetation changes have been rapid, and these changes have further influenced animal populations and behavior.

Another example is sphagnum moss (*Sphagnum* sp.), which appears to be the trigger to switch northern hardwood swamps and open water bodies to bogs. Highly efficient at controlling nutrient availability, moisture availability, and water chemistry, sphagnum leads to the development of a unique suite of species and ecosystem.

The dominant and more abundant species of a community are not necessarily keystone species. For example, the American elm (*Ulmus americana*) was a dominant species of the floodplain forests of the eastern United States. From the 1950s through the 1970s, this species was largely eliminated from the canopy in most floodplain forests due to Dutch elm disease. Other canopy trees were quick to fill in the gaps left by the elm, and the floodplain forests persist in a similar structure to that of the past. Any influence on the community resulting from the reduced prominence of elm in the ecosystem is still unclear. You may need to use adaptive restoration to determine the keystone species within a restoration site you are planning.

Species Functions: Pioneers or Rapid Colonizers

Small-scale disturbances that expose bare ground to potential colonists (pioneers) are common in most communities (see Chapter 2). These disturbances are similar to what occurs when a restoration is initially planted. It is often important to include native colonists that will be able to quickly expand into these disturbances. If such natives are absent, the opportunity for colonization by undesired species greatly increases.

Annuals and many biennials produce large quantities of seed and can quickly colonize open ground, but without additional future disturbances, they are unlikely to remain over time. Some perennial species also provide similar functions to those of "pioneer" species, and although they are often slower than annuals to establish, they have the advantage of persisting for much longer periods once established. Pioneer species are desirable for colonizing disturbed areas, such as gopher mounds, wild boar scrapes, abandoned wallows, or wave-scoured banks.

Species Functions: Phenology

Phenology is the timing of different events that regularly occur throughout the year within a community. These include production of foliage, flowers, fruits, and seeds for plants, and periods of migration, nesting, and foraging for animals. Many plant species flower for one to several weeks and then produce mature fruit a few weeks later. Communities differ in the times when species are in bloom and fruit. In temperate zone forests, most understory species flower in the spring, with a smaller number coming into bloom in summer and fall. In midwestern tallgrass prairie, about one-quarter of the species flower in spring, one-half in summer, and one-quarter in fall.

By developing a desired species composition with phenology in mind, you can match the seasonal dynamics of a community. You can create phenology tables, from your own observations or published observations of others. Items to include are the time periods for flowering, fruiting, seeding, and foliage growth; or the periods in which a plant species provides cover, nesting, and food for specific wildlife species. Examples of phenology tables for plants and wildlife are provided in Tables 7.4 and 7.5. In recent decades, scientists and naturalists have begun to see a change in some of the regional phenological events in plants, such as blooming times. Climate change is one likely cause, but the degree and direction of its effect on plant phenology will depend on the geographic location in which you work.

Species Functions: Interactions

When determining the species composition of your restoration, the species–species interactions and species–environment interactions must be taken into account (see Table 2.1). These interactions are site-dependent and community-dependent. They can be placed into various groupings, including interactions of organisms with their physical environment and surrounding ecosystem processes, interactions between individuals of the same species,

Table 7.4 • Plant Species Phenology (Northern Hemisphere–Eastern Deciduous Forest of the United States)

	J	F	M	A	M	J	J	A	S	O	N	D
Plant Species A												
Foliage			▮	▮	▮	▮	▮	▮	▮	▮		
Flower				▮								
Fruit					▮							
Seed								▮	▮			
Aboveground cover	▮	▮	▮	▮	▮	▮	▮	▮	▮	▮		▮
Plant Species B												
Foliage					▮	▮	▮	▮	▮			
Flower							▮					
Fruit								▮				
Seed										▮	▮	
Aboveground cover	▮				▮	▮	▮	▮	▮		▮	▮
Plant Species C												
Foliage					▮	▮	▮	▮	▮			
Flower							▮					
Fruit									▮	▮	▮	▮
Seed										▮	▮	
Aboveground cover					▮	▮	▮	▮	▮	▮	▮	▮
Plant Species D												
Foliage				▮	▮	▮	▮	▮	▮			
Flower					▮							
Fruit						▮						
Seed								▮	▮			
Aboveground cover				▮	▮	▮	▮	▮	▮			
Plant Species ...												
Foliage					▮	▮	▮					
Flower							▮					
Fruit							▮					
Seed							▮					
Aboveground Cover					▮	▮	▮					

and interactions between different species. All types of interactions influence overall plant composition, as well as particular species that might be targets for restoration.

Periodic events, such as drought, fire, floods, and wind throw, are common occurrences in some parts of the world, acting almost as a community stabilizer; in places where they are less common, they act as a disturbance (see Chapter 2). For example, in the African plains, periods of drought, rain, and fire are part of an annual cycle, and the plants and animals are adapted to these periodic events. Species that tolerate frequent fire, spring floods, and/or drought, or that are able to colonize areas after wind throws, should be included in the desired composition mix when circumstances warrant.

However, infrequent or catastrophic events—canopy fires, floods, and multiyear droughts—can result in major compositional changes to a community, or can even change the community type itself. Hurricane Katrina in 2005 caused major changes to water depth

Table 7.5 • Phenology for Plant Species for Wildlife Support of Wildlife Species A

	J	F	M	A	M	J	J	A	S	O	N	D
Plant Species 1												
Cover	■	■	■	■	■	■	■	■	■	■	■	■
Food	■						■	■				
Nesting					■	■						
Plant Species 2												
Cover				■	■	■						
Food							■	■	■	■	■	
Nesting												
Plant Species 3												
Cover	■	■	■	■	■	■	■	■	■	■	■	■
Food									■	■		
Nesting					■	■						
Plant Species 4												
Cover				■	■	■	■	■				
Food												
Nesting								■	■			
Plant Species 5												
Cover							■	■	■	■		
Food												
Nesting												
Plant Species ...												

and salinity levels in the bayous and swamps of coastal Louisiana that have resulted in the decline of large areas of bald cypress. Ecologists are currently conducting research on how to restore these swamps for both ecological values and functional values, such as protecting interior lands from flooding.

In developing your desired restoration species composition, you need to consider and adjust for species dependencies, relationships, and competitiveness. Some species have narrow habitat requirements, have specific (obligate) food sources, or, for some plants, depend upon a single pollinator. The larval phase of the regal fritillary butterfly (*Speyeria idalia*) only feeds on violet leaves. To restore habitat for this species, violets must be included in the desired species composition. Giant pandas (*Ailuropoda melanoleuca*) have a narrow diet limited to bamboo and few other species. These species would need to be included in a panda habitat restoration. Predator-prey relationships, pollinators and plants that rely on each other, and herbivores or grazing relationships (including those involving insects) are additional factors that should be considered when developing the species composition outcomes or guidelines for a given restoration (see Sidebar).

The criteria used to develop the desired species composition—one that supports diversity, functionality, and adaptability to changing environmental conditions—are key elements of site planning. To find the specific information for a community of interest, explore biology, ecology, and botany texts and journals; relevant web pages (i.e., government and nonprofit conservation organizations); and herbaria. In addition, consult with experienced restorationists.

| Sidebar | Providing for Wildlife Populations |

Although this text does not address relocating wildlife species to new habitats, wildlife habitat requirements are inherently considered in the development of goals and outcomes (see Chapters 4 and 5). Habitat needs should be addressed for all life forms relevant to the project, including invertebrates, fish, herptiles, birds, mammals, and soil organisms.

Many of the criteria used for determining plant composition also apply to wildlife species, that is, the inclusion of keystone species, life form diversity, and habitat interactions. Additional criteria include:

- Populations of (targeted) species can survive within the scale of the site.
- Populations of species exist in habitats within dispersal distances of the restoration.
- Populations or individuals are available for transplant to the site.
- The landscape encompasses a variety of slopes, uplands, and lowlands that support species' needs in all seasons and during extreme climatic events.
- The proposed vegetation provides food, nesting opportunities, and cover.
- Both water quality and water quantity are sufficient for aquatic and terrestrial species.
- Social outcomes for the site are compatible with species requirements.

Animal species reintroductions can be approached directly (introducing the desired species) or indirectly (creating the habitat to encourage colonization by the desired animal species). The first approach requires an understanding of habitat requirements (size and context of the restoration, access to food and cover), reproduction needs, animal tolerance to relocation, and political/public support. And, introductions of larger prey animals and native herbivores can be highly controversial, particularly in ranching and farming communities. The second approach assumes that providing habitat will result in habitation by the targeted species, and yet this is often not the case.

Habitat suitability indices (HSI) are one way of quantifying the quality of habitat for wildlife species populations, while looking at the potential effects of human activities, such as management, restoration, and area development (Larson et al. 2003). The HSI score is mathematically calculated from the strength of hypothesized relationships between wildlife species and habitat variables, such as shrub density, canopy cover, patch size, and water availability. How critical each variable is to supplying quality habitat is species-dependent and may be supported by empirical data, or based on expert opinion.

7.1.2 Physical Conditions

In writing the site outcomes, you want to address not only the vision for the biological community, but the visions for the physical landscape and desired user infrastructure as well. A focus only on vegetation is unlikely to be successful if the physical landscape necessary to support the vegetation, and provide for the dynamic processes associated with the community, is not present. In many instances, the physical landscape will require modification, particularly in cases where past land use has significantly modified its original character. Here we are addressing changes in topography and hydrology, which, for example, occur with streambank channelization, mining, and damming of rivers. The latter can alter gravel shoals used for spawning beds, as well as the flooding regime on which many plant and animals depend. The Kootenai River Restoration case study in Chapter 14 presents a number of outcomes that address the restoration of landscapes where such changes have occurred.

Soils, too, are often modified under agriculture or industry. If soils have had fertilizers or pesticides applied to them, outcomes may list the levels of nutrients, for example, that are targeted or acceptable. Brownfield sites, or former industrial sites that have severe soil and groundwater contamination from the storage and use of hazardous materials, are examples of extreme conditions where outcomes for soils and hydrology could dominate over those focused on plants and wildlife. Site outcomes only provide the targets that we may want to obtain once restoration is complete; they do not spell out how they are to be achieved. Those recommendations will be located in the implementation plan (see Chapter 8).

7.1.3 People Activities and Uses

Service resources and visitor uses and needs are additional areas that you will need to address in the outcomes. All restorations will require some level of service resources, ranging from smaller infrastructure components, such as trails, fences, or firebreaks for management, to larger components, such as service areas or equipment storage and drives to access the various areas of a site. For sites where visitors are allowed or encouraged, trails, parking, visitor centers, educational centers, signs (directional, interpretive, warning), wildlife viewing areas and overlooks, and restroom facilities may have to be integrated. If the restored site is to have an educational or public purpose, then you need to decide how that component can best be integrated into the landscape. Zoning and other land use regulatory rules will also dictate what uses can and cannot occur and what can be built on the land. These rules might indicate setback locations for restorations and buildings relative to roads and property lines. Outcomes would state the levels of infrastructure that are needed based on the agreed-upon use-policy, but not how they are to be implemented.

Here are some of the questions to consider in developing the site outcomes, relative to possible people activities and uses:

- What are the existing uses and intended public uses of the site?
- What are public expectations for the land?
- If a society is dependent on these lands for resources, can its needs be incorporated into the restoration?
- Will there be periods of high use and low use?
- What will the major attractants of the restored landscape be to potential users?
- Will the future land use surrounding the site be compatible with the project goals?
- What management or service needs will the site have?

▶ 7.2 PLAN LAYOUT: CONFIGURING THE SITE PLAN

Taking its cues from the master plan, the site plan gives detail to the shape and size of each community that is planned for the site, and provides for natural transitions between each one. Transitions between community types can be abrupt, as found along rapid elevation changes such as those that often occur when moving from wet meadows to drier upland ridges. Or transitions can be almost imperceptible, as when moving from grassland to savanna to woodland. General locations for community placement in response to the physical environment will have been delineated in the master plan. Some reassessing of a community's final placement may be needed after additional site study and consideration of its location relative to that of other nearby communities or adjacent land covers and uses. The final shape should incorporate the extent of edge desired, the site's orientation to prevailing wind direction and solar access, and the site's physical characteristics.

The concepts of shape and edge, and inputs and outputs between site and adjacent lands, were discussed in Chapters 2 and 4. The site plan must address the shape and edge of each restored community in relation to the adjacent land cover. If similar land cover or remnants exist next to the restoration site, then maximizing edge to enhance interaction with this landscape can be beneficial. In this case, as the restoration develops, the hope would be that the edge would eventually blend and disappear. But when the adjacent land cover is composed of species (plant or wildlife) or uses that are not compatible with the restoration outcomes, and these have the capacity to disperse onto the restoration site, the edge should be minimized. The major consideration is that, as the interior space is reduced and edge is increased, the influence of adjacent landscapes will be magnified.

Although the legal boundaries of a site are usually a given, the site design should take advantage of opportunities to expand size and connectivity to other sites on a regional scale, through either direct ownership or cooperation with landowners with similar land cover. For most restoration situations, the area in which the restoration will occur will seldom be adequate to reintroduce all former ecosystem processes. Project area size must relate scale to the project goals, and if this is not possible, the project goals may need to be revised. For example, a sedentary insect will require a much different restored habitat area than a larger mammal, such as a grizzly bear, or an anadromous fish species, such as salmon, which migrates between fresh water and salt water.

The size of a planned project may be larger than the current restoration site, and the planners might hope additional lands can be acquired over time. Examples of this situation are found in several ongoing projects by The Nature Conservancy to protect and restore the watersheds of high-quality streams and rivers.

Another site size consideration is major landscape disturbances, such as a catastrophic fire, major flood, or climate change. These are disturbances that can significantly change the communities. The implications of future climate change, for example, reinforce the need to incorporate high levels of flexibility into restoration plans by once again considering the regional landscape in which the site exists.

One option is to provide opportunities, whenever possible, for species to migrate to acceptable habitats as the regional climate changes. The widespread arrangement of designated core conservancy lands over substantially large land areas, and the establishment of species migration corridors between these areas, may provide opportunities for species to shift habitats in response to climate and vegetation changes across continents. Large conservancy areas that cover more than one community type may also allow some plant species to shift ranges, or for the transition of one community type to another when the climate changes. Restoration, land protection, and private landowner agreements, are ways to help create and preserve these large conservancy areas. Whenever possible, enhancing the opportunity for plant and wildlife migration is one way to mitigate the potential impacts that catastrophic events have on species survival.

7.2.1 Physical Features

The site plan will show physical changes recommended for the existing site in order to accommodate vegetation, wildlife, or any environmental process upon which the restoration may depend. Although physical alterations are common to all types of restorations, they are probably most needed in highly degraded sites such as former mine sites, and in wetland and aquatic restorations where landforms are critical to hydrology, particularly water flow, volume, and velocity. A few examples from wetland restoration follow.

Stream Channels

Historically, streams were often channelized or straightened by landowners to expand the land area for agriculture or other uses, as well as to increase the rate that water is drained from the land. Channelization had several undesirable impacts, including an increase in downstream volume and speed. These increased water volumes and flow rates resulted in a higher incidence of flooding, the degrading and incising of the channel, or its accelerated erosion both vertically and horizontally. The result was that banks changed from being vegetated with gradual slopes to barren with steeply dropping slopes, thereby positioning stream levels well below their original elevations. Channelization also caused a loss of wildlife access to water, the disappearance of pools and riffles, head cutting of the streambed, and increased water temperatures.

Reintroducing meanders to a stream, along with restoring former slope geometry, can result in a slower rate of flow, less flooding, and increased bank vegetation diversity. These outcomes are in part due to the increased length the water now travels and the obstacles that vegetation and reconstructed land peninsulas provide (Riley 1998). Restoration site plans will often show relocation of stream channels or the addition of meanders to existing channelized streams (see Figure 4.12).

Shorelands, Ponds, and Basins

One of the first steps in the restoration of prairie wetlands is the configuring of shorelands, ponds, and basins (Galatowitsch and van der Valk 1994). Shoreland configuration has been found to have significant influence on waterfowl behavior, particularly molting or breeding pairs of ducks. The visual barriers of irregular shorelines provide isolated protected areas for seasonal periods, when these and other species may be most vulnerable to predation. The restoration site plan will often show new shoreland configurations.

Similar to shorelands, the composition and structure of ponds and pools are significant in their ability to create diverse habitats. For example, the physical and vegetation variety of a series of small pools compared with a single large pool often results in the former having much greater waterfowl diversity. Waterfowl also benefit when the pools are a mix of temporary, seasonal, and semipermanent wetlands.

The shape, depth, and placement of basins within a watershed also influence the types of wildlife and plants that may colonize or establish in an area. As with streambanks, shallow basins with gradual slopes have expansive vegetation zones that provide for a diversity of wildlife, whereas steeply sloped and deep basins have narrow vegetation zones and limited access by wildlife to water. Success in restoring communities to wetland basins also depends on the quality of water that flows into them. For example, where basins are situated to receive surface runoff from agricultural lands, water quality is often high in nutrients and organics and low in oxygen. This can result in a habitat of low plant and wildlife diversity, leading to failed restoration goals.

The above physical changes would be detailed in the implementation plan (see Chapter 8). The site plan notes the desired state of topography, soils, or hydrology that are needed and locates where stream meanders might be placed or the distribution of a series of pools, but it would not explain how these are to be implemented.

7.2.2 Plant Communities, Wildlife Habitat, Species Placement, and Species Patterns

Along with establishing outcomes for the species composition, some site plans specify where to locate the species mixes, and possibly individual species for each community type identified on the site plan. The level of location specificity for individual species or species groups can follow one of two approaches: one that allows substantial flexibility and one that does not.

The first approach locates the communities to be restored and simply specifies the desired plant species. It is assumed that the majority will establish and successfully reproduce, redistributing themselves into the most appropriate environments given both abiotic and biotic interactions. Such a planting would, in theory, allow for natural growing patterns. One problem with this approach is that many species may be slow to redistribute themselves, may not have been located where they grow best, and may actually have been located where the environment is incompatible. In this case, resources are lost, some of which are not abundant to begin with. This strategy is best used when a site is rather homogeneous in its physical characteristics.

The second approach specifies the location of individual species and species groups in the plan. This is done for several reasons, the most important one being to match species of interest to changes in the physical characteristics of the landscape at a much more detailed level than that used in the first approach. In some instances, species are placed into mixes that are matched to microtopographic changes or variances in soil moisture and lighting. Many sites are not homogenous and can have distinct zones, such as moisture level gradients that extend from the center of a marsh or wetland to upland or shade gradients that occur from forest to open plains. In these situations, it is helpful to prepare separate mixes for each zone in order to achieve the best species-environment match. Where the environmental gradients that separate two zones are gradual, it is also useful to prepare three seed mixes—a mix for the interior of Zone A, a mix for the interior of Zone B, and a third mix that contains species from each of the others, assigned to Zone AB, a region along the edge where the two zones meet. This planting pattern allows for a gradual species transition from one community to the next.

There are instances where the separate placement of individual species in specific locations rather than including them in a mix is desirable, despite the added time and cost this requires. Some species have a narrow environmental tolerance and are quite specific to habitat conditions. These species are often rare; the planting of their seeds or seedlings, which may be limited in supply, in areas where survival is questionable should not occur. Assigning species to specific locations is also done for aesthetic reasons.

Species frequency is one measure for deciding how an individual species might be distributed across a site. For our purposes, frequency provides an indication of how widely distributed a species should be in the community to be restored. As defined in Chapter 2, frequency is the number or percentage of quadrats or points where a species can be found when sampling a site. The higher the frequency, the more likely the species will be found when walking through an area. However, frequency is not a measure of a species' abundance; species with small populations are often scattered throughout an area.

How you show the location of woody species or trees and shrubs on the graphic plan depends on the community type and the method of planting, as well as the scale of the site (Figure 7.3). For savanna systems,

FIGURE 7.3 Species Placement. This graphic shows the approximate tree placement from the student plan for the CPIANSR west subsection. The west-facing slopes on the left show a scattering of trees in savanna-like groupings, and the trees on the east-facing slope on the right are at a density representative of woodland or forest. (Adapted from Read et al. 2004.)

generalized individual tree placements are helpful for indicating approximate tree densities and the desired distribution. For continuous forest canopies, outlines of the tree canopy are sufficient. Written specifications can then provide the actual tree planting density per unit area. In some situations, a large number of tree species can exist in a community, but only a few species represent the majority of trees to be planted. The plan has to show or describe how the less populous tree species are to be located on the site. Shrubs are most easily located in groupings with projected densities (number of plants per area unit). On small sites where greater detail is feasible, individual shrubs or shrub areas can be located on the plan.

For large acreages displayed at fine scales on the plan, it is impractical to show individual tree and shrub placements or even herbaceous mixes. Where placements have to be specific, portions of the site plan must be enlarged. Text and tables that elaborate on the desired densities and distribution characteristics accompany most plans.

Although aesthetics would not alter the plant species selection for desired ecological components or processes, aesthetic touches can increase people's enjoyment and understanding of sites, particularly when they are located in botanical gardens, arboreta, and other public educational and interpretation areas. In such situations, aesthetics might be used in deciding on plant composition along certain trail junctions and near buildings, for framing vistas, or in creating species patterns within the landscape. Fast-establishing flowering species that bring color to a site and lessen the visual impact of weedy species, and that can establish during a restoration's early stages as slower-growing plants develop, might be included in mixes for highly visible sites.

One additional human-oriented issue is the use of species with armaments or toxic oils and chemicals. Although ecological arguments can easily be made to include such species in plant mixes, their ability to affect human health and safety has often kept some of them off restoration lists. While we do not advocate eliminating species that meet other selection criteria for the site, the placement of these species requires considerable thought for the safety and health of onsite workers and managers, as well as the public.

Figure 7.4 summarizes the species criteria discussed in this chapter that can be used in creating a list of species for a project setting.

7.2.3 Landscape Infrastructure

As noted earlier, all project sites require some level of infrastructure to support site uses such as access for service staff and vehicles. In situations where the restoration purpose is solely for restoration or research, the infrastructure, such as trails and/or service roads, can be rather primitive and limited. On the other hand, when the projected community type requires ongoing management or when site use is for education, programmed interpretation, and other public functions, the need for service access infrastructure can be extensive. Infrastructure to support site access and/or human services can include pedestrian and vehicle access and parking, visitor centers or gathering spaces, water sources, and signage. The level of infrastructure will vary with the community type and planned outcomes.

Trails, for example, can be designated as primary, secondary, and service-related. Primary trails are typically the easiest to negotiate. In some countries, primary trails are legally required to meet standards for accessibility, such as maximum percent slope and wheelchair-accessible trail surface, width, and height (distance to the lowest branches extending over the trail). Many government and nonprofit agencies have guidelines for trail planning

Appropriateness to the Abiotic Conditions of the Site

Climate

Soils

Topography

Hydrology

Geography

Composition

Family representation in community

Keystone species

Life form

Life cycle (annual, biennial, perennial)

Pioneer species

Climax species (which may include pioneering species)

Cover/companion crop

Diversity and Structure

Representation in community examples for which data are available: commonness, rareness (density, frequency, dominance, importance values)

Phenology (seasonal attributes)

Reproductive strategies (vegetative, wind-pollinated, insect-pollinated)

Ecotype/genotype

Functions

Event reactions (fire, grazing, drought, flooding, wind)

Species interactions (competitiveness)

Nutrient cycling

Wildlife supporters (food, cover, nesting)

Aesthetics and Safety

Flower color

Plant height or size

Fragrance

Armament

Toxicities

Wildlife Species

Food, nesting, and cover

Seasonal support

FIGURE 7.4 **Plant Species Composition Criteria.**

and construction (see the Resources at the end of the chapter). Such guidelines include trail widths, clearance heights, slope ranges, surfaces, and water diversion features.

The design of a trail and/or road network is specific to the site. For example, trail placement can be tied to experiencing site features (landscape and wildlife viewing, geologic features, community representations) and used for negotiating the site (topography, hydrology, and soils). Therefore, one of the best ways to locate trails (roads) is to begin with the trail purpose, and then map the specific areas on the site the trail needs to reach. Next, you analyze soil and topographic maps to identify locations where the trail will have minimal site impacts and still support the desired purpose. Once a rough trail is plotted, the proposed location can be walked and modified as opportunities for enhancement arise, or constraints to the proposed layout become apparent. If possible, use a GPS unit to walk and digitize the route onsite, and then download the data points into a computerized mapping program, such as a GIS, in order to create maps showing trail placement relative to topographic contours and other site features.

Trails are best located to avoid increasing erosion and rates of stormwater flow, both of which are influenced by soil characteristics, slope steepness, and slope length. Some general considerations for erosion control are positioning trails to gently ascend and descend steep slopes, avoiding slopes steeper than 8% when possible, and not locating trails perpendicular to the slope. Minimizing the trail length before switching direction to move up or down a slope reduces the ability of water flowing down the trail to gain speed and volume and carry soil. These "switchbacks," when used, should be gradual, with wide bends, to minimize the temptation for users to go off the trail and cut corners where the trail switches back. Avoiding the use of impermeable surfaces, such as asphalt, on slopes will help limit erosion along path edges. In wetland locations, raised boardwalks allow for access with minimal damage to vegetation. Boardwalks in wetland settings can be subject to high water, wave action, and ice action. A floating boardwalk may be necessary on deep and unconsolidated soils (Figure 7.5).

FIGURE 7.5 **Trails and Boardwalks.** The boardwalk in the foreground is only wide enough for one person but provides access to a wet-mesic prairie restoration at the University of Wisconsin–Madison Arboretum. The boardwalk is not anchored at this position, thus allowing for movement during periods of wetness or flooding and when soils freeze. Note in the background the wires and housing that may interrupt, for some people, the experience of the vast prairie that once dominated the landscape in this region. (Photo reprinted by permission from John Harrington.)

Additional trail considerations will depend on the project's use purpose. But the ability for trails to traverse most community sites, pass highlights, and avoid sensitive areas and species that require special protection are major considerations. For educational purposes, bump-outs, or areas along trails that are widened for small group discussion or presentations in the field, are beneficial.

Parking and road networks should be located where they will require minimal earthwork, soil disturbance, and impervious surfaces. Parking and roads have specific soil requirements (i.e., bulk density and drainage) and topographic requirements (i.e., slope, aspect) for siting. When possible, they should be constructed not only to avoid creating additional stormwater runoff but to minimize the expansion of site edges and effects on wildlife. In situations where parking occurs only occasionally, grass can often be used for the parking surface. When site conditions and resources permit, road surfaces should be constructed for permeability, whether they are grass, gravel, or another type of porous surface.

7.2.4 Rationale for the Site Plan

The site plan includes a written rationale explaining the outcomes, plan, and solution. The rationale will help future managers of the restoration understand the original decisions and outcomes. The rationale synthesizes the data used in making your decisions, including those derived from the site inventory and analysis, reference models, background review of previous projects and documentation, and discussions with other restorationists and experts, as well as your clientele. The rationale explains why this solution, why these targets, and why these outcomes. It supports your work in two ways: by explaining to the client the logic of your solution, and by helping you understand and reevaluate your decisions.

▶ 7.3 CREATING THE LANDSCAPE EXPERIENCE

Many restoration projects are planned to include such activities as education, hiking, birdwatching, and nature observation. Kiosks, trails, and visitor facilities are often constructed to support these activities. Another way to enhance visitation is to create an experience where the user feels fully immersed in the community, free of outside influences.

In small settings and in open communities where adjacent lands and distant horizons are apparent—often with utility lines, buildings, and highways interrupting the horizon (see Figure 7.5)—creating a sense of true immersion into a landscape can be difficult. Yet several mechanisms, each working with landforms or vegetation, can help with the illusion of a continued space.

Hills and elevated landforms that protrude into an open landscape interrupt views; the site boundaries become less obvious, and the views disappear into valleys and around bends. The illusion of expanding size is also accomplished when open communities disappear into a savanna-like setting among scattered trees, when a forested peninsula juts out into the open landscape, or when a river meanders around a forested bend and allows the open ground plane to disappear from view without a definitive visual terminus (Figure 7.6).

Open vistas that terminate on a rise can also provide the appearance of continuance, as can open communities that border open landscapes in the distance. These adjacent lands often have an agricultural use, such as pasture, but help extend the restoration's spatial character of an open landscape. The difficulty with these landscapes, of course, is that the restoration manager has little control over changes in the bordering or "borrowed" landscape (Figure 7.7).

FIGURE 7.6 The Illusion of Continued Space. The Greene Prairie restoration at the University of Wisconsin–Madison Arboretum uses peninsulas of oaks to interrupt views. The result is that the prairie appears to extend much farther than it actually does. (Photo reprinted by permission from John Harrington.)

FIGURE 7.7 An Open Landscape with Invisible Boundaries. By orienting the locations of views with topography and background land cover, the restored landscape appears to extend outward beyond its actual boundaries, blending with its context. This is the Swamp Lovers restoration, Dane County, Wisconsin. (Photo reprinted by permission from John Harrington.)

Trails that move across varied topography and through several communities often provide the illusion of having traveled greater distances than a straight trail over level ground and through only one community. Trail placement can also be strategically planned to expose visitors to prominent landforms, overlooks, and unique vegetation or wildlife, as well as to avoid the less desirable features of a site. Where and how trails are placed, however, is a function of the use-policy, the desired visitor experience, and service and management requirements (Figure 7.8).

Aesthetic guidelines to trail placement can entice the public to learn more about a landscape and, in doing so, become a supporter of the project. By understanding what people find attractive and appealing in a landscape, onsite educators can often find ways to connect with their audiences. In many situations, the restoration of the community itself may accomplish this goal.

▶ 7.4 DEVELOPING A SITE PLAN FOR THE CROSS PLAINS ICE AGE NATIONAL SCIENTIFIC RESERVE

A 151-acre (61-ha) unit of the proposed 1600-acre (647-ha) Cross Plains Ice Age National Scientific Reserve (CPIANSR) is a good example of how a site plan can be developed. Our restoration class has developed site plans for the CPIANSR over several different years. This section summarizes the student group reports (Acker et al. 2007; Bouressa et al., 2007; K. Olson et al., 2009; Read et al. 2004). The CPIANSR is located near Black Earth in Dane County, Wisconsin, a rapidly growing metropolitan area extending from Madison.

The master plan proposes seven community types within the Cross Plains Unit of the CPIANSR (Figure 7.9). These community types and their locations were determined after a review and analysis of the U.S. federal land office survey conducted in the 1830s, physical characteristics of the land, and the existing vegetation. The federal land office surveys

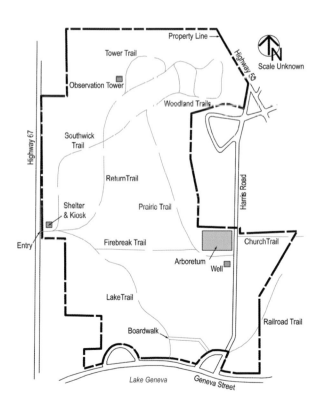

FIGURE 7.8 The Trail System for the Kishwauketoe Conservancy, Wisconsin. (Adapted from Carlson et al. 2004.)

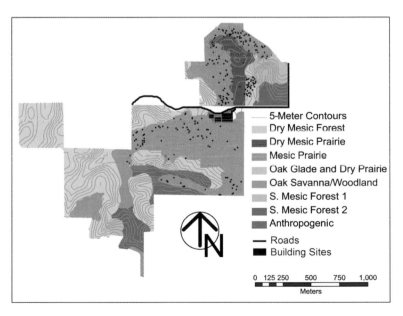

FIGURE 7.9 A Proposed Master Plan for the Cross Plains Unit of the CPIANSR. The plan shows conceptual plant community placements; suggested tree densities (black dots) for savanna, glades, and prairies; and infrastructure locations (roads and general location for visitor center and service buildings). Tree densities are not provided for forested areas. (Harrington, J.A. 2008.)

found the area of the reserve to consist of oak savanna and prairie. The landform was characterized as a series of ridgetops with steep slopes to the east, north, and west and rolling topography to the south. Limestone cliffs were exposed by glacial meltwaters along many of the slopes, but no lakes or streams occur on the site today. Soils are well-drained silt loams and vary from thin on the slopes to deep on the flat ridgetops. In 2000, the vegetation consisted of crops, old-field dominated by bromegrass, and 200-plus-year-old oaks surrounded by younger aged oak woodlands. The older oaks were of a structure or habit that suggested they once grew in the open instead of among the dense shrubs and younger trees that surrounded them now. In addition, small pockets of prairie survived on some of the limestone cliffs, and a sugar maple forest was developing in a deep gorge in the eastern half of the site. Interviews with long-time landowners were conducted about the recent history of the site. The selection and location of community types were also based on ecosystem models.

After the class analysis, the communities selected and located included:

- Dry-mesic prairie: Level to slight rolling topography with well-drained, moderately shallow, low-fertility soils; no canopy.
- Mesic prairie: Level fertile deep soils, moist but well drained, no canopy.
- Oak savanna: West- and south-facing slopes steeper than 8% on average, upper ridges of some north-facing slopes where open canopy exists, shallow depth to limestone bedrock.
- Oak glade: Level topography, shallow depth to limestone bedrock, low fertility, existing scattered and semi-open white oak canopy.
- Dry-mesic forest: Existing oak canopy on gently rolling to steep south- and west-facing slopes, well-drained soils with moderate depth and fertility. Dominant tree species: white, bur, and red oak; shagbark hickory (*Carya ovata*).
- Mesic forest 1: Small areas of steep slopes where light is limited, moist soils; sugar maple (*Acer saccharum*) dominates.
- Mesic forest 2: North- and east-facing slopes with deep soils. Red oak dominates canopy.

The proposed restoration site plan (Figure 7.10) results in a landscape composed of dry-mesic and mesic prairie, southern dry-mesic and mesic forests, oak savanna, and oak glade ecosystems (see Sidebar). The restored oak savanna is an example of the "natural" transition between the eastern deciduous forest and the tallgrass prairie zones of the U.S. and fits within The Nature Conservancy and U.S. government designation of the Prairie-Forest Border Ecoregion. The dry-mesic prairie occurs on the gently rolling plateau, while the mesic prairie occurs in the lower historic Black Earth Creek Valley. The oak savanna habitat extends along the western edge and upper northern and northeastern edges of the prairie. The oak glade extends east of the prairie along an exposed limestone ridge. The dry-mesic forest occupies eastern slopes along the western boundaries of this site and above the gorge. The mesic forest occupies the gorge and some of the upland areas east of the gorge. Hiking trails and interpretive signage are located throughout the site.

The CPIANSR provides opportunities for interpretation and education on the glacial history and geology of the site, nature recreation (hiking, birdwatching, photography, and nature study), and volunteer activities for public visitors.

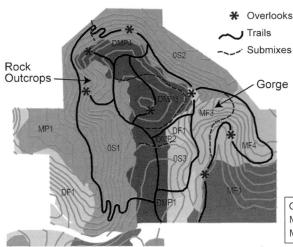

* Overlooks
〜 Trails
-〜-' Submixes

Rock Outcrops

Gorge

OS = Oak savanna; DMP = Dry-mesic prairie;
MP = Mesic prairie; DF = Dry forest;
MF = Mesic forest; Numbers identify species mix.

FIGURE 7.10 The Proposed Restoration Site Plan for the Cross Plains Unit of the CPIANSR. This site plan shows placements of vegetation types, species mixes, trails and user facilities. Parking and interpretive center zones, as well as sign locations, would be enlarged to present detail. Asterisks indicate overlooks. (Adapted from Acker et al. 2007; Bouressa et al. 2007; K. Olson et al. 2009; Read et al. 2004.)

Sidebar | Setting Targets for the CPIANSR

Developing a site plan has many challenges—particularly when resolving the goals and outcomes for a site where the original native vegetation structure is only generally defined, and where the site's physical conditions could support several community types that would reestablish perhaps different but desired environmental processes.

The National Park Service CPIANSR soils can support prairie, oak opening, or oak forest, and in some areas, maple woods. In a few locations, the existing vegetation has traces of the former native vegetation—former savanna trees now surrounded by younger saplings, or native forbs and grasses growing among a tangle of brush that has grown up where cattle used to graze. In other areas, plowing and disking to favor crops have annually disturbed the soil, and there are no native species remaining. In addition, project funds and other resources are limited. The goals for the site are to reestablish native species associations that may have been here in the past, but a specific community type is not cited. The goals also look to having a land cover that ensures the site's geologic features remain visible for interpretation.

The students set goals to develop a grassland-oak landscape composed of seven community types, yet the total site is relatively small. One question the students dealt with was: Would fewer larger-sized communities or many smaller-sized communities be more appropriate as a restoration goal? During the ensuing discussion, students considered additional questions, including:

- How diverse are the soils and topography? What site attributes (soil ph, nutrient levels, elevation, etc.) are most important in contributing to that diversity? At what level of differences in these attributes are distinctions in community types most likely to occur?

- Would a single community of larger size be more beneficial to wildlife, or would wildlife be better served by a diversity of habitats regardless of size?

- Under which configuration of landscape would visitors be better served with a use-policy focused on landscape interpretation?

FOOD FOR THOUGHT _____

1. How would you address these questions?

2. Are there additional questions to ask before making a final decision?

3. What would you have recommended in regard to the number of community types, given what you know about the site and ecology?

4. What additional information would help you with this decision?

7.4.1 Site Goals and Outcomes for the CPIANSR

A set of goals with their outcomes taken from the CPIANSR restoration report appears in Figure 7.11. These examples are abbreviated for space economy; many more outcomes could be written under each goal. Several sets of outcomes under a series of goals are shown so that you can see how the former are written in a time-neutral format and as measurable targets. Several of these outcomes describe major species groups desired for the restoration, and provide the minimum number of species to be selected from a prescribed species list.

7.4.2 Plan Layout: Configuring the CPIANSR Site Plan

Species were selected for each community type based on the above criteria and outcomes, and the species existing on the site. Table 7.6 has species lists for oak savanna that correspond with the site plan in Figure 7.10. The list is abbreviated; an actual species listing would have many more species. The site plan shows the placement of species and species

---

Goal: Maintain a complex of southern dry-mesic and mesic forest, oak savanna, and prairie communities common to the Upper Midwest.

  Outcome 1: Oak savanna occurs on at least 25% of the CPIANSR and 40% of the Cross Plains Unit within the CPIANSR.

  Outcome 2: Canopy coverage ranges 20–50%, represented by species from Table 7.6.

  Outcome 3: Trees are present as scattered individuals or in groves of no more than 47 individuals per hectare. At least 15% of the trees have a dbh (diameter at breast height) of 46 cm (18″) or greater.

  Outcome 4: Canopy dominants are bur and white oak (*Quercus macrocarpa* and *Q. alba*) representing at least 70% of the canopy trees in terms of numbers and biomass. At least 95% of the remaining tree species are native to the oak savanna.

Goal: Preserve and highlight the geologic effects left by the Green Bay lobe of the most recent period of glaciation.

  Outcome 1: Kiosks explain the history of glaciation in the area and highlight the influence it had on this landscape in terms of its geology, topography, and soils; and how that history influenced land use (agriculture, quarry) and subsequent vegetation since human settlement in the region.

  Outcome 2: Vegetation is sufficiently open (low stem and foliage densities) and limited in height so as not to obscure the landforms of the hanging valley, Black Earth Creek Valley, and the rock outcrops.

Goal: Allow people to use the site for nonconsumptive recreation, education, and research purposes.

  Outcome 1: Passive recreation permitted, excluding dogs, horses, motorized vehicles, or forms of consumptive recreation.

  Outcome 2: Trail locations highlight the site's physiography and natural aesthetics and pass through each restored vegetative community type.

  Outcome 3: Trails follow universal design principles and range in access levels from easy to difficult. At least one path allows for maintenance vehicle access and at least one major path is ADA (Americans with Disabilities Act) compliant.

  Outcome 4: Interpretive signage indicates the site's glacial history, significant landscape and ecosystem features, vegetation, and wildlife communities.

Goal: Supply parking for visitors on site

  Outcome 1: Parking is provided for 5–10 vehicles.

  Outcome 2: Parking occurs off the CPIANSR and is located on land currently occupied by farm buildings on the National Park Service unit.

  Outcome 3: Parking surface is permeable to allow for infiltration of precipitation.

FIGURE 7.11 Goals and Outcomes for the CPIANSR Restoration.

Table 7.6 • Species Mixes for Oak Savanna Restoration on the Cross Plains Unit of the CPIANSR

Oak Savanna Species	Mix 1 (<25% Canopy) Relative Abundance	Mix 2 (50–80% Canopy) Relative Abundance	Mix 3 (25–50% Canopy) Relative Abundance
Trees			
Carya ovata	Low	Low	Moderate
Quercus alba	High	Moderate	High
Quercus macrocarpa	Moderate	Low	—
Quercus rubra	—	High	Moderate
Shrubs			
Ceanothus americanus	Low	Low	—
Cornus racemosa	Low	Low	Low
Corylus americana	Moderate	—	Low
Viburnum lentago	Low	Moderate	Low
Grasses and Sedges			
Bouteloua curtipendula	High	Moderate	Low
Carex pensylvanica	Moderate	High	High
Dichanthelium leibergii	Moderate	Low	Low
Dichanthelium perlongum	Moderate	Low	—
Elymus hystrix	Moderate	—	Low
Hesperostipa spartea	Moderate	—	—
Schizachyrium scoparium	Moderate	Low	—
Sporobolus heterolepis	Moderate	Low	Low
Forbs			
Asteraceae			
Coreopsis palmata	Low	—	—
Helianthus stumosus	Moderate	Moderate	—
Liatris aspera	Low	—	—
Antennaria neglecta	Moderate	Low	Moderate
Solidago ulmifolia	—	Low	—
Symphyotrichum laeve	Moderate	Moderate	Low
Symphyotrichum oolentangiense	Moderate	Moderate	Low
Legumes			
Amorpha canescens	Low	—	—
Desmodium glutinosum	—	Moderate	—
Desmodium illinoense	Moderate	Low	—
Ranunculaceae			
Anemone patens	Moderate	—	—
Anemone virginiana	Moderate	Low	—
Aquilegia canadensis	Moderate	Moderate	Low
Prenanthes alba	Low	Moderate	—
Santalaceae			
Comandra umbellata	Moderate	Low	Low
Primulaceae			
Dodecatheon meadia	Moderate	Moderate	High
Euphorbiaceae			
Euphorbia corollata	Low	—	—

(table continues)

Table 7.6 • Species Mixes for Oak Savanna Restoration on the Cross Plains Unit of the CPIANSR (cont.)

Oak Savanna	Mix 1 (<25% Canopy)	Mix 2 (50–80% Canopy)	Mix 3 (25–50% Canopy)
Species	Relative Abundance	Relative Abundance	Relative Abundance
Geraniaceae			
Geranium maculatum	—	Moderate	Low
Boraginaceae			
Lithospermum canescens	Low	—	Low
Lamiaceae			
Monarda fistulosa	Low	Moderate	—
Scrophulariacea			
Pedicularis canadensis	Low	—.	—
Polemoniacea			
Polemonium reptans	Moderate	Moderate	Low
Polygalaceae			
Polygala senega	Low	—	Low
Polygala verticillata	Low	—	—
Liliaceae			
Maianthemum racemosa	Moderate	Moderate	Low
Commeliaceae			
Tradescantia ohiensis	Moderate	Low	Low
Caprifoliacea			
Triosteum perfoliatum	Low	Moderate	Moderate
Violaceae			
Viola pedatifida	Moderate	—	—
Viola sagittata			

mixes in the dry-mesic prairie and adjacent oak savanna and glade. Each of the community types is divided into subareas, each with its own species mix. Species are assigned to each mix, and many of the species are found in several, if not all, mixes. Every species has a relative proportion—high, moderate, or low—assigned to it within each mix in which it occurs. All species are to be collected from plants growing within 40 miles of the project area. Sufficient remnants exist within this radius to provide necessary seed for propagation and planting.

Species are assigned to subarea based on subtle changes in the site's physical and biological environment. Species proportions are also based on these differences. For example, the dry-mesic prairie has one mix for the base of the hanging valley, a mix for the small cone of elevation to the north, and two mixes for areas with level topography. The overlap in species between mixes recognizes that most landscape differences are subtle gradations and not discrete environments. This is also true with the oak savanna mixes and the overlap between the savanna and prairie mixes. Some species are found in both community types, while others are found in only one of the two types. The proportions are relative and qualitative. The implementation plan will specify amounts of seed or transplants, but the site plan, particularly for herbaceous species, does not need to be overly specific. The precise proportions of herbaceous species to be planted depend on numerous variables, over which planners have little control. Moving beyond the level of relative proportions at this stage wastes time and resources.

The site plan could also show specific placement of trees to be planted or cite the densities at which they occur. The plan can also show specific locations for plant species that are rare, have a narrow range of environmental conditions under which they will grow, and are being reintroduced. For extremely large areas, such a placement scheme is impractical, and single species allocations over broad areas will be required.

A site plan's level of detail is determined by the expertise of the individuals responsible for the restoration. However, most restorations remain in place long after the original author of the plan is no longer associated it with. By providing documentation at the level of detail as expressed above, individuals who follow the original plan will have a clearer understanding of what was being attempted. In addition, detailed documentation that can be made available to many will enhance the knowledge of restoration among a broader range of practitioners.

7.4.3 Rationale for the CPIANSR Site Plan

Along with the CPIANSR site's glacial history, this site plan was developed to preserve the presence of multiple large (dbh >70 cm), open-grown bur and white oak trees located throughout the project site and to restore the site to an oak savanna/prairie/forest landscape common to the area prior to settlement by cultures of European descent in the 1800s. The wide-spreading tree form is indicative of a historical environment with substantial spacing between individual trees to allow for uninhibited and extensive horizontal branching. Aerial photographs dating to 1937 reveal these randomly scattered, open-grown trees populating a considerably more open area than is currently present (see Figure 4.15).

The U.S. Federal Land Office surveyors' records indicate that the site (T: 7N, R: 7E, Sections 13 and 14) was not densely forested in the 1830s. It supported an open community type with large bur and white oaks and prairie, suggestive of oak savanna (see Figure 4.14). Restoring the site to oak savanna, prairie, and oak woodland is also contextually appropriate, because the neighboring reserve unit supports this same community type. This solution also provides for a contiguous habitat corridor between differing units of the CPIANSR and the surrounding area. Wildlife diversity can also be augmented through restoration. For example, the red-headed woodpecker, identified as a "special concern" species in Wisconsin and historically recorded on this site, responds positively to oak savanna restoration.

The one exception to converting the entire project site to oak woodlands, savanna, and prairie occurs within a narrow gorge and the lands east of it. These areas currently contain mesic forest dominated by red oak forest that is slowly converting to sugar maple dominance. Unlike the remainder of the property, this area's steep physiography and associated cooler microsite conditions, due to sun blockage and moist soils, best suit it to support a nutrient- and moisture-demanding plant community. The presence of this community provides a second vegetative community type for park visitors to experience, and an opportunity to educate visitors about the way fire and topography shaped this landscape.

KEY CONCEPTS

- The site plan is the detailed description of what the restoration intends to achieve. It adds the detail to the master plan vision through quantifiable outcomes, in written and graphic form.

- Depending on the restoration purpose, the site plan guides the site's physical environment, plant species composition and location, wildlife habitat requirements, and accommodations for human use through placement and specifications of infrastructure components.

- Plant and wildlife composition outcomes consider the roles that species play in community processes and random events, such as fire or drought, and build in species redundancy for these roles.

- Species composition and structure are dynamic, and the site design must allow for maximum flexibility in its species composition.

- The site design considers potential connections to the external landscape in order to assist species adjustments to ecosystem dynamics or change, including climate change.

- Infrastructure components (buildings, roads, trails, signs, etc.) meet multipurpose requirements for economizing on resources and space.

- Aesthetic principles can enhance the user experience of the site through the location of trails and viewpoints and, where appropriate, visitor centers. For smaller sites, design principles can be used to fully immerse visitors in the restored community.

FOOD FOR THOUGHT

1. List several reasons why a wildlife species may not always be able to recolonize a restored community, even if its habitat needs seem to have been restored. What strategies would you add to the site plan to resolve this?

2. Consider the ways in which wildlife or vegetation that occur outside the physical project area affect your success for restoration. After listing each one, discuss how you might plan for these potential influences when writing the site plan.

3. Select a restoration project occurring on public lands in your area. What are the public expectations for it? How are the public expectations and uses incorporated into this landscape? Is the restoration successful in terms of its use-policy and public satisfaction? How would you determine whether the restoration is successful?

4. Select a community type that occurs within 100 miles of where you live. Assume you have the opportunity to develop a restoration project for this community. What specific outcomes would you establish in terms of species composition and structure? Defend each outcome.

5. Establish and defend a set of criteria for selecting specific plant and animal species that meet the outcomes for the restoration in question 4. In this defense, explain how you would use each criterion to select species.

6. Why do some large landscapes have no more plant species than smaller landscapes? Do you think this would also be true for animal species? Explain how your responses could influence your outcomes for species composition at a particular restoration site.

7. What other criteria would you use to develop a proposed restoration's species composition, in addition to those listed in Table 7.6? How might these criteria change under different outcomes or environmental circumstances?

Resources for Further Study

1. For a good discussion of the importance of understanding the restoration site as determined by the results of site inventory and analysis, see:

   Burns, C. J., and A. Kahn. 2005. "Why Site Matters." In *Site Matters: Design Concepts, Histories, and Strategies*, edited by C. J. Burns and A. Kahn, vii–xxix. New York: Routledge.

2. Two sources we have found to be very helpful in thinking about how to establish site plan outcomes are these:

   Hobbs, R. J., and D. A. Norton. 2005. "Ecological Filters, Thresholds, and Gradients in Resistance to Ecosystem Reassembly." In *Assembly Rules and Restoration Ecology: Bridging the Gap Between Theory and Practice*, edited by V. M. Temperton, R. J. Hobbs, T. Nuttle, and S. Halle, 72–95. Washington, DC: Island Press.

   Micheli, F., K. L. Cottingham, J. Bascompte, O. N. Bjornstad, G. L. Eckert, J. M. Fischer, T. H. Keitt, B. E. Kendall, J. L. Klug, and J. A. Rusak. 1999. The dual nature of community variability. *Oikos* 85:161–169.

3. For more information on the case studies discusses in this chapter, check out the following documents:

   Brockway, D. G., K. W. Outcalt, D. J. Tomczak, and E. E. Everett. 2005. "Restoration of Longleaf Pine Ecosystems." General Technical Report SRS-83. Asheville, NC: U.S. Department of Agriculture, Forest Service, Southern Research Station.

   Harrington, J. A. 2008. "Vegetation Patterns and Land Cover Change for the Cross Plains Ice Age National Scientific Reserve, 1937–2007." A report submitted to the Ice Age National Scenic Trail Office, National Park Service, Madison, WI.

4. Several agencies in the U.S. federal government have produced helpful guides to trail design and construction for a variety of uses, including hiking, skiing, and the use of motorized recreational vehicles. These are readily available online. For example:

   U.S. Federal Highway Administration and Forest Service. 2004. "Wetland Trail Design and Construction." http://www.fhwa.dot.gov/environment/fspubs/01232833/toc.htm

   U.S. Federal Highway Administration and Forest Service. 2011. "Manuals and Guidelines for Trail Design, Maintenance, and Operation, and for Signs." http://www.fhwa.dot.gov/environment/rectrails/manuals.htm

   U.S. National Park Service. 2011. "Handbook for Trail Design, Construction, and Maintenance." 2011. http://www.nps.gov/iatr/parkmgmt/trail_handbook.htm

The Implementation Plan

Site Preparation and Installation

LEARNING OBJECTIVES

After reading this chapter, you will be able to:

- Differentiate between the site plan and the implementation plan.

- Describe five steps in the development of the implementation plan.

- Recognize and discuss the various ways a site's physical and biological characteristics, as well as its social and landscape context, influence plans for site preparation and installation.

- Differentiate between goals, objectives, and performance standards, and describe
where in the restoration process each should be used.

- Describe the variety of ways that propagules are introduced to a site, and how to
evaluate which ones would work best for a particular situation.

- Compose a timeline of actions that need to occur during the implementation
phase of a restoration.

The restoration implementation plan is the road map for moving a site from what it is like at the start of a project to achieving the vision set forth in the site plan (Figure 8.1). It is the set of strategies that explain what is to be done—how, where, when, and by whom. Implementation can include, among other things, removing or adding plants, animals, and microorganisms; changing landforms; removing or adding soil; altering hydrology; and adding

FIGURE 8.1. The Restoration Process: Implementation.

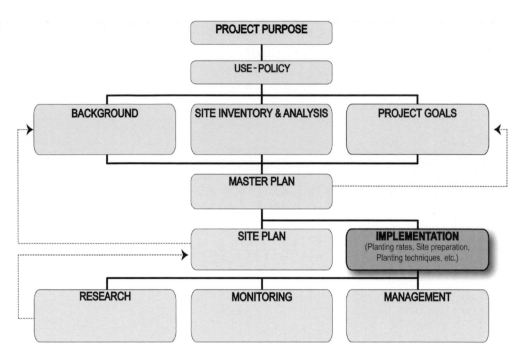

trails or other amenities for people. Planners consider a number of factors when determining implementation strategies, including the restoration goals, the presence or absence of natural disturbances, the potential for the site's natural recovery, the land's productivity, existing and desired ecosystem services, and long- and short-term resources available to the project.

Many of the approaches used to implement restorations have been borrowed from other fields, such as agriculture and engineering, and many of these techniques have not been evaluated to determine how successful they are for restoration activities. Since every site is unique in both space and time, implementation procedures that rely solely on the past anecdotal experience of others are limiting. For situations where past techniques do not seem applicable, adaptive restoration or experiments with several techniques can be tried. For maximum benefit to the profession and discipline, it is important that the project be documented and the results be shared (see Chapter 3). Because restoration is relatively new, trying new ideas that are based on sound ecological understanding may well result in improved techniques.

Implementation involves five steps: (1) defining the implementation units and boundaries, (2) assessing existing site conditions in relation to those required by the site plan, (3) developing implementation strategies for site preparation, (4) developing implementation strategies to reintroduce biotic components, and (5) resolving logistics, such as resource needs and permits. As is true for the restoration process as a whole, several of these steps may occur simultaneously.

▶ 8.1 DEFINING THE IMPLEMENTATION UNITS AND BOUNDARIES

The first step in planning the implementation is to define and locate the implementation units, each of which is addressed separately in the implementation plan (Figure 8.2).

FIGURE 8.2. Implementation Units for the Shea Prairie Preserve Restoration. The left image shows implementation units throughout the area, but avoids several remnant prairies that would be part of the site plan. The right image shows implementation units for tree and shrub removal. The units are part of the site preparation that would occur prior to seeding the prairie. (Adapted from Fondow et al. 2006.)

The implementation units and the plant community units shown on the site plan will not coincide in all situations. The units depicted on the site plan are representative of the locations for the communities once the restoration is complete. Not all land areas, even within a community unit, have the same implementation needs, however. Consider, for example, the restoration of a large coastal dune system. The site has been impacted by nearby development, runoff, pest species, and human use; none of the disturbances extends the entire length of the site. The existing vegetation varies along the coast as a result. In this case, different implementation strategies would be required in the different sections to achieve he common restoration goal. Therefore, the coast would be divided into several implementation units.

▶ 8.2 ASSESSING EXISTING SITE CONDITIONS

The second step involves determining how closely the existing site conditions match the restoration objectives of the site plan. In most cases the site inventory will contain this information, but sometimes it is necessary to collect additional data (see Chapter 4). It is useful to develop a table summarizing the site conditions that differ from the restoration goals, including a brief statement about the action needed to resolve each of the issues. Table 8.1 is an example of such an assessment for the Kootenai River restoration project (see Chapter 14). In this case, as is true for many restorations, the required actions involve modifying the abiotic environment, adding community/ecosystem components, and removing or modifying human infrastructure.

If the desired biotic components are not present, you will determine why and attempt to correct the problem. Often this is due to altered soil or hydrologic components, as well as site context. If species are missing from a site, they can be added directly by you or your restoration team, by planting vegetation or transplanting wildlife, or indirectly through colonization from surrounding areas. In degraded sites where the soil surface has not been disturbed or where some native species yet remain, the soil seed bank should be investigated before restoration or any activities that result in soil disturbance occur.

In all cases, in order to increase the chances for success, it is necessary to begin by making the site conditions as hospitable as possible for the establishment of the new species. This is accomplished in the next step: site preparation.

► 8.3 DEVELOPING IMPLEMENTATION STRATEGIES FOR SITE PREPARATION

Site preparation is the careful manipulation of site elements in order to enhance the establishment of desired organisms, populations, and processes, as well as trails, roads, and facilities. The quality of site preparation is one of the greatest determinants of a restoration's success or failure. Site preparation plans address this question: What aspects of the site and its context currently prevent the community from existing? Under the best circumstances, the site will have had minimal development, minimal hydrologic changes and soil modification, and will have an existing vegetation component that is compatible with your goals. Often, however, this will not be the case, and significant site preparation will be required. Site preparation and the next step, reintroducing the biotic components, are not done without knowledge of the other, as how one prepares the site will direct the reintroduction options, and vice versa.

8.3.1 Modifying Incompatible Human Uses

Often the sites proposed for restoration have been visited and enjoyed by people in ways that are incompatible with the restoration goals. For example, people may have been able to ride all-terrain vehicles or horses, let pet dogs run free, hunt animals, gather plants, or collect firewood for decades. Modifying such uses to meet restoration goals can be difficult, because longtime users now consider the site to be "theirs." Hopefully, during the master planning and site planning processes, the issue of past use has been discussed and resolved. However, if, during the site preparation phase, people persist in engaging in activities that are no longer accepted, you and your team will need to determine how to overcome this hurdle. Fencing and signage are two means, but these can tend to discourage compatible activities as well. Education is preferable. Chapters 12 and 13 address these human interactions with restoration in more detail.

Table 8.1 • Limiting Factors, Restoration Strategy Components, and Restoration Treatments for the Kootenai River Habitat Restoration Project

Limiting Factors	Restoration Strategy Components	Restoration Treatments
Morphological Limiting Factors		
River and floodplain response to altered flow regime and altered hydraulics	Establish channel dimensions that are sustainable given the morphological setting and governing flow and sediment regimes	• Excavate or dredge the river to modify the channel geometry • Install bank structures • Install instream structures
River and floodplain response to altered sediment supply and sediment-transport conditions	Gradually reduce sediment supply and transport competence in downstream direction in order to promote deposition of sediment on the floodplain in the braided reaches and reduce deposition of sediment on the channel bed in downstream reaches	• Excavate or dredge the river to modify the channel geometry • Construct floodplain surfaces • Install bank structures • Excavate floodplain adjacent to the river • Manage the backwater from Kootenay Lake
Loss of floodplain connections	Establish channel and floodplain connection at mean annual peak flow where feasible given constraints from river and floodplain	• Excavate or dredge the river to modify the channel geometry • Excavate floodplain adjacent to the river
Bank erosion and reduced boundary roughness	Establish bank vegetation Increase channel roughness	• Revegetate the floodplain • Revegetate the riparian corridor and establish a riparian buffer • Install bank structures
Riparian Vegetation Limiting Factors		
Lack of surfaces that support riparian recruitment	Increase floodplain areas with suitable substrate and elevation relative to the water table in order to support riparian vegetation recruitment and establishment	• Construct floodplain surfaces • Construct or enhance wetlands • Construct or enhance secondary channels • Excavate floodplain adjacent to the river
Lack of outer bank vegetation	Establish bank vegetation	• Revegetate the riparian corridor and establish a riparian buffer • Install bank structures • Revegetate the floodplain
Frequent scour/deposition of floodplain surfaces	Increase stability/longevity of floodplain surfaces	• Construct or enhance wetlands • Revegetate the floodplain • Install bank structures
Altered hydroperiod	Increase floodplain areas with appropriate elevation ranges relative to the water table to support native tree and shrub species	• Construct or enhance wetlands • Construct or enhance secondary channels • Construct or enhance wetlands adjacent to the river • Construct or enhance wetlands behind the levees and connect to the river • Construct or enhance secondary channels adjacent to the river • Construct or enhance secondary channels behind the levees and connect to the river

(table continues)

Table 8.1 · Limiting Factors, Restoration Strategy Components, and Restoration Treatments for the Kootenai River Habitat Restoration Project (continued)

Limiting Factors	Restoration Strategy Components	Restoration Treatments
Invasive weeds	Reduce weed cover so that weeds do not limit recruitment and establishment of native plant species	• Construct or enhance wetlands • Revegetate the floodplain • Revegetate the riparian corridor and establish a riparian buffer • Revegetate the floodplain adjacent to the river • Manage land use practices
Lack of native plants and seed sources	Establish nodes of diverse, native vegetation within the Straight Reach and Meander Reaches	• Revegetate the riparian corridor and establish a riparian buffer • Construct or enhance wetlands adjacent to the river • Construct or enhance wetlands behind the levees and connect to the river • Revegetate the floodplain adjacent to the river
Lack of nutrient sources for primary productivity and limited carbon storage (reduced primary productivity)	Increase amount and diversity of native vegetation and wetlands within the Meander Reaches	• Revegetate the riparian corridor and establish a riparian buffer • Construct or enhance wetlands behind the levees and connect to the river • Revegetate the floodplain adjacent to the river
Aquatic Habitat Limiting Factors		
Insufficient depth for Kootenai sturgeon migration preference	Provide depth conditions for normal Kootenai sturgeon migration and spawning behavior in sturgeon migration reaches	• Excavate or dredge the river to modify the channel geometry
Insufficient velocity for Kootenai sturgeon spawning preference	Provide velocity conditions for Kootenai sturgeon spawning and embryo/free-embryo incubation and rearing in sturgeon spawning reaches	• Excavate or dredge the river to modify the channel geometry • Install instream structures
Lack of coarse substrate for Kootenai sturgeon egg attachment	Provide substrate conditions for Kootenai sturgeon embryo/free-embryo incubation and rearing in sturgeon spawning reaches	• Excavate or dredge the river to modify the channel geometry • Install instream structures
Lack of cover for juvenile fish	Increase instream and bank cover by constructing instream structures and establishing bank vegetation	• Construct or enhance secondary channels • Install bank structures • Revegetate the riparian corridor and establish a riparian buffer • Install instream structures • Revegetate the floodplain adjacent to the river
Lack of pool-riffle complexity	Increase hydraulic habitat complexity by establishing ratios of pool and riffle habitat that are appropriate for the morphological setting	• Excavate or dredge the river to modify the channel geometry • Install instream structures
Simplified food web from lack of nutrients	Increase nutrient availability	• Construct or enhance wetlands • Revegetate the floodplain • Construct or enhance wetlands adjacent to the river • Construct or enhance wetlands behind the levees and connect to the river
Insufficient pool frequency	Establish pool frequency that is appropriate for the morphological setting	• Excavate or dredge the river to modify the channel geometry • Install instream structures

Table 8.1 • Limiting Factors, Restoration Strategy Components, and Restoration Treatments for the Kootenai River Habitat Restoration Project (continued)

Limiting Factors	Restoration Strategy Components	Restoration Treatments
Lack of fish passage into tributaries	Establish fish passage at known barriers on tributaries within the project area	• Tributary restoration
Lack of off-channel habitat for rearing	Increase availability of off-channel habitat for native aquatic species	• Construct or enhance secondary channels adjacent to the river • Construct or enhance secondary channels behind the levees and connect to the river • Tributary restoration
Altered water quality	Identify and reduce point source pollutant inputs into Kootenai River and tributaries	• Revegetate the riparian corridor and establish a riparian buffer • Construct or enhance wetlands adjacent to the river • Construct or enhance wetlands behind the levees and connect to the river • Revegetate the floodplain adjacent to the river
Constraints on Restoration Due to River and Floodplain Management		
Dam controlled flow, sediment, thermal, and nutrient regimes	Develop habitat actions that are compatible with modified regimes and work with Libby Dam managers so that operations support habit restoration efforts	• Manage the discharge from Libby Dam
Floodplain land use	Coordinate with landowners and grazing leasees to explore development of grazing management plans that allow floodplain vegetation to develop	• Develop land use practices that support habitat restoration
Bank armoring	Coordinate with appropriate parties to maintain, modify, or remove bank armoring to support channel, riparian, and floodplain ecological processes according to specific habitat actions	• Mitigate for impacts from or to infrastructure
Levees and diking districts	Coordinate with diking districts and other affected parties to maintain and/or modify levees to support channel, riparian, and floodplain ecological processes according to specific habitat actions	• Mitigate for impacts from or to infrastructure
Transportation corridors	Develop habitat actions that are compatible with existing infrastructure; and work with owners to mitigate for potential impacts to infrastructure from project actions	• Mitigate for impacts from or to infrastructure
Floodplain draining/pumping	Work with diking districts, NRCS, SCD, and landowners as appropriate to identify areas where floodplain draining and pumping can be modified to restore floodplain hydrology	• Mitigate for impacts from or to infrastructure
Backwater influence from Kootenay Lake	Reduce the negative effects of the backwater influence in the Braided Reaches	• Manage the backwater from Kootenay Lake
Urban development adjacent to river	Design habitat actions that do not place urban infrastructure at risk, and create riparian buffers to separate Bonners Ferry from the river where possible by working with the City of Bonners Ferry and landowners	• Mitigate for impacts from or to infrastructure • Revegetate the riparian corridor and establish a riparian buffer

Adopted from Kootenai Tribe of Idaho, "Kootenai River Habitat Restoration Project Master Plan: A Conceptual Feasibility Analysis and Design Framework." Bonners Ferry, ID, 2009 (with funding provided by Bonneville Power Administration)

8.3.2 Removing Structures

Site preparation often requires the removal of such structures as earthen dams, buildings, drainage tiles, ditches, and even abandoned cars. In some situations, infrastructure components, including industrial buildings, roads, and miscellaneous human-built structures, may have to be removed. In the case of industrial sites, or any other former uses involving hazardous materials, it is important to check the site for the presence of such materials during the site inventory. If such materials are found, you should consult with specialists to determine how best to remove the danger posed. Physical site modifications may include altering the topography, and rebuilding banks and former channels of degraded rivers and streams. Two examples from wetland restorations follow, to help illustrate such actions.

Many wetlands have been drained for conversion to agriculture either by ditching or by tiling (the creation of a network of underground drainage pipes). These actions cause water to move offsite, leaving dry and (temporarily) fertile soil. The dewatering of wetlands and their conversion to cropland can have long-lasting effects on soil fertility and chemistry, and these can contribute to the loss of soil organisms that may be important to wetland plant establishment. A first step in restoring wetlands to such a site is to fill the ditches or break the tiles, thus increasing water levels once again. In some situations, this may be all that is necessary to return a landscape to its former wetland community, as wetland plants and animals will colonize the site. But in many cases, additional site preparation (e.g., involving the removal of competing vegetation) along with the active additions of species will be necessary.

Rivers and streams meander naturally due to the energy and dynamics of water. The meanders lengthen watercourses and widen their land coverage, resulting in a gentler flow. Channelization—the straightening of meandering river channels—was often done to claim floodplains for agriculture, improve navigation, or move storm water rapidly through an area (Figure 8.3). After watercourses are straightened, if the banks are not reinforced, the volume and rapid flow of water of during storms can cause severe flooding, and the erosion and deepening (or incision) of banks. As a result, some wildlife become isolated from the water, the vegetation composition along the watercourse changes, and the diversity of aquatic species in the channel decreases. Many watercourses flowing through agricultural land face a different impact: deposits of soil that has eroded from the fields (Figure 8.4).

Site preparation in these situations involves digging new channels, recontouring of the slopes of the banks, using erosion control matting or even rocks to stabilize the new banks, and removing excess soil and sediment from the site. (In some cases, farmers are willing to place nutrient-rich sediments back onto their fields.)

Infrastructure can also interfere with wildlife behavior, and sometimes, particularly where the restoration goal is to restore habitat, it might be necessary to remove infrastructure components. For example, Europe's wild reindeer occur in several disjunct fragments of their original range, and efforts to link these require the restoration of migration routes across developed landscapes (Nellemann et al. 2010). Based on surveys of reindeer migration routes and behavior, these researchers make several recommendations that include reducing resort development and expansion near winter ranges and migration corridors, and removing cabins and trails that currently exist near these historic routes.

FIGURE 8.3. **Stream Channelization.** Ditching or channelization of former streams is done to drain wetlands and expand land areas for agricultural purposes. (Photo reprinted by permission from John Harrington.)

FIGURE 8.4. **A Stream Restoration Project in Southwest Wisconsin.** The stream is in the basin of a large rural watershed where farming has occurred for at least 150 years. Prior to restoration, the stream had been straightened for agricultural purposes. Several feet of sediment deposits eroded from the surrounding uplands were removed and stockpiled for use by rural landowners and agencies. The streambank was then recontoured and planted to native species. (Photo reprinted by permission from John Harrington.)

8.3.3 Reducing Competition from Existing Plants and Animals

A common focus of site preparation is the removal of existing foreign plant or animal species that compete with, slow, or prevent the establishment of species to be planted. In some cases, however, species need to be reintroduced to help reestablish the community structure and interactions that help control the overpopulations of other species. Wildlife has a significant influence on the success of new plantings, and prior to taking any action, restoration ecologists must understand how the presence or absence of wildlife can affect the success or failure of a restoration. Site preparation may involve reducing wildlife populations or creating a temporary habitat that might be conducive to beneficial wildlife.

Rodents such as rabbits and squirrels, insects, and ungulates can decimate young shoots of many herbaceous species and increase the difficulty of woody plant establishment. A number of animals will also locate and eat newly planted seeds. Waterfowl will eat plugs planted along waterways. Wild pigs will dig new plantings and quickly disturb plantings and ground, thereby increasing the potential for pest species to establish.

Wildlife populations are also influenced by government policies and social conditions. The loss of large predators due to habitat change and hunting has allowed the populations of many herbivores to increase, resulting in the overbrowsing of some species and thus leading to significant structural changes in the landscape's vegetative component and increased difficulty in its restoration. On the other hand, some predator populations are increasing in countries where laws protect their habitat and where the percentage of the population that hunts has declined. This has resulted in a behavioral change for prey species, many of which are herbivores, and with that another change in vegetation composition and structure. In many developing countries where human populations are increasing and hunting is still a necessity for obtaining meat, a different set of interactions is occurring, as both prey and predator may be hunted. In all of these situations, the interactions of both animal species and animals with vegetation shape the landscape and will create unique challenges for each restoration.

Solutions that successfully control wildlife disturbances are difficult to implement. In many regions, the trapping or hunting of overpopulated animals is restricted, certain problem-causing animals have very efficient reproduction, and for some of the more charismatic species, public outcry against their removal is intense. Attempted solutions to keep specific wildlife species from restorations under development include fencing, placing netting over especially susceptible plants, using chemical and organic repellants and/or noise-making devices, and the use of dogs to run the problem wildlife off the site. None of these solutions is 100% effective, particularly as the size of a restoration project expands.

Where wildlife species are missing but are deemed important to the restoration, reintroductions can be attempted, but this will often require permits, public input, and lengthy environmental assessment studies. Wildlife in the role of seed dispersers can be extremely important to the success of a restoration, particularly in cases where funding requires the natural recolonization of sites. Additional services that wildlife provide for plant establishment include selective grazing of dense grasses and brush, which releases desired species by the removal or thinning of competing vegetation; soil aeration and the addition of nutrient and organic matter to the soil; and pollination.

Simply put, understanding animal behavior—foraging, travel patterns, cover and nesting needs—within specific settings can be used to predict potential influences on restoration efforts (Fink et al. 2009).

With globalization, exotic plant species have become a major problem for most restorations. Those that are highly competitive or have pest tendencies can prevent restorations from establishing. Because of their tremendous impact on restoration success, we devote a separate chapter to pest species (see Chapter 11).

Similar to exotic pest species, however, some native species can also conflict with restoration goals. Many fire-adapted communities, for example, occurred throughout North America prior to European settlement. By the early 1900s, naturally recurring fires, which maintained the openness of these communities, were stopped by European settlers. With the cessation of fire, these partially open landscapes rapidly became enclosed with shrubs and saplings, both exotic and native. In order to restore these communities, the thinning of native trees, saplings, and shrubs may be required along with eradication of non-native species. Such scenarios exist for the longleaf pine (*Pinus palustris*)–wiregrass community of the southeastern U.S., the ponderosa pine (*Pinus ponderosa*) ecosystem in the western U.S., and the oak savanna of the Midwest, as well as other natural communities around the world.

For areas that are to be planted as part of the restoration process, eliminating the existing problem species will reduce competition for new seedlings. The numerous techniques for removing existing vegetation include mechanical and chemical, as well as flooding, grazing, and fire. The choice of methods depends on the existing site conditions, the existing vegetation, and the proposed vegetation. Consider the following scenarios.

Open Sites

Open lands are often former pastures or croplands with a cover of sod-forming or rhizomatous grasses. Where grass cover is dense, the germination and seedling growth of new plantings are suppressed. In Uganda, Africa, for example, researchers found that slow recovery of the forest from seeds and nuts on former agricultural lands was due to suppression from elephant grass (*Pennisetum purpureum*) and the herb *Acanthus pubescens*, as well as high seedling mortality from seed-eating rodents that were abundant in the grass cover of these open lands (Chapman and Chapman 1999).

Under this scenario, the team mechanically or manually removes the undesirable vegetation and prepares the soil surface for seeding of the desired species. One approach is to plow, particularly when the land to be restored is covered with dense sod-forming grass. Plowing turns over the upper 30 cm of the soil, breaks up the grass sod, and cuts through roots, exposing many to the air where they will dry and die. Disking or loosening the upper 5–8 cm of soil after plowing will dislodge any rhizomes that may have survived, thereby cultivating the soil into an evenly textured seedbed.

In addition, plowing, disking, or tilling brings buried dormant seeds to the soil surface, and the combination of light and moisture will often promote their germination. Many of these seeds are from opportunistic species that grow quickly, sometimes quite robustly, and are competitive with native perennials that are slow to grow and establish. To reduce this seed bank, disking and tilling begin one or more growing seasons before the restoration planting. Each tilling brings new seeds to the surface, and many of them will germinate. The next tilling will eliminate these seedlings and bring new seeds to the surface. In theory, after tilling several times, the number of dormant seeds (as well as rhizomes) in the soil will decrease and no longer be of concern. In reality, such seed is seldom eliminated entirely, and wind, birds, and other dispersal vectors are constantly reintroducing new seed.

One method for reducing the need to disk lands repeatedly over several years is to plant the land to crops. This method accomplishes the site preparation of annual disking and tilling, removes the weed bank that might compete with restoration plantings, and provides income for the restoration efforts from harvesting the crop. In some cases, herbicide-resistant crops are planted, which provides another opportunity to control weeds.

Herbicides are an alternative technique for removing undesired vegetation, but they can require several repeated applications before the regrowth of rhizomatous species no longer occurs. For the removal of sod-forming grasses, herbicides would likely be broadcast sprayed, a method that can have unintended environmental consequences due to drift and runoff. A second approach is to wick apply herbicides to vegetation, thereby reducing the potential for herbicide drift and damage to desired vegetation. Chapter 10 discusses herbicide use in greater detail.

Canopied Sites

If the goal is to restore forest and the desired canopy exists, cultivation is not an option. Cultivation in almost any form is likely to damage the roots of the occuring species and, in some cases, may lead to their death. Examples include forests where the canopy is in place but a cover of weedy non-native species has replaced the native herbaceous layer, forests where the canopy has been invaded by non-native tree species, and savanna where the desired trees and shrubs exist but at a much greater density than is typical for the community. In the latter two examples, sunlight becomes reduced below the levels needed by the desired mid- and ground layer species for survival. For some regions, a common cause of the third example is fire suppression.

In these situations, prescription burns are one possible solution for thinning and removing undesired species (see Chapter 10). If the unwanted herbaceous, shrub, or tree species are fire-sensitive (and the desired existing canopy trees are not), repeated burns over several years will often reduce them to a density at which the planting of desirable species can successfully occur. Sometimes, however, fuels (litter) under fire-sensitive species are scarce, and burning is difficult without first reducing canopy density in order for more light to reach the ground. The desired result is an increase in foliage production, resulting in larger volumes of litter or fuels. The greater sun exposure also helps dry the litter, so it is more likely to ignite and burn.

If fire is not practical, possible, or desired for the community of interest, mechanical thinning of trees and shrubs to the appropriate density is another solution. Another option is not to cut, but to girdle the trees that resprout or form large clones on one root system (see Chapter 10). If the tree or shrub is a species that will resprout, the cut can be painted with an appropriate herbicide. Brush cutters, flail mowers, and backhoes are machines that can remove trees and shrubs from large areas, but their use often damages the soil.

Where the existing canopy is composed of large stands of undesirable species, several mechanical methods for removing trees can be used. Bulldozing is one method when all vegetation across an area needs to be removed. For thinning a canopy, several kinds of timber harvesting equipment are available, including Feller bunchers, whole-tree chippers, roller drum choppers, and skid steers; they are all machine-driven devices, some which can grab and cut standing trees up to 24″ dbh and chip them. These machines are highly efficient, reducing days of labor to hours. However, they will cause significant soil disruption and can produce considerable debris and slash that can be a fire hazard.

Selective grazing has also been used to remove or reduce shrub cover when it is so dense that fire or manual cutting is impractical. Grazing, if continuous or uncontrolled, has significant negative impacts on the world's landscapes, including soil compaction and erosion, altered soil nutrient levels, reduced vegetative cover and biodiversity, and, in arid regions, desertification. Past experiences with unmanaged grazing have led some land managers to avoid its use. However, recent studies suggest that managed grazing or rotational grazing, where ungulates are frequently moved from one paddock (grazing unit) to another, can result in a landscape where shrubs and saplings are reduced without significant impact on the herbaceous vegetation or soil. Beef cattle, goats, and sheep are three ungulates for which managed grazing shows promise. However, woody species react uniquely to grazing regimes; not all shrubs will display the same response. Grazers are also selective in the species they graze; therefore, not all species will decline. And species will be grazed differently depending on the context in which they grow. Some shrubby species, for example, will be removed by grazing under canopies where little grass is produced. In sun-exposed areas where grasses are plentiful, these same species may receive little attention from grazers (Harrington and Kathol 2009).

On land that was degraded from recent overgrazing and whose ecosystem at one time included natural grazers, a lighter grazing regime may be tried to see if at least some of the desired species can be restored before further restoration activities occur. The recovery of native species under such treatment may be particularly feasible for lands where only the biotic functions have been damaged and the abiotic functions are still intact (Papanastasis 2009).

8.3.4 Sites Without Competition

There are scenarios where the removal of existing vegetation is not needed. These include areas that are currently being cropped or have a vegetation type that is not highly competitive. In such instances, seeding or planting into the bare ground or existing vegetation can be successful. Fire or some types of machinery can also be effective at preparing a site for planting when the goal is to remove litter and expose the soil surface. This approach mimics, for example, the jack pine (*Pinus banksiana*) barrens of northern regions. The seeds of jack pine require mineral soil to germinate, so litter removal is important. The pine often holds onto its cones and seeds until after a fire, when the litter layer has been removed and the soil exposed. The scales of the cone open with the fire's heat, releasing seed onto the exposed soil surface. Where propagules are not present on site, there will be a need to interseed or plant within days after the fire.

8.3.5 Soil and Hydrologic Properties

The success or failure of restorations is linked to achieving a good site-to-species match. Two environmental components of particular importance are soils and water. The ability to alter soil and hydrologic properties for a restoration is limited by resources and budget costs, the soil or water characteristics that require alteration, and current understanding of soil-water-plant relationships. In many situations, goals would not list a target community for a site where the existing soils or hydrology are inappropriate. However, modification of physical site properties is sometimes necessary in order for plant propagules to establish.

Soils

Many restoration projects are undertaken on sites where agriculture has been practiced for many years, sometimes for centuries. Western agricultural practices, characterized by repeated plowing and disking together with frequent additions of fertilizers and herbicides, result in soil properties that are significantly different from those under native cover. In such cases, soil properties may need to be modified for the restoration of native plant communities. The soil characteristics for which modifications may be needed include soil moisture, nutrient status, compaction, microbial community composition, and chemical residuals from past use (e.g., farming, industry). Soils in many restorations, even after several decades, have been found to be more similar to an agricultural field than a relatively undisturbed ecosystem in terms of their microbial populations.

Highly fertile agricultural soils present challenges in controlling the growth of undesired pest species, or the rapid establishment of some desired species that then inhibit others that are slow to establish. Some species are excellent competitors where nutrients and moisture are plentiful. When resources such as nutrients and moisture are limited, many of these weedy species become less competitive and may fail to establish or survive. In contrast, many native species will establish, grow, and create a diverse restoration under low-fertility conditions.

More than one solution has been attempted to lower fertility, particularly in terms of nitrogen levels, prior to planting a restoration. Growing nutrient-demanding crop species without applying fertilizers, and then harvesting the plants and removing all litter, can lower soil nutrient levels if done over several years. High carbon-to-nitrogen ratios (25:1 and greater) can immobilize nitrogen, temporarily stopping its conversion by soil bacteria to a form of nitrogen that plants can utilize (see Chapter 2). A research study in Minnesota found high soil carbon additions (fresh organic matter, such as sawdust or sucrose) to immobilize plant-available nitrogen resulted in a sevenfold increase in prairie biomass and a 54% decrease in weed biomass (Blumenthal et al. 2003). A second study in western North America found that the reintroduction of the microbial community by soil inoculum, along with a reduction of available nitrogen, resulted in an increase in native perennial cover and a decrease in nonnative cover (Rowe et al. 2009). However, similar attempts by others have had mixed success.

Not all regions suffer from areas of high fertility. For many tropical forests and arid lands, low rather than high soil fertility is more common. Nutrients in many tropical forests reside in the vegetation and leaf litter and not in the soil. When these forests are logged and then farmed, nutrients are rapidly lost from the ecosystem. Under these conditions, restoration can be slow.

Chemical residues from pesticides or herbicides in former agricultural lands, and manufacturing waste on industrial sites, are also problematic for restorations. Toxic chemicals frequently occur in older industrial and mining sites and may require extensive cleanup prior to restoration. Elevated salinity levels and pH, common to some arid regions, can also result in restoration complications.

Soils that have been exposed to extensive grazing or subjected to heavy machinery are often highly compacted and absent of ground layer vegetation. Seeding into these areas without opening up or decompacting the soil is unlikely to be successful. One method of decompaction is to use a chisel plow consisting of blades with ripping teeth that break up the soil to depths of 30 cm. Simply decompacting a soil may not lead to the reestablishment of a suite of native plants, however. One study conducted on African rangelands found that tilling alone resulted in annual plant establishment, while the presence of perennial grasses only occurred when seed was applied after tilling (Kinyua et al. 2010).

Hydrology

Because hydrology usually affects a larger region than a single restoration site, and because changes to hydrology can have physical and legal ramifications beyond the restoration site, it is one of the more difficult and complex aspects of site preparation to resolve (see Chapter 4). Restorations that involve streams, rivers, lakes, and wetlands will benefit from the planners doing a broad review of the site and the watershed, to find ways to accomplish the restoration goals without impacting lands under other ownership.

Some wetland systems depend on seasonal water cycles, in which periods of flooding and other periods of low soil water levels alternate with some regularity. Often, these cycles have been altered by human development. The channelization of rivers and the drainage of wetlands, discussed above, are two examples. Another is the installation of stormwater basins for holding back the high waters of floods during a storm or snowmelt, and then releasing them at lower, less damaging flood levels. This constant, instead of intermittent, surge of water results in a landscape that is constantly wet instead of one that is flooded in the spring (or wet season) and then proceeds to dry down in the summer (or dry season). The results for floodplain forests, wet grasslands, and some aquatic communities from these constant artificially high levels of water include loss of species, invasion by undesired species, changes in water chemistry, and a shifting of the community structure and functions. In order to restore such communities, site preparation involves replacing floodplains and water cycles.

Strategies for water management on drier sites can include improving water capture, storage, and infiltration. Scarifying, roughing up, distributing coarse woody debris, and pitting soils, as well as applying organic mulches and establishing vegetation, are all ways to enhance water capture and infiltration. In extreme arid environments, drip irrigation and the use of buried clay pots that can be filled with water have been used in restoration (Bainbridge 2007).

Riparian systems may require the restoration of lands throughout the watershed to control the rate and volume of surface runoff, as well as the sediments, nutrients, chemicals, and elevated water temperatures it can contain. In some instances, removing dams or restructuring their operation may be needed. In others, converting uplands from row crop agriculture to native grasses has improved water quality by lowering nitrate concentrations in creek sub-basins.

When restoring hydrology is feasible, the impact on adjacent lands must be considered. Filling ditches, for example, may return desired water levels to a site, but it can also increase water levels on adjoining lands, which may not be desired. In such cases, careful engineering and discussion with adjacent landowners will be required, and sometimes filling ditches and restoring natural water levels may not be realistic. Where municipalities and industries use aquifers for their water needs, groundwater levels may be depressed. In this situation, solutions for restoring wetlands that depend on groundwater are limited, without considerable expense and the agreement of all affected parties. Many times, when evaluating site hydrology (or, for that matter, any other aspect of restoration), it is best to first ask what the system was previously like and what changed it. If the answers can be determined, the goals for site preparation can be established and strategies for accomplishing them can be explored.

▶ 8.4 DEVELOPING IMPLEMENTATION STRATEGIES TO REINTRODUCE BIOTIC COMPONENTS

Once the site preparation phase is complete, procedures for implementing the restoration focus on how to introduce both plant and animal species, any infrastructure or programs required for accommodating the use-policy, and, for some communities, disturbance cycles. The reintroduction of plant species involves answering numerous questions (many which also apply to wildlife):

1. Which species from the site plan list will need reintroduction?
2. Which species should be included in the final species mix?
3. What should the origin of the planting propagules be?
4. How are the seeding and planting rates determined?
5. How are seeds and plants prepared for planting?
6. How should seeds, seed mixes, and transplants be located on the site?
7. Is there a need to sequence the installation of species?
8. What are the alternative planting methods?
9. How will the finished restoration be evaluated?

Each of these questions is discussed and answered in the sections that follow.

8.4.1 Which Species from the Site Plan List Will Need Reintroduction?

This question could be rephrased as: Which species that are slated to be in the restored community now exist onsite, have the potential to colonize the site without human assistance, or will need to be artificially reintroduced? Recognizing how the different species reproduce and disperse will help you to determine which species you will want to reintroduce artificially. Plants reproduce sexually and vegetatively, the latter typically through stolons and rhizomes. Species that reproduce vegetatively are often clonal and can quickly colonize areas that have had the soil surface disturbed, areas that would otherwise be colonized by pest species. Some of these species can also be rather aggressive in a restoration, particularly where competition is low or where other species have not become fully established.

Plant species also have different strategies for pollination and dispersal of seed. Some plants self-pollinate, but many are pollinated by animal species, including insects and bats, or by wind-dispersed pollen. Some plant species are pollinated by several animal species, and certain plant species depend on a single animal species for pollination.

Many species rely on wind, water, or animals for seed dispersal. Ants, primates, ungulates, and birds, for example, are major dispersal agents for many herbaceous and woody species. In restorations where a particular dispersal agent does not occur, colonization by woodland herbaceous species is slow to nonexistent. Understanding a plant species' strategy for reproduction and dispersal of propagules can help you determine whether that species needs to be planted or will come in on its own, how competitive a species is in a new planting and therefore when and how much to plant, and whether specific pollinators need to be introduced.

Onsite Sources

In some cases, the source for native vegetation is on the restoration site. Once the site is prepared so that it is conducive to the reestablishment of a community, many of the associated species may appear without planting. In former fire-adapted communities, restorationists will often burn the site and wait to see what species appear before proceeding with additional planting. Seed can then be collected from plants growing on the site for additional spreading. Since seed is not being introduced from offsite, the success of this approach is dependent on the soil having a sufficient seed or root bank of desirable species. However, for sites that have been under cultivation or used as pasture for a long period, the soil may hold few propagules of past native species.

Colonization from Offsite Sources

Another option is to prepare the site and then wait to see what species colonize the restoration. If remnants are nearby, local ecotypes of native species that are well adapted to the setting may establish. However, this scenario can also result in the colonization of undesirable species, some of which can be aggressive, competitive, and difficult to constrain.

This "hands-off" method can result in the establishment of pioneering and disturbance-adapted native species, with few conservative or slow-to-establish species. Species that drop seed near their origin may have less success at colonizing new areas relative to seeds that are windblown or animal-dispersed. Windblown or bird-dispersed seeds are more likely to cross barriers such as roads and fields than seeds carried by small animals that travel on the ground. Taller species also seem to have an advantage at dispersing seed more broadly. As distance to the next remnant or seed source increases, the likelihood of colonization decreases. Most often, if colonization does occur, only a partial representation of the possible community species composition will establish. Supplementing with seed from additional species will be required.

Several studies have been conducted exploring the recolonization of former forests in East Africa and South America (Duncan and Chapman 1999; Parrotta et al. 1997). In many of these tropical regions, the majority of seeds are animal-dispersed, in contrast to temperate species whose seeds are often wind-dispersed. These methods of seed dispersal have implications for restorations that depend on colonization from adjacent lands for some of their species composition. For example, lightly disturbed sites in Uganda had a much slower recovery than more heavily degraded sites in South America. Several reasons were suggested. Studies of frugivore visitation to Ugandan forestlands placed into cultivation and then abandoned found that birds and bats were frequent visitors, but large mammals rarely used these areas. Seed dispersal from adjacent lands therefore was restricted to species that were distributed by birds and some fruit eating bats. Unlike Amazonian forests, midelevation African forests have few small-seeded colonizer genera and only a few trees that rely on wind dispersal. Larger-seeded forest species were less likely to be dispersed to degraded sites than smaller-seeded species unless larger frugivores, such as primates, were present. Recruitment into the fields of Uganda was also slow due to the existing grass cover and high seedling mortality from high populations of seed-eating rodents.

Within Mediterranean mountain ecosystems, large mammals are responsible for long-distance dispersal of fleshy-fruited woody species; smaller seeds are dispersed shorter distances by birds. Here, mammals tended to visit a variety of habitats, including pine plantations, whereas birds tended to be more particular about the habitats they visited (Matias et al. 2010). Fink et al. (2009) found seed dispersal by birds in Costa Rica to be influenced by tree canopy size and patch size among tropical forests. As previously noted, by knowing the behavior of wildlife species and their habitat preferences, you can use the presence of mammals and birds as indicators of potential offsite colonization.

8.4.2 Which Species Should Be Included in the Final Species Mix?

In our site plan discussion, the objectives did not specify an exact species mix; instead, they provided a list of species from which you or your restoration team can choose, based on a set of criteria or objectives. At some point, you have to decide on the species to be planted. Here are three possible approaches:

1. Plant more species than the minimum suggested in the site plan, or as many species as are available.

2. Plant a specific set of species expected to occur on the site.

3. Plant a base mix, perhaps of representative species, and allow additional species from the list to enter the site from offsite remnants.

The first approach assumes that not all species that are planted will establish and be sustained. But if they were, the objectives would still be met. The assumption is that the interactions between site and propagule will determine which species succeed and which do not. This approach allows species to sort out based on the habitat, the weather conditions at the time of planting, and other natural events, such as seed predation. The more species that are planted, the more species the system will have to select from. This approach can also result in the waste of species propagules that will not establish, and it provides less control over the final composition and structure.

The second approach lists the specific species you wish to have restored for structural, functional, or even aesthetic purposes, and that will be sufficient to result in the communities desired. This strategy assumes it is possible to predict which species matching the restoration goals will grow on the site. If predictions are correct, this method provides the greatest control over which species will occur on the site and reduces propagule waste. However, it is important to be aware that predictions are usually probabilities, since dynamics in the system will result in a species composition that is largely, but not exactly, derived from the species planted.

The third approach relies on offsite colonists arriving at the site on their own after an initial base planting. A core set of specific species are restored for compositional, structural, functional, and aesthetic purposes. Once this framework is established, other species that are adapted to the environment are expected to slowly colonize. This method reduces costs because fewer propagules need be purchased or collected and planted, but it also leaves the final composition of the restoration to chance. Before deciding on this approach, you should have an understanding of what species are likely to invade from the surrounding area and their chances of survival over time.

Cover and Companion Crops

In addition to including the suite of species that are desired in the final composition of the restoration, you may wish to include two other species groups, cover and companion crops. These species groups are sometimes planted in the early stages of implementation to support the establishment of the final species mix.

Cover crops are used in situations where bare ground exists that will not to be planted for several months or years. The cover crops are planted, prior to the restoration planting, to compete with weed species and to help reduce wind and water erosion. When it is time for the actual restoration planting, the cover crop may or may not be removed. When they are harvested and removed, cover crops can be used to reduce high levels of fertility, pesticides, and herbicides in former agricultural lands. In some situations the cover species will diminish as the desired species establish, in which case the restoration planting can be interseeded into the cover crop planting. In other cases, the cover crops can be used to provide shade and shelter to forest seedlings and saplings.

If the restoration will not be planted for several years, placing the land into crops can help control erosion, reduce weed establishment, and, with tillage, reduce the potential seed bank for weeds. This strategy also provides income that can be used for implementing the restoration in the future.

Companion crops are sometimes planted at the same time seed or transplants are planted in a restoration. The idea is that these crops will facilitate the survival of the desired species by controlling erosion and limiting competition, as well as partially sheltering new seedlings from sun, wind, and intense rains (facilitation; see Chapter 2). Species used for companion crops may be native or commercially developed, often as agronomic crops, and are chosen for quick germination and establishment. Some restorationists, however, question the effectiveness of these species and view them as competition for the newly planted species. If cover or companion crops are used, they must be species that are poor competitors with the targeted and planted species.

8.4.3 What Should the Origin of the Planting Propagules Be?

Here, we consider origin in two ways: the geographic location where propagules originate, and the place where the propagules are acquired, usually through collection or purchase from a native plant nursery. The acquisition of species can come from local, regional, or national scales depending on the species' natural range. However, acquiring seed from geographic regions that are distant from the intended planting location can have implications for the restoration's success.

Ecotypes and Genotypes

An ecotype is a population of a species that is morphologically and/or physiologically different from other populations. The difference has a genetic basis and often reflects adaptations for survival within a particular habitat. A genotype is the genetic composition of an organism, and those individuals that occur within a geographic region are often referred to as the local genotype. Ecotypes may exist among populations of some (but not all) species and can affect a species' ability to compete and survive in a restoration.

Many ecotypes result from geographic distance or isolation of one population from another over a period of time, and they can influence environmental adaptations. One example from North America is red maple (*Acer rubrum*). It is native to the 48 contiguous states of the U.S. as well as Canada; however, plants from populations in the extreme south are unique enough that they will not necessarily survive when grown in northern states. Similarity of the environment (temperature, precipitation) between the source site and the recipient site is highly important to new planting success.

When tallgrass prairie restoration was in its infancy, much of the grass seed used throughout the Upper Midwest came from western sources where seed was being produced for rangeland use. When a switchgrass (*Panicum virgatum*) ecotype grown in the west for its rapid recovery under grazing was included in restorations in the eastern region of the tallgrass prairie, it quickly reduced the populations of other grass species that normally would have been much more abundant than switchgrass. In some cases, a nearly monotypic stand of switchgrass resulted. Introduced ecotypes can sometimes interbreed with local populations and, if restorations are large enough, can replace the local population, thus resulting in the loss of genetic diversity.

In order to avoid ecotype issues, restorationists will restrict the distance from the project site where seeds or plants can be collected or grown. The difficulty with this approach is that distance per se does not always differentiate ecotypes, and some species do not contain obvious examples. Rather, soil or other site differences and methods of reproduction and pollination mean that the separation of ecotypes is unique to each species.

The cost of determining ecotypes or local genotypes for all species that might be specified for a restoration would be prohibitive. Therefore, restorationists will often identify seed and plant sources within a region and compare the success of species of different origins within past restorations to help them determine the extent of the region from which they are willing to utilize seed and plants.

To counter the above issues, a number of states have moved to producing foundation seed from selected ecoregions. For example, Iowa has been divided into three latitudinal regions and Wisconsin into seven ecoregions. Seeds are harvested from each of the regions, and then this base stock is grown separately within its region of origin to produce seed unique to a defined ecoregion. The seed is planted only within the ecoregion in which it was collected.

Our discussion thus far leads to a focus on restoring species populations of a certain origin or ecotype; however, climate change suggests that we may want to be restoring with a focus on ecosystem services, and a look to the future as to what species populations may be best suited to a site. Under this reasoning, we may wish to select for species populations that have the greatest potential to survive climate change effects as predicted by models for a geographic area. Here, when species in our restoration are known to have a wide geographic range, we may want to add ecotypes that may not currently exist within the geographic area but would fit within the predicted model for climate change. The thinking is that plant groupings or ecotypes that are limited to one landscape may actually become less resilient and more likely to fall apart or collapse under climate change. Without a broad gene pool, the chances that species and ecosystems can adapt to large changes in habitat resulting from climate change are diminished (Harris et al. 2006).

Collecting or Purchasing Propagules of Species

You can obtain plant materials for projects from two main sources: (1) by collecting seeds from remnants or established restorations, and (2) by purchasing seeds or plants from nurseries, either established on the project site or through commercial nurseries. Removing plants from remnants is seldom warranted or desired unless the remnant is going to be destroyed. For many species, transplant survival is poor due to root loss, physical site differences, and storage issues prior to planting. These actions also seriously impact the collection site (see Sidebar).

Sidebar | Are Seeds or Transplants More Likely to Establish Successfully?

The question of whether to use seeds or transplants is highly dependent on the plant community, the species morphology, and the setting. The use of seed mimics the natural dispersal processes but has drawbacks, including the time required for germination and plant establishment, as well as an often high rate of loss due to predation or weather conditions. In arid environments, for example, the presence of drought, wind, limited rain, erosion, and predation limits the periods in which seeds establish. Transplanted seedlings and plants require care in terms of watering and protection from browsers, and they are more expensive than seed. Transplants of herbaceous and small woody seedlings suffer from frost heave in colder climates, which is not an issue for seedlings arising from seeds planted on site. Transplants do, however, have advantages. Individuals can be placed where desired, are a more efficient use of species that produce seed in limited quantities, and, for some species, establishment is much higher. Seeds, on the other hand, tend to cost less, which is particularly important for large projects requiring lots of plants.

Sod from remnants or large preserves can be transplanted to restoration sites, but this method has had mixed success. Shallow, fibrous-rooted species tend to have greater success than deep-rooted species, which lose a large portion of their roots during removal. In some cases, weeds quickly establish in cracks and edges between two pieces of sods. These areas have greater exposure to air and tend to dry out, increasing the stress on the transplanted species. In most instances, sods are procured from salvage areas or remnants that are going to be destroyed to make way for an alternative land use.

Live stakes are often used in stream and riverbank plantings or along water bodies. Live stakes are cuttings from select species of woody plants that have the ability to quickly establish roots when inserted into the ground. Species such as willows or dogwoods are highly effective for this. The use of live stakes along with erosion control fabrics, such as jute or straw mats, increases slope stability and reduces soil loss. Live stakes are cut in lengths of 1.0–1.25 m, with diameters of 5–10 cm, and are best installed during dormant seasons. Approximately two-thirds to four-fifths of the stake should be inserted into the ground (Figure 8.5).

FIGURE 8.5. Live Staking for Establishing Bank Plantings. Left: Live stakes are inserted into the bank and anchored. Stones are placed at the toe of the opposite outer bank where water forces are high. Right: The live stakes root and begin to resprout during the next growing season. (Photos reprinted by permission from John Harrington.)

Collecting seeds from remnants near the restoration site has several advantages, the most significant one being that the origin of the propagules is known. Given the proximity to the restoration site, the plants and their seeds are likely to be of a local ecotype that is adapted to the site conditions. However, some disadvantages accompany collection from remnants. First, due to the rarity of remnants in many regions of the world, chances are that a sufficient number of nearby stands will not be found. Second, sites in less than pristine condition are likely to have lower species diversity and may not contain all the desired species.

Even if species are found on a remnant, seed availability may be limited during the year the restoration is being implemented. Seed availability is limited to species that have flowered, been pollinated, and not parasitized that year. Species react differently to weather conditions that precede flowering and seed set. For example drought and wet years can affect seed production among different species. Some plant species produce large seed sets in alternate years or at some other frequency, and small seed sets in the intervening years. The extent of seed predation is influenced by the population size of the predators, which is also variable from year to year.

Removing seed from remnants may harm local populations. Ecologists do not know how much seed can be collected from a site before damage is done to a population, nor the extent to which collecting seed from several remnants may have an impact on regional populations. A rule of thumb used by many is to "collect no more than 25% of the seed within an area," but there is no definitive research to support this threshold. Whether any seed should be collected from rare species is an even larger question. When sufficient seed quantities from a single species are difficult to procure through collection and the restoration is planned to continue for several years, temporary nurseries can be established on the restoration site. Access to rare or difficult to grow species is particularly benefitted by such an arrangement.

Collecting seeds from past restorations is not advised, unless you know the origin of the seeds. Many early restorations were done when few native plant nurseries existed, and the propagules used may have been obtained from nurseries or sites that could be hundreds of miles away.

Purchasing seed or plants from nurseries has advantages. In particular, the immediate availability of many species is generally more reliable. You need not wait an entire year or two to collect seed from a diversity of species that have different seed production periods—the nursery may have done that. Many plants that are difficult to establish when seeded onsite are grown by nurseries for transplant onto restoration sites. However, some species propagules are quite expensive to purchase, and budget considerations may mean leaving out some species until a future date or not planted at all. Other species are difficult for nurseries to grow. As a business, nurseries may focus on those species that are easiest to grow and in greatest demand. Nurseries may or may not be producing or supplying local ecotypes. Where this is an issue, you should ask about the geographic origin of a species' seed. Last, commercial nurseries are not found in all areas where restorations are conducted; nurseries that specialize in native species are even less common.

8.4.4 How Are Seeding and Planting Rates Determined?

Seeding and planting rates are based on what you, as the restoration planner, believes is required to represent the plant composition and structure found in a typical stand of the community. If every seed planted would germinate and every transplant survive and grow into an adult, this task would be simpler. However, most undomesticated species are highly variable in annual seed viability and germination success, and seedlings are browsed and often succumb to diseases and competition.

Seed can be assessed for its viability and potential for germination. There are multiple methods to test the germination rates for seeds. One of the least expensive is to germinate a subset of the seeds, by placing seeds in soil flats or petri dishes with wet filter paper that remains moist but not soaked for 2–3 weeks. These tests should have a predetermined end date, as seeds for many species will not germinate at the same time and instead may germinate over a period of months or even years. For reliability, use a minimum of 200 seeds for each species of interest, divided into groups of 10–20 seeds distributed among the petri dishes.

A second test using staining with tetrazolium dye can identify whether or not a seed has living tissue, by determining if respiratory activity is present. A third less-reliable test involves squeezing the seed. If the seed is firm, even when pinched strongly, it is likely to be viable. If the seed collapses or feels hollow, it is not viable.

Many nurseries will test their seed for germination and pure live seed (PLS) rates prior to sale. If this is not automatically done, the service can often be requested for an additional fee. Pure live seed is based on purity and germination rate. Purity is the volume or weight that is seed and not chaff and other inert material. For example, 45 kg of bulk seed that has a 70% germination rate and is 90% pure provides 28 kg PLS. Under this scenario, if you were to order 28 kg of PLS from a nursery, the seed received would weigh approximately 45 kg.

Pure live seed (weight) = Bulk (weight) × % germination × % purity

To determine planting or seeding rates, you begin by estimating the average density of plants in the reference models. Then you adjust the rate based on experience with the system to be restored. Where pest species are likely to cause problems, some restorationists may increase the seeding rate, particularly for species that germinate quickly and are competitive. A higher planting rate will result in the quick and dense establishment of a suite of native species that will then restrict undesired species from establishing in the planting bed. The downside is that higher planting rates may also keep slow-to-germinate species from establishing, and the site will then lose desired diversity and patchiness.

Reduced seeding rates, although they potentially allow for the establishment of undesirable species, potentially provide more opportunity for the establishment of conservative or slow-to-germinate-and-establish species, resulting in a much more diverse mix in the end. Many studies that have compared seeding rates find differences in plant density and diversity in the first few years of the planting, but in some cases these differences disappear over time, as long as the high or low seeding rates are not too extreme. Some of these studies also suggest that, although there is a minimum planting rate for a successful restoration, there is also a limit to the maximum planting rate necessary for success. In other words, after reaching a hypothetical planting rate, planting a greater density of plants or seeds does not necessarily result in greater diversity and success for the restoration (Harrington, unpublished data; Williams et al. 2002).

One problem with specifying planting rates for different species is that the weather during the current and past years, as well as that immediately preceding and following the planting, can have a tremendous influence on germination and establishment. When preparing a species mix, individual seeding rates should be adjusted by a species' known ease of establishment, reproduction capabilities, and overall ability to compete.

When ordering materials from nurseries and contractors, seed numbers will often be converted to weight. It is much easier to work with weight than with seed numbers when ordering or mixing and measuring seed to be planted. The seed from different species is, of course, of different sizes and weights. A mix with 10% of its weight from Species A and 10% from Species B will seldom result in two species equally represented in the mix or either species composing 10% of the total mix. Species A could feasibly have 800,000 seeds per ounce and Species B 20,000 seeds per ounce (see Sidebar). Most nurseries will convert seed numbers to an estimated weight.

Sidebar | **Converting Seed Numbers to Weight**

Acquiring seed by weight is much easier than trying to count the seeds for an entire project. However, we cannot proportion seed abundance by weight, because of the weight variations of seeds from different species. Let's look at how to convert between seed numbers and seed weight.

In this example, there are 4 ha (10 ac) and we want to plant at 440 seeds/m² (40 seeds/ft²). The weight of the seed needed for the restoration can be quickly calculated. We begin by calculating the total number of seeds needed for all 4 ha. One hectare requires 4,400,000 seeds, or 440 x 10,000 m² (40 x 43,560

ft² [ac] equals 1,742,400 seeds/ac). Then we set proportions of the seed mix for each species based on its typical abundance in the community. Let's say we have a mix of four species. We calculate the average weight of the seed of each of the species. This is done by taking a small amount of seed, weighing it, and counting the number of seeds. After repeating this several times (about 10 repetitions), we average the number of seeds from all repetitions and then extrapolate the average number of seeds per weight used to ounces or pounds (Table 8.2).

Table 8.2 • Converting Seed Numbers to Weight

Conversion of Seed Numbers to Pounds Per Hectare

Species	Percentage of Mix	Seeds/Hectare	Seeds/Pound	Pounds/Hectare
A	30	1,320,000	200,000	6.60
B	10	440,000	67,200	6.50
C	50	2,200,000	428,000	5.10
D	10	440,000	320,000	1.40
Total		4,400,000		19.6

Conversion of Seed Numbers to Pounds Per Acre

Species	Percentage of Mix	Seeds/Acre	Seeds/Pound	Pounds/Acre
A	30	522,720	200,000	2.60
B	10	174,240	67,200	2.60
C	50	871,200	428,000	2.00
D	10	174,240	320,000	0.50
Total		1,742,400		7.70

1 hectare equals 10,000 square meters.	1 acre equals 43,560 square feet.
440 seeds per square meter.	40 seeds per square foot.
Seeds desired per hectare: 10,000 x 440 = 4,400,000.	Seeds desired per acre: 43,560 x 40 = 1,742,400.

8.4.5 How Are Seeds and Plants Prepared for Planting?

Many seeds have natural barriers to germination, and breaking them requires specific environmental actions. The barriers presumably help ensure that germination occurs when environmental conditions are most likely to support plant establishment. For example, in climates with cold winters, many seeds produced in late summer and fall require a period of cold before they will germinate, in essence inhibiting germination until spring when growing conditions are more favorable.

Seeds, with hard or thick seed coats, are often impervious to water and can take one or more years to germinate; as the seed coat softens or is nicked or scratched (scarified), it begins to imbibe water. Under natural conditions, the seed coat barrier can also be overcome when the seeds pass through the gut of an animal, blow across abrasive surfaces such as sand, expand and contract in freezing and thawing cycles, are charred in a fire, or are digested by microbes. In preparing for a restoration, seed can be artificially scarified by scratching the seed coat using sandpaper, by submersion in boiling water, and sometimes by using acid baths.

Stratification of seeds to overcome physiological dormancy uses several techniques, depending on the species. For some species, storage for a period of time in dry conditions is sufficient. Seeds of other species, particularly those native to temperate and boreal climates, require cold stratification that involves storing the seed under cool moist or cool dry conditions at 1°C–3°C (34°F–38°F) for a species-specific period of time.

Many species benefit from interactions with soil microorganisms. For example, legumes fix nitrogen through a relationship with *Rhizobium* bacteria, which colonize their root nodules. Although in many situations soils will have the necessary *Rhizobium* for nitrogen fixation, nurseries can inoculate the seed mix to ensure the presence of bacteria if you are working in areas where their occurrence may be doubtful.

8.4.6 How Should Seeds, Seed Mixes, and Transplants Be Located on the Site?

As noted in Chapter 7, several methods are used to determine where a species or species mix will be placed in the restoration planting. One method is to plant each species mix at designated rates evenly throughout an implementation unit, and allow the species to establish in response to the microvariations in the environment. This method is relatively inexpensive and requires the least amount of planning. The result is a diverse mix of species, self-sorted to the site so that location and species are well suited. Since most restoration units will have environmental variations, this method will result in seed waste if the seeds for many species fall in less than desirable conditions.

An alternative method is to divide the community implementation unit into subunits based on soils, topography, hydrology, and, if present, vegetation. Each subunit is seeded with mixes developed for the specific environment. Seedlings and transplants can be individually matched to microenvironments at a predetermined density. In theory, this planting method should result in greater species-to-environment compatibility and less loss of seed and transplant. On the downside, this method takes additional time and expense for site inventory, planning, and implementation, and, due to constant environmental variation, the final result may not be much different from using the previous method of species location.

A third alternative is a hybrid of the first two. A single mix of representative species is planted throughout the area, and then separate mixes of individuals of more conservative or less widely distributed species are planted in targeted areas over the first planting. The submixes can also be used to provide for transitions from one community type to another, to match zones that correspond to different water depths or shade gradients, and or to emulate natural species aggregations for both ecological and aesthetic advantage.

The site plan presented in Figure 7.10 and Table 8.3 illustrates this third option. Notice that several of the mixes used between community types contain overlapping species, but that the proportions change as one community transitions to another. The restoration team could have planted one mix for each community type and allowed the species to sort out over time, but as noted previously, this would waste seed or seedlings, which for many species is consistently in short supply.

8.4.7 Is There a Need to Sequence the Installation of Species?

Another decision that needs to be made during implementation is whether to plant all species at once or in stages. In many situations, this decision is a function of who is doing the restoration, the time frame for completing the installation, and the existing site conditions. If this is a contracted installation or one that needs completion in a given budget year, there is little leeway for sequencing a planting. However, in situations where the restoration can be completed over several years or more, sequencing the restoration may result in a higher-quality product. The answers to the following questions can help determine the planting sequence.

- Is the landscape to be planted entirely bare of vegetation? If so, are some species on the list less able to withstand exposure to the elements and best planted once other species have grown sufficiently to provide some degree of shelter?
- For forested or savanna restorations, is the canopy structure present, or does it need to be introduced prior to the ground layer species?

Table 8.3 • Hypothetical Seed Mix Arrangement for the Cross Plains Unit of the CPIANSR

(For the site plan, see Figure 7.10.)

Grass Species	Base Mix	DMP 1	Percentages DMP 2	DMP 3	DMP 4
Andropogon gerardii	5	15	70	5	20
Bouteloua curtipendula	25	0	0	10	0
Elymus canadensis	10	0	0	0	0
Hesperostipa spartea	0	30	0	40	20
Schizachyrium scoparium	40	0	0	0	0
Sorghastrum nutans	10	55	30	20	40
Sporobolus heterolepsis	10	0	0	30	20
Forb Species					
Allium cernuum	0	5	0	5	10
Amorpha canescens	5	5	0	5	5
Anemone canadensis	0	0	10	0	0
Asclepias tuberosa	0	5	0	10	10
Baptisia alba	0	10	0	5	10
Dalea purpurea	5	5	0	5	5
Desmodium illinoense	0	10	10	0	5
Dodecatheon meadia	0	0	0	5	0
Echinacea pallida	0	10	0	10	10
Eryngium yuccifolium	10	5	5	0	0
Heliopsis helianthoides	5	5	10	5	5
Heuchera richardsonii	0	0	0	5	0
Lespedeza capitata	10	0	0	0	0
Liatris aspera	10	0	0	5	5
Liatris pycnostachya	0	0	10	0	0
Monarda fistulosa	0	10	10	5	5
Pedicularis canadensis	0	0	0	5	0
Ratibida pinnata	10	0	10	0	0
Rudbeckia hirta	10	0	10	0	0
Silphium integrifolium	10	0	0	0	5
Silphium laciniatum	0	10	0	10	10
Sisyrinchium campestre	0	5	0	5	5
Solidago rigida	10	0	0	0	5
Symphyotrichum laeve	5	5	5	5	0
Symphyotrichum oolentangiense	0	0	0	10	10
Tradescantia ohiensis	10	0	0	0	0
Veronicastrum virginicum	0	5	10	0	5
Zizia aurea	0	5	10	0	0

Base mix is planted at 12.4 lb/ha (5 lb/acre); all other mixes are planted at 5 lb/ha (2 lb/ac). Forb-to-grass ratio is 60:40. DMP = dry mesic prairie.

- Are desirable species still present on the site, and if so, how might these influence the establishment of introduced species?

- Are there seasonal events or weather conditions that may favor the planting of some species at different times of the year—for instance, planting some species in the spring and others in the fall?

- Which species germinate best immediately after seed matures, and which ones require a period of dormancy before germination?

- Are some species dependent on the presence of others for productivity and survival (e.g., hemiparasitic species)?

Succession theory and the concepts of facilitation and inhibition (discussed in Chapter 2) help us understand the importance of recognizing the physical and biological features on the landscape that may be of greatest importance to the establishment of a particular species. Some species have a very wide environmental range of tolerance in which they can survive and grow successfully. Other species have very specific environmental requirements, and until these conditions occur onsite, there is little reason to plant these species. For example, the ability of sphagnum moss to enter a landscape, establish a population, and create the conditions for an acidic bog community often depends on a suite of shrub species colonizing wet basins first. The sphagnum invades the setting by climbing over and supporting itself on the branches and roots of woody plants that have already colonized the site and extend out into open water. Bog conditions then follow that facilitate colonization by a new suite of plant species, but inhibit the colonization of others.

Anticipating how plant species will respond when planted into a setting with existing species is difficult. One challenge is recognizing those species that are poor competitors compared with species that effectively compete with and limit the establishment of others. Why is this difficult? The level of competitiveness a species exhibits is often related to the site's physical and biological environment. Species that are not aggressive or highly competitive in remnants can become so in new plantings where resource levels can be quite different.

The planting environment in many restoration projects is unlike that found on sites largely undisturbed by humans. In a restoration started from scratch, the environment can lack plant cover, have moisture levels that fluctuate rapidly to uncommon extremes, and have extremely high or low nutrient levels. Sites that still have some elements of the former community may have beneficial soil microorganisms, which will often be missing from highly degraded sites. How species will respond to these various site conditions is difficult to determine and will only be learned by working in the same region for many years.

Being highly competitive should not exclude a species from being planted into a restoration. But it may mean delaying its introduction until other less competitive species are established and reproducing; or it may mean that fewer individuals of this species are introduced than what is specified for final species composition when the restoration is completed. The assumption is that over time, its numbers will increase without additional planting.

In addition, the ability of a plant to establish and compete with other plant species in a given physical environment can be a factor in how the site is managed. Management practices such as grazing, mowing, and fire will give certain species a competitive advantage.

Many restoration situations begin with site conditions that are not suitable to all species proposed to compose the targeted community. When working with open exposed ground, for example, a base slate of species from the community mix can be planted. Periodically, additional species can be interseeded or interplanted, as the initially planted species become established and the environment changes to accommodate the introduction of newer species. This technique lets you tailor the restoration over time, based on what establishes during the initial seeding or planting. The method is often used for forested or savanna situations where no initial canopy exists.

Environmental Considerations

Deciding when to plant depends on many factors, and more research is still needed to determine the best time to plant in order to restore the overall community, species groups, or individual species. The primary factors include (1) the growth and reproduction requirements of specific species, (2) weather and other stimulants for germination, (3) the site characteristics, and (4) the planting method.

If using seed, plants need to be installed with sufficient time for germination and establishment prior to the occurrence of seasonal conditions that are detrimental to growth and survival. In regions that have regular wet and dry seasons or warm periods followed by freezing temperatures, planting must be done when seed germination or root growth is most likely to occur, with sufficient time to establish before conditions become too harsh for young plants to survive.

In temperate grasslands, for instance, one suite of species flowers and produces seed by early summer. This seed will germinate and establish during that first year if planted immediately after it is produced. Seed produced in mid to late summer is either stratified over winter and then planted in the spring, or seeded during the winter months. The latter situation has several advantages; during planting, the distribution of seed is visible on top of any existing snow, the seed migrates into the soil during freezing and thawing events, and the seed is planted prior to melting snow or early spring rains. Seed is not planted one or two months before winter conditions, as young seedlings will lack sufficient time to harden off prior to freezing temperatures. In some tropical climates, the production or release of seeds is linked to seasonal periods of rain.

Predictable cyclical events, such as flooding, which might carry recently planted seed offsite, or drought, which would prevent germination or result in the death of young seedlings, will also dictate times to plant or avoid doing so. Some planting techniques are also weather-dependent, such as avoiding high wind days for broadcasting light seed that can be blown away from its desired planting location, or wet periods for drilling seed that can result in compacted soils, as well as soil and seed being picked up by the drill's tires.

8.4.8 What Are the Alternative Planting Methods?

Once the site is prepared in a manner that is conducive to meeting your restoration goals and seeds and/or transplants are procured, planting takes place. As noted before, the manner in which the site was prepared must be compatible with the chosen planting techniques. There are numerous ways to plant a site, and each method has its advantages and disadvantages.

Many methods in use today have been adopted through trial and error, not necessarily out of research or long-term monitoring. The result is a lack of uniformity in techniques used by restorationists that makes it difficult to compare the success of different methods.

To overcome the common dilemma of having to apply implementation strategies to a new set of landscape variables or the need to try an untested strategy, the concept of adaptive restoration has been adopted. Adaptive restoration asks questions about the various ways to restore a site or community, and then tests them as the restoration progresses in stages. As you learn from one test or experiment, you can modify or adapt the next phase of the restoration (see Chapter 3). Some of the implementation techniques often used in three major ecosystem categories— grassland, forest, and wetlands—are discussed next.

Grassland and Open Systems

Grassland systems are found around the globe, particularly in the interior portions of large landmasses where the environment is arid and fire and grazing have historically occurred. These systems are dominated by grass in terms of biomass, but they often have many more species of blooming forbs or herbaceous broad-leaved plants. The primary causes for much of the loss of native grasslands are conversion to agriculture, overgrazing, and the cessation of fire.

Four common site conditions for restoring grasslands are pasture, cropland, overgrazed natural grasslands, and old-field. Maintained pastures are typically composed of one or a blend of grass species that serve as feed for livestock. These landscapes are relatively weed-free compared with old-fields and abandoned pastures, but the pasture grasses themselves are often rhizomatous and highly competitive. Old-fields are landscapes that were probably pasture at one time but are no longer being maintained as such. This lack of maintenance results in the invasion by woody plant species, as well as many generalist weed species. These landscapes often contain highly competitive species, making restoration difficult. Croplands are generally one of the easier landscapes to prepare for seeding grassland mixes. The vegetation, in most cases, is easily removed, and these fields are also relatively weed-free. However, depending on farming practices, the soils can be extremely fertile, allowing for quick establishment by undesirable and highly competitive species once cropping ceases. Croplands and pastures can also be compacted, contain pesticide or herbicide residues that require multiple years to break down, and have a lower level of microfauna that are beneficial to the establishment of native species.

In many situations, grasslands are planted using seed because the high diversity of species, high plant densities, and focus on herbaceous species make the planting of seedlings impractical. Transplants, however, have several advantages over seed in that they provide a more efficient use of seed particularly for species that are scarce, and are perhaps the best chance for the establishment of many species that germinate poorly from seed. The planting of seeds has evolved over the years, with most techniques originating from traditional agricultural practices. For grassland plantings, seed drills, no-till seed drills, and broadcast seeding are three of the more commonly used planting techniques.

Seed drills used for native plantings are agricultural range or crop drills that have been modified to accommodate light, chaffy seed and a mixture of seed with a wide range of sizes and weights. Two or three boxes sit behind the driver: one box for small seed, a second box for large or fluffy seeds, and an optional box for cover crop seeds. Tubes drop the seed from each box into furrows opened in the soil by disks. The furrows are subsequently closed by a set of press wheels.

Seed drills are labor efficient and waste less seed than nearly any other planting method. However, all seed is planted at the same depth, which favors some species over others. The rows in which the seed are distributed are visible for many years. Seed drills often make several passes from different angles to cross rows and lessen their visual impact. The chaff, awns, bracts, and other appendages surrounding the seed can still clog the seed tubes of the drill and are best removed. Fine seed drops more quickly than larger seed, resulting in an uneven planting, with the smaller-seeded species being dispersed early in the process.

No-till seeding is similar to using a seed drill except for slight modifications that allow the drill to plant seeds into a shallow groove that it opens in the soil without tilling (Figure 8.6, left). Because the site is not tilled, the use of a no-till seed drill requires a site that has little competition from undesirable species. If the site has existing vegetation, usually several applications of herbicide will be needed to remove it. The advantage of no-till seeding is the absence of an open seedbed where undesirable species can establish. It also creates less opportunity for wind and water erosion. The disadvantages are similar to those found for traditional seed drills.

Broadcast seeding, by either hand or machine, is another preferred method. This method is frequently used on open ground with little litter or vegetative cover. In addition to the growing season, broadcast seeding is often done successfully on frozen or snow-covered ground in northern climates. Broadcast seeding does not require high-end equipment. If done by hand, labor is the most expensive aspect, and if volunteer groups participate, this cost can be quite low. Mechanical broadcast seeding requires relatively inexpensive equipment as well.

A broadcast seeder is attached to a tractor or similar drivable equipment; it has a vault to hold seed and a controlled arm that casts seed outward (Figure 8.6, right). Seed to be broadcast is often mixed with crushed vermiculite, silica sand, or other low-weight inert material to add bulk so that the seed is not distributed too quickly. A single handful of seed could contain hundreds of thousands of individual seeds. The added inert ingredients are also light-colored and will show on a dark surface so that the driver can see where seed has been distributed.

FIGURE 8.6. Planting Seeds. Left: This no-till seed drill has two seed boxes, one for fine and one for larger fluffier seeds. The drill has 12 coulters or disks that cut through litter and into the soil. These are followed by openers, which provide a furrow for seed to drop into. The depth of the furrow is controlled with depth bands. (Photo reprinted by permission from John Harrington.) Right: In this broadcast seeder, each hopper can hold about 30–40 pounds of seed. Each arm waves back and forth and scatters seed 15–16 ft. Planting on snow, where available, allows you to see where seed has been cast. (Photo reprinted by permission of McHenry County Conservation District, Illinois.)

The advantages of broadcast seeding are relatively low cost, the ability to use seed that has not been entirely cleaned of chaff, a more natural appearance of plant distribution than the straight rows that drilling produces, and, for hand broadcasting, the ability to reach spaces that many machines drills cannot.

Hydroseeding is a fourth seeding method that is occasionally used in restoration efforts. This method allows seed to be distributed on slopes and areas that are difficult for machines to access directly, as long as such areas have an open seedbed. Hydroseeding involves placing seed into a large water tank typically attached to a truck cab. An agitator in the tank keeps the seed distributed throughout the water. Water and seed are shot up to 200 ft (61 m) from a hose at no less than 100 pounds per square inch (psi) or 100 gallons per minute (gpm). Compared with the other seeding methods discussed, hydroseeding appears to require considerably higher seeding rates. Some restorationists find that hydroseeding results in poor seed contact with the soil. For some species, the seed when held in contact with water for more than 30 minutes will begin the germination process, regardless of the climatic conditions, once it is sprayed onto the ground.

Native hay, cut and taken from nearby remnants or restorations, can be distributed across an open restoration seedbed with the hope that it will introduce seed of desirable species. Hay has limitations in that the seed it carries is time-dependent and varies according to morphology. Not all species produce seed at the same time, seed maturation varies tremendously between species, and not all seeds persists on stalk but rather fall off the plant as it matures. Most remnants can only be cut once or twice a year, limiting the ability to use hay to collect seed from different periods throughout the growing season unless many remnants are available for this use. For these reasons, hay should be considered as one of several seed sources, rather than the sole seed source.

Forest and Savanna Systems

Unlike grasslands, forested and savanna systems are composed of three main vertical vegetation layers: ground, mid, and canopy. The stage at which a forest restoration can begin often depends on whether a canopy exists and the state the canopy is in. A site can have a canopy composed of the desired species, a canopy of species that are not in the long-term targeted composition but are acceptable in the short term, or a canopy of species that are not desirable and could be problematic with the establishment of a desired canopy; or it may have no canopy.

In settings where no canopy exists, the restoration process will continue until the targeted canopy can be established and has grown sufficiently that the appropriate understory and ground layer plants can establish as well. In most situations, seeding or planting seedlings of the desired canopy species will be necessary; the recovery of forest tree species in areas that have been farmed or under slash-and-burn agriculture will be slow or nonexistent without such aid. The desired canopy species for the restoration may not be those that are initially planted. In situations where the desired mature canopy will be composed of shade-adapted forest species, you can begin with species that are better adapted to being planted in full sun and unprotected from wind. Once these species grow sufficiently to develop the beginnings of a canopy and shelter, species adapted to shade and closed canopy conditions can be underplanted. The expectation is that as the underplanted canopy trees develop, the initial cover crop of trees will slowly weaken with increasing shade, fail to reproduce, and die out over time.

Other settings may have a canopy that does not contain the desired species, but neither is it problematic and, in fact, may be useful to the growth of the desired canopy species.

In some situations, the targeted forest canopy already exists, and it is the shrubs and ground layer that need to be restored. In this case, you will want to determine why these layers are absent and eliminate the existing elements or functions that are hindering their development, perhaps reintroduce the missing species. Often a site like this will be invaded by undesirable species, and these will need to be removed.

Unlike many forest communities, most savanna communities have a ground layer dominated by grasses, sedges, or low ground-hugging shrubs and are highly adapted to fire. Worldwide, savannas have been used for grazing, often being overgrazed, and this can result in bare and compacted soils, as well as erosion. Others have experienced fire suppression, and have subsequently transformed into dense forestlike communities. The absence of fire for several years allows a number of shrubs and herbaceous plants to invade, increasing shade and thus limiting the production of the grassy fuels needed for future fires (Figure 8.7) The savanna trees accustomed to high levels of light and wind slowly decline as the tree density increases and the community shifts to a more shaded and less windy condition. Removing grazers or altering the grazing regime and reintroducing fire may not always solve the problem. Any methods that are considered for reducing soil compaction, healing erosion, reopening the canopy, or removing undesirable species must recognize the impact such treatments could have on existing tree species that are to be maintained in the restoration.

Finally, there are settings where the existing canopy is best removed prior to any planting. Typically this occurs where the existing canopy is unwanted and would be highly competitive with the desired canopy and ground layer species. For example, in situations where the goal is to establish forests of largely shade-intolerant trees the existing trees will need to be removed.

FIGURE 8.7. A Changing Landscape. Formerly open-grown oak is now surrounded by shade-tolerant shrubs and saplings. Thousands of acres of oak savanna have been engulfed by shrubs and forest trees with the suppression of fire that began in the late 1800s in North America. Many of the species that composed the previous savanna herbaceous layer are no longer present due to light and litter suppression. A few of the species that tolerate shade remain, but most do not flower and produce seed. Over time, the surrounding plants will increasingly shade the oak, causing its demise as well. (Photo reprinted by permission from John Harrington.)

The Wisconsin program of The Nature Conservancy (TNC) has undertaken an oak savanna restoration project at a property where a pine plantation was started in the 1930s. The pines have reached a mature height and are shading the former savanna oaks. The savanna ground layer is gone due to the shade and litter the pines produce. To restore the savanna structure, TNC has been cutting the pine. But part of the restoration site is on steep slopes where machinery cannot be used to remove the fallen trees without creating significant damage. Instead of machines, TNC has used oxen to drag the pines out during winter months when the ground is frozen and unlikely to be disturbed by the dragging of logs. The logs are then cleaned of branches, and any timber of commercial value is sold to support further restoration efforts.

Trees are often planted with bare root seedling stock, but container-grown trees and seeds can also be used. The number of individuals to plant in order to reach the desired tree density requires experience with the species being planted. Typically, bare root seedlings will be planted in much greater numbers and more tightly spaced than container trees. Seeds will be planted at even greater densities to account for individuals that do not germinate and establish or are preyed upon.

The site to be planted must be free of competition, such as dense grass. Preemergent herbicides can keep weed seeds from germinating and establishing new plants. A small crew can plant a large number of seedlings in one day. Whether tree, shrub, or herbaceous seedlings, making a slit in the soil and opening it with a spade, locating the plant, and closing the slit back up will suffice for many plantings, particularly for well-drained, loamy soils. Seedlings can also be machine planted. This is advantageous where thousands of seedlings are to be planted and labor is limited. The disadvantage is the distribution of plants in rows, similar to that in grassland plantings.

Forest clearing and agricultural practices have resulted in the loss of forests and land productivity in many developing countries. Such practices, including repeated burning, intense grazing, and cutting, lead to limited stump sprouting, depletion of the soil seed bank, soil erosion, nutrient loss, and loss of habitat for seed-dispersing wildlife. In countries where resources are highly limited, restoration relies on the abandonment of the land and hoped-for forest regeneration. To increase chances of colonization by indigenous species, tree plantations are sometimes used to facilitate succession in their understories. The plantations help facilitate forest succession through changes in microclimatic conditions, increased vegetation complexity, the development of litter and humus layers, the suppression of grasses, seed dispersal by canopy-dwelling wildlife, modifications to light levels and surface temperatures, and increased soil moisture. At the same time, the plantations gain public support by providing economic and social benefits to neighboring landowners.

Wetlands

Wetlands cover a wide range of ecological communities, each of which has its own restoration needs. Many of the ideas presented under grasslands and forested systems will apply to one or more wetlands systems—sedge meadows, hardwood swamps, fens, and so on. However, wetland systems are often difficult to repair because of their unique position in the landscape, that of being a collector for most surface waters originating in uplands, their dependence on specific hydrologic systems, and, in many countries, significant government regulations to address.

CASE STUDY   Bog Restoration in the Chiemsee Lake Region of Bavaria

Hydrology is the key parameter for reestablishing plants and processes in wetlands; hydrologic processes must be restored before planting. Influences on plant establishment include maximum and minimum water levels, duration of seasonal flooding periods, water quality and volume, and water flow rate. One example is the restoration of ombrotrophic peatlands or raised bogs in the Chiemsee Lake region of Bavaria in a long-term study undertaken by faculty at the Technische Universitat Munchen (Figure 8.8).

In the early 1990s, approximately 220,000 ha of peatlands, both fen and bog, existed near the base of the Alps in the main basin of the Chiemsee glacier. The bogs were drained for forestry and mined for peat dating back at least to the early 1800s. Since then, 900 ha of the area were placed into the Kendlmuhl-

FIGURE 8.8. **Bog Restoration.** This long-term study, conducted by faculty at Technisceh Universitat Munchen, in Germany, explores the physical, chemical, and biological variables that can affect the success of bog restoration. Top row: The ditches are plugged and the peat surface is scraped to raise water levels. Middle row: The land is flooded for a period of time before water levels are lowered. Bottom row: Ericaceous shrub and sedge species establish on the flooded peat, followed by species of sphagnum. (Photos reprinted by permission from John Harrington.)

filzen Nature Preserve, although peat mining still occurred into the 2000s. The southern half of this area remained nearly intact, with bog peat thickness of 4 m. The northern half was heavily disturbed by peat extraction and drainage. After drainage, the bog vegetation developed into dry heathland and open woodlands dominated by downy birch (*Betula pubescens*), *Pinus rotundata*, and heather (*Callua vulgaris*).

Following the concept of adaptive restoration, experiments began in 1987 to determine the most efficient way to rewet and reestablish vegetation on the former bog lands (Sliva and Pfadenhauer 1999). These included recontouring the basin floors, experimenting with remaining peat depth, and creating "dams" to impede the flow of water. Once areas were rewetted to a targeted level, the water level was lowered to the peat surface, and sedges (*Carex rostata, Eriophorum* sp.) and pioneer

bog shrub species were reintroduced. After 3–4 years, pioneering vascular species were established and began to provide a stabilized the peat surface, the water level was raised 10–20 cm, and sphagnum moss was introduced. As sphagnum mats formed, the water levels were raised even further. Additional nutrient studies found that fertilization with phosphorus enhanced bryophyte establishment.

The success of this and other bog and fen restoration projects is uncertain because of the long-term maturation such ecosystems require (Poschlod et al. 2007). However, the above research suggests that without the thorough and permanent rewetting of the peat substrate, these restorations are not possible. Even then, the restoration success depends on the presence of propagules, nutrients, and favorable climatic conditions. ■

CASE STUDY   The Nygren Wetland Preserve

The Nygren Wetland Preserve is a 292-ha (721-ac) restoration project of the Illinois Natural Land Institute, located northwest of Rockford, Illinois. The first project goal is the restoration of the original Raccoon Creek channel and its confluence with the Rock and Pecatonica Rivers' channels as they come together. A second goal is to return the land to its original vegetation of wetland and lowland to upland prairie and forest. Thirteen restoration zones that were former agricultural fields are being restored to prairie, wetlands, and savanna. An additional seven zones are being restored to bottomland forest and savanna. A third goal is to restore habitat for raptors, migratory waterfowl, and passerines, as well as terrestrial and aquatic species (see Sidebar).

Raccoon Creek was ditched and channelized in the 1970s to drain the wetlands and rush floodwater downstream to the Pecatonica River. To restore its original route, ditches that drained and channelized the creek were filled, plugged, or diverted back to the original channel, berms were removed to reflood areas or rebuilt if needed to hold water back, and the original river and creek channels had

excess sediments removed and were reshaped as needed. Stream banks were recontoured, stabilized, and then replanted with plugs and willow live stakes (Figure 8.9). Rock weirs were added when necessary to prevent eroding stream bottom elevations, and a vegetated cobble was added to the toes or bottoms of some banks to protect them from large volumes of fast-moving water. More than 10,000 live willow stakes were installed to complement the restored channel seed mix. Onsite nurseries were established to grow seedlings and plants for seed collection. The removal of pest species and the environmental causes for their presence have been major hurdles (see Chapter 13). However, over a span of several years, the flow of Raccoon Creek has been returned to many of its natural meanders and restored to its original mouth on the Pecatonica River.

In addition, a series of wetlands and upland prairies are being established using contractors and volunteers. The main ditches through the wetlands were plugged in order to raise the water table for recolonization by wetland plants and wildlife. The upland prairies were disked and then seeded. Some areas

FIGURE 8.9. Stream Channel Restoration, Nygren Wetlands, Rockton, Illinois. Top row: An original stream channel of the Raccoon Creek is reestablished. Formerly straightened channels are blocked so that water backs up into the newly formed channels. Middle row: New banks are stabilized with straw, erosion fabric, and willow stakes. Bottom row: Banks are revegetated with prairie and cutting of lowland shrub species. In the background of the right image is the adjacent floodplain forest, which is also being restored. (Photos reprinted by permission from John Harrington.)

that planted in rhizome-producing grasses, such as brome, were sprayed with the herbicide Roundup prior to disking; others were burned and then sprayed and lightly disked. Although volunteers collected considerable seed to insure seed availability, several native plant nurseries for forb, grass, and shrub production were established on the property. Prior to planting, pest species were treated with herbicides (the type dependent on the species being treated), hand pulled, or had their seed heads removed.

Prior to seeding of the final mix within several restoration zones, a cover of quick-to-establish native species was planted to suppress weed competition. Initial seeding of the final mix began the following year and continued over the next 2 years.

Bottomland forest restoration has also been ongoing at the Nygren Wetland Preserve. Container trees are planted at 100/ha (40/ac) and with an average spacing of 9 m (30 ft) and shrubs at 279/ha (690/ac). Savanna zones were planted at a rate of 10 trees per hectare (25 trees per acre). Bare root seedlings are planted at 865/ha (350/ac) on a 3-m (10-ft) spacing. Seeds are planted at 1,200–1,800/ha (3,000–4,500/ac). Expectations are that the container-grown trees will have a lower mortality than bare-root trees, and that not all seeds will germinate; for those that do, the first-year seedlings will have high mortality.

The restoration has been monitored for plant and animal species. Overall, an increase in species diversity has been recorded. Well monitoring established that the water control structure has raised the water table in the lower wetlands. ■

Sidebar Evaluating the Habitat Value of Preexisting Vegetation

Prior to implementing the restoration plan for the Nygren Wetland Preserve, a survey of the existing vegetation was conducted. The survey was constructed so that it could be used in the future to evaluate the progress of the restoration plantings against prerestoration vegetation. Specific goals were to document how the vegetative cover progressed during the first decade after seeding, to measure seeding guarantees by the contractor (percentage of seed that produced established plants), and to document wildlife habitat conditions and use. Data were collected for the different restoration zones during the surveys; they included total species richness, planted species richness, percent cover of grasses, sedges, forbs, and non-native woody and herbaceous species. Other surveys were conducted for birds, mammals, amphibians, and invertebrates. The survey found the presence of streamside bird species among dense shrub thickets along the Raccoon Creek channel. A discussion centered on whether these shrubs should be removed, thinned, or left as they were. As the dense shrubury was not natural but an artifact of agriculture and the channelization of the creek, some members argued for its removal. Others, although agreeing that the stands were not a representation of what would likely have occurred along the creek, argued that the shrubs did serve as valuable habitat for a number of songbirds and therefore should not be taken out.

FOOD FOR THOUGHT

1. In order to make this decision, what would you want to know about the existing habitat prior to implementation, the songbirds, and the projected future vegetation cover if these stands of shrubs were to be taken out?

8.4.9 How Will the Finished Restoration Be Evaluated?

Performance standards define how a finished product will appear and function upon acceptance or completion; they are supplemental to the site and implementation plans. These standards provide for broad flexibility in methods, as long as the final product meets the project objectives. A performance standard is similar to an objective with a time frame added to it. These time frames may be fixed, but by applying temporal trends, we account for the dynamic nature of the restoration process and the environment, as the examples in Figure 8.10 show.

All performance standards must provide a clear, concise description of what is to be accomplished; provide a measurable target for the restoration or management activity; and ensure that the variable to be evaluated measures what it is purported to—that is, nests per acre reflects successful breeding. In this example, the standard lacks validity; nesting does not equate to successful breeding. The standard should have a threshold of acceptance (a minimum of 100 nests per acre of which at least half produce fledglings) that needs to be met for approval, and it should provide a time period over which the target is to be accomplished.

Performance standards may be fixed:

- 80% of the species planted from list A will be established at densities of at least 40 plants per square meter within 3 years of planting.
- Within 5 years of the restoration, undesirable herbaceous species will represent less than 5% of the ground cover.

Or performance standards may look at temporal trends:

Composition:

- For each year through year 5 there is a minimum species (planted) increase of 10% from the previous year.
- There is a continuous decline in invasive species beginning by or before year 4 and continuing until less than 5% of species remain in this category.

or

- 25% of species are established (present) by year 3.
- 50% of species are established by year 5.
- 50% of remaining species are established by year 10.

Structure:

- An annual measurable increase in native cover occurs by year 3 with a continuous decline in exotic cover. This decline shall continue until there is less than 5% cover of exotic species or the applicable objectives are met.

Disturbance regimes:

- Be able to carry at least one fire during the first 5 years of establishment.
- Be able to carry a minimum fire frequency of every 3 years during years 5–10.
- Be able to carry a minimum fire frequency of every 2 years from year 10 on.

Standards should be within 10% of the above thresholds.

FIGURE 8.10. **Performance Standards.**

Performance standards require an understanding of the restoration process and the technology available. They focus on measurements whose values or trends best reflect the status and dynamics of the community or system.

The use of performance standards overlaps with monitoring and asks similar questions:

- What ecological measures will be used to judge the success of a restoration or management technique?
- What are the threshold acceptance and rejection values for selected measures?
- At what stage of the restoration is it appropriate to apply standards? When do you know that a restoration is progressing toward its target?

▶ 8.5 RESOLVING LOGISTICS

Planning logistics, or the coordination of implementation activities, includes the following tasks, several of which occur throughout or at different stages of the implementation process:

1. Developing a timetable or phasing plan for the implementation planning and installation.
2. Obtaining the necessary labor and equipment resources.
3. Obtaining permits; notifying public agencies, neighbors, and other pertinent parties.
4. Constructing a budget (see Chapter 6).

8.5.1 Developing an Implementation Timetable

An implementation or action timetable lays out what needs to be done and when to establish the restoration. You begin by listing all the activities that must be conducted to accomplish each goal and its objectives. To assist you, review the implementation timetables from similar projects, noting the steps taken and the time required for their completion. Organize the activities for your project into a logical sequence of steps, to be conducted over the anticipated time frame. Activities listed in your timetable may or may not refer to specific calendar dates. Often the sequence of steps follows the development of the restoration; for example, "When the canopy cover is more than 40%, begin to interseed the partial shade mix." Timetables should be flexible, in order to accommodate weather fluctuations, the availability of propagules, site preparation success, and the inherent flexibility and variation of natural systems.

CASE STUDY | The Hog Island and Newton Creek Restoration Master Plan

The Hog Island, Hog Island Inlet, and Newton Creek restoration is part of a large brownfield remediation project near Superior, Wisconsin, at the western edge of Lake Superior (Biohabitats, Inc. 2007). Newton Creek drains into Superior Harbor near Hog Island, passing through an area contaminated by industrial discharges and sewage overflow (Figure 8.11). Remediation of the contaminated sediments in the creek and inlet occurred from 1997 to 2005. Restoration planning began at this time. The project is part of a much larger endeavor to preserve and repair areas of concern, or severely degraded geographic areas, within the Great Lakes ecosystem.

The purpose of the restoration project is to provide a vision along with specific goals, objectives, and actions for restoring the natural communities and ecosystem processes of the area. The project use-purpose is to increase environmental awareness through recreational, educational, and stewardship opportunities and through a process of stakeholder engagement and collaboration.

Hog Island Inlet occupies 17 acres supporting shallow wetlands and mudflats in Superior Harbor. Hog Island was created from dredge spoils taken from the harbor between 1919 and 1935 and hosts a wide diversity of migratory birds and fish populations. Prior to restoration, the area had several habitat types and species assemblages, including estuarine aquatic habitats, swamp, meadow, and lowland forest. Much of the forest was discontinuous, however, and disconnected from its surroundings. The hydrology in Newton Creek was controlled with little

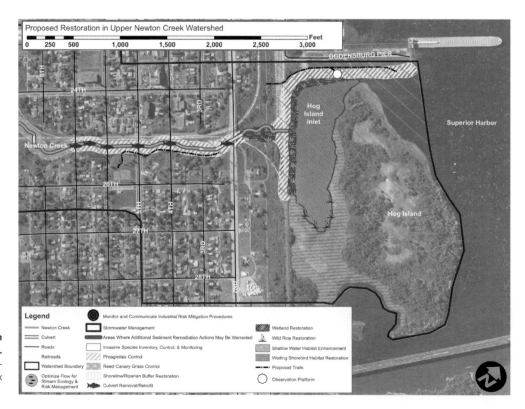

FIGURE 8.11. Proposed Restoration in Upper Newton Creek and Hog Island. (Reprinted by permission from Biohabitats, Inc., "Hog Island and Newton Creek Restoration Master Plan," 2007.)

natural variation in flow, and the waters contained high levels of sediment and pest species.

Chapter 6 introduced the master plan components for this project. The master plan has four major goals:

1. Improve water and sediment quality conditions in Newton Creek and Hog Island Inlet and reduce the threat of future contamination.

2. Conserve and protect ecologically sensitive habitats.

3. Restore selected ecosystem components in a manner that is consistent with the ecological restoration guiding principles.

4. In conjunction with restoration activities, create recreational, educational, and environmental stewardship activities for City of Superior and Douglas County residents.

In addition, the plan sets forth several guiding principles that are similar to the time-neutral format we suggest for goals and objectives (Figure 8.12). Each of the project goals has a set of objectives established as part of the site plan, and each objective has a set of actions or implementation steps. Table 8.4 provides a small portion of these actions and how they are sequenced in the phasing plan for the Hog Island and Newton Creek Ecological Restoration Master Plan by Biohabitats, Inc. Although the objectives are not stated in the table, each action is coded to the objective in which it occurs. Table 8.5 similarly presents a portion of a timetable for the same project. The time targets are provided in relative years, since specific dates for the master/site plan adoption and for plan installation were unknown. ∎

- Functional groups are present, or they have the ability to successfully colonize.
- Reproducing populations of target species are present.
- Characteristic assemblages of species/communities, as in reference ecosystems, are present.
- Indigenous species are present.
- Self-sustaining natural communities are present.
- Potential ecosystem threats are eliminated or reduced.
- Ecosystems are resilient to normal ranges of ecological stress.
- The restoration site is integrated into a larger ecological landscape.
- Habitat diversity is maximized.
- The goals of the Lower Saint Louis River Habitat Plan are integrated.
- Sensitive ecological areas are placed under permanent protection.
- Restoration and resources management occurs according to watershed planning principles.
- Educational and volunteering opportunities are integrated.
- Human uses, which compromise long-term ecological sustainability, are restricted.
- The plan is flexible, allowing for the integration of new ideas and stakeholders.

Adapted from Biohabitats, Inc. 2007.

FIGURE 8.12. Guiding Principles for the Hog Island and Newton Creek Restoration Master Plan.

Table 8.4 · Phasing of Restoration Activities for Hog Island and Newton Creek

Phase 1: 0–1 Year from Master Plan Adoption
- Initiate ecological flow regime determination and feasibility assessment (A1:1, A2:2).
- Initiate risk assessment of industrial water contamination (A3:1).
- Initiate surveys and research to determine the extent of residual sediment contamination (A4:1).
- Initiate public and private property conservation and acquisition efforts (B).
- Initiate invasive species surveys and control efforts (C1:1, C1:2, C1:3).
- Initiate SAV restoration in Hog Island Inlet (C4:2).

Phase 2: 2–4 Years from Master Plan Adoption
- Implement ecological flows band begin monitoring (A1:2, A1:3).
- Initiate stormwater management in Newton Creek watershed (A2:1).
- Continue construction of additional industrial pollution control facilities as needed (A3:1).
- Complete residual sediment contamination surveys and research (A4:1).
- Establish and monitor phytoremediation pilot projects (A4:2).
- Complete public property conservation efforts, and acquisition/conservation of key private parcels (B1:1, B2:2, B2:3).
- Continue conservation/acquisition of private parcels in upper watershed (B2:1).
- Complete invasive species inventories (C1:1) and begin invasive species monitoring (C1:4).
- Complete *Phragmites* control (C1:3), continue reed canary grass control efforts (C1:2).
- Establish riparian and shoreline buffers, begin culvert removal efforts (C2).
- Initiate wetland restoration and expansion efforts (C3).
- Complete restoration of open water habitats in Hog Island Inlet (C4:1, C4:2).
- Restore piping plover habitats and begin monitoring (C5:1).
- Initiate post-project monitoring of any restoration projects that have been completed (C6).
- Begin construction of footpaths and birdwatching platforms (D1).

Phase 3: 5–10 Years from Master Plan Adoption
- Continue monitoring ecological flows in Newton Creek (A1:3).
- Continue stormwater management in Newton Creek watershed (A2:1).
- Continue construction of additional industrial pollution control facilities as needed (A3:1).
- Complete monitoring of phytoremediation pilot projects (A4:2) and expand phytoremediation plots to other areas (A4:3).
- Continue conservation/acquisition of private parcels in upper watershed (B2:1).
- Complete reed canary grass control efforts (C1:2).
- Continue and complete invasive species monitoring efforts (C1:4).
- Continue and complete culvert removal efforts (C2:3).
- Complete wetland restoration and expansion efforts (C3).
- Continue monitoring of completed ecosystem restoration efforts (C6).

Phase 4: 11 Years to Completion of Ecosystem Restoration Efforts
- Continue phytoremediation plot expansion and monitoring (A4:3).
- Continue monitoring of completed ecosystem restoration efforts (C6).

Letter-number codes (i.e., A1:1) represent objectives. The code links the action statement to specific plan objectives.
Adapted from Biohabitats, Inc. 2007.

Table 8.5 • Timetable for Hog Island Inlet and Newton Creek: Initial 22 Years After Planting																							
	Years from Master Plan Adoption																						
Action	0	1	2	3	4	5	6	7	8	9	10	11	12	13	14	15	16	17	18	19	20	21	22
A1:1—Determine ecologically optimal flow regime for Newton Creek																							
A1:2—Determine release schedule compatible with plant operations																							
A1:3—Monitor Newton Creek to refine release schedule and establish in-stream and riparian habitat benefits																							
A2:1—Develop recommendations for appropriate stormwater best management practices (BMPs) in the watershed																							
A3:1—Perform a risk assessment of industrial contamination in the project area																							
A4:1—Determine if contaminated sediments remain along the shoreline of Hog Island inlet and along the Newton Creek watershed																							
A4:2—Establish a series of phytoremediation test plots and monitor for success																							
A:3—Expand phytoremediation plots to other areas demonstrated to contain unacceptable levels of sediment contamination																							
B1:1—Work with the City of Superior and Douglas County to protect remaining vacant public lands on Hog Island and within the watershed																							
B2:1—Permanently protect privately held upland and riparian habitats within the Newton Creek watershed																							
B2:2—Acquire/protect Ogdensburg Pier properties on the northwestern end of Hog Island inlet																							
B2:3—Work with Burlington Northern Santa Fe railroad to acquire the railroad berms running parallel to the shoreline																							
C1:1—Perform a comprehensive invasive plant species inventory and map throughout ecologically sensitive areas; control invasive vegetation																							

(table continues)

Table 8.5 • Timetable for Hog Island Inlet and Newton Creek: Initial 22 Years After Planting (continued)

Action	0	1	2	3	4	5	6	7	8	9	10	11	12	13	14	15	16	17	18	19	20	21	22
C1:2—Control reed canary grass along Newton Creek		▓	▓	▓	▓	▓																	
C1:3—Control *Phragmites australis* along the Hog Island shoreline areas		▓	▓	▓																			
C1:4—Actively monitor for migration of exotic invasive plants from the adjacent landscape				▓	▓	▓	▓	▓	▓	▓	▓												
C2:1—Establish a 75-ft-wide riparian buffer along Newton Creek between 7th Street and 2nd Street			▓	▓	▓	▓																	
C:2-2—Establish a 100-ft vegetative buffer along the southwestern and northwestern shorelines of Hog Island Inlet			▓	▓	▓	▓																	
C2:3—Remove/retrofit culverts at road and sanitary sewer line crossing along Newton Creek			▓	▓	▓	▓	▓	▓	▓	▓	▓												
C3:1—Restore sustainable, reproducing communities of wild rice in the Hog Island Inlet and along the shoreline			▓	▓	▓	▓	▓																
C3:2—Expand areas of emergent wetland vegetation into the northern and western areas of the inlet			▓	▓	▓	▓																	
C3:3—Expand areas of wetland vegetation into the seiche-influenced areas of Newton Creek			▓	▓	▓	▓	▓																
C4:1—Use natural features such as logs and rocks in the open waters of Hog Island to promote cover for aquatic organisms and bird habitat			▓	▓	▓	▓																	
C4:2—Restore populations of SAV in the open-water areas of Hog Island Inlet		▓	▓																				
C5:1—Establish nesting habitats for piping plover on Hog Island Inlet			▓	▓	▓	▓																	
D1:1—Extend existing trail system to include limited access to Newton Creek and Hog Island and improve preexisting informal footpaths			▓	▓	▓	▓	▓																
D1:2—Establish additional birdwatching platforms on Ogdensburg Pier and Hog Island			▓	▓	▓	▓	▓																

(table continues)

Table 8.5 • Timetable for Hog Island Inlet and Newton Creek: Initial 22 Years After Planting (continued)

Action	Years from Master Plan Adoption																						
	0	1	2	3	4	5	6	7	8	9	10	11	12	13	14	15	16	17	18	19	20	21	22
D2:1—Recruit an entity to direct public outreach efforts and advocated for ecosystem restoration in the watershed	X	X	X	X	X	X	X	X	X	X	X	X	X	X	X	X	X	X	X	X	X	X	X
D2:2—Create an "adopt a stream" or ecological education program in local schools that focuses on the ecosystems and restoration processes	X	X	X	X	X	X	X	X	X	X	X	X	X	X	X	X	X	X	X	X	X	X	X
D2:3—Use the ecosystem restoration efforts at Hog Island and Newton Creek as a focus for student research projects at local universities	X	X	X	X	X	X	X	X	X	X	X	X	X	X	X	X	X	X	X	X	X		
D2:4—Create and maintain a project website to keep stakeholders, watershed residents, and citizens informed of the restoration process	X	X	X	X	X	X	X	X	X	X	X												

Letter-number codes (i.e., A1:1) represent objectives. The code links the action statement to specific plan objectives. The shaded cells link action to years.

Adapted from Biohabitats. 2007.

8.5.2 Obtaining the Necessary Labor and Equipment Resources

Many different kinds of equipment are used in restoration activities. Specific needs will vary depending on what is to be restored, the existing land conditions, and the restoration's location. Decisions on equipment use depend on what has to be accomplished and on the preferences and experience of the user. Prior to finalizing the implementation plans, labor requirements must to be anticipated and evaluated as to whether they can be met with the staff at hand. If not, you have to determine where the extra labor will come from (volunteers, hired temporary staff, private consultants, etc.).

By recognizing potentially limiting resources (labor, equipment, propagules, water, etc.), you can be prepared with contingency plans if the essential resources cannot be found. As with the site plan, before the actual implementation process begins, goals and objectives must be clearly understood and communicated to others who have a direct or indirect role in the process. Review the potential site preparation and planting strategies and budget with those responsible for the actual onsite implementation, and determine whether specialists will be needed for any portion of the implementation.

8.5.3 Obtaining Permits; Reviewing Plans with Jurisdictional Powers; Notifying Public Agencies, Neighbors, and Other Pertinent Parties

In many situations, it will be necessary to obtain construction permits and provide notification to public agencies, neighbors, and other pertinent parties of the activities being conducted. Even when not required, these are desirable actions, particularly in situations where activities will physically alter a site's use, remove large swaths of vegetation, and impact its aesthetics, such as altering public views or creating noise and dust. This is the time to revisit all parties of interest and remind them of the goals, explain in detail any updates since the master planning process, and review who is responsible for which aspects of the project and for supplying which resources.

This evaluation step is particularly important for large-scale projects in which several agencies are involved and for which public dollars are being used. The restoration of the Everglades in Florida is a good example of a project whose restoration goals and plans for restoring the biological systems are in place, but where delays have occurred due to multiagency and private sector interactions. Lack of funding has resulted in the scope of the restoration being significantly decreased and the short-term goals, at least, being altered.

KEY CONCEPTS

- Implementation is a five-step process: (1) defining the implementation units and boundaries, (2) assessing existing site conditions, (3) developing and applying implementation strategies for site preparation, (4) developing and applying implementation strategies to reintroduce biotic components, and (5) resolving logistics.

- Implementation tests our understanding of the environmental processes, both internal and external to the landscape site, that influence the restoration's success.

- Implementation begins with evaluating the existing site and determining where it does and does not meet the restoration site plan objectives, and for those aspects that do not, why.

- Site preparation is one of the most important steps in implementing a site plan—if not the most important step. Site preparation addresses the question: What aspect of the site and its context currently prevent the targeted restoration community or ecosystem from existing?

- The past, present, and predicted future uses of the site by people all have the potential to influence implementation.

- The information you have about a species' method of reproduction, its ability to disperse across landscapes, and its competitiveness in the setting where it will be initially placed strongly influences the manner in which you introduce the species to the site.

- Performance standards add an operational time frame to the restoration goals and objectives, and are used to evaluate whether the progress of a project meets expectations. They are clear, concise, and measurable. Although they are time driven, performance standards often involve temporal trends, not fixed dates.

- Implementation occurs in stages (site preparation, trail building, planting, etc.), and implementation schedules list the activities to be conducted under each stage. Activities are scheduled according to weather and logistical needs, and in relation to the completion of preceding stages. Because many variables that affect implementation are unpredictable, schedules must be flexible.

- Implementation plans are unique to each setting, regardless of the similarities in goals between sites.

FOOD FOR THOUGHT

1. Both the Kissimmee River and the Everglades ecosystem in Florida have been undergoing major restoration efforts during the past decade. Using journal articles, internet sources, and government agency reports, discuss who the interest groups are for each project, the goals for the restorations and their time frames, and any conflicts or problems that have surfaced in the restoration efforts.

2. Select a possible restoration project from your geographic region, establish goals, and discuss the variety of site preparation needs that would be required.

3. Select a second ecological community found within your region for a hypothetical restoration project. Put together your desired species composition, including the proportions of each species, and defend these. Then consider whether you would keep these same proportions in the actual planting mix, or whether you would adjust the rates for specific species based on establishment considerations. For each species, explain and defend any adjustments.

4. Consider the actual planting of the above species. Would you plant them all at once or sequence the planting over time? Explain and defend your decision.

5. What ecological measures would you use to judge the success of the restoration?

6. Develop a timeline for the implementation, beginning with site plan review.

Resources for Further Study

1. A number of journals publish articles on the pros and cons of restoration implementation techniques. *Restoration Ecology* is a bimonthly scholarly journal published on behalf of the Society for Ecological Restoration (SER) that serves as a means of exchanging both technical and theoretical information. *Ecological Restoration* is published quarterly and contains short articles, notes about ongoing projects, book reviews, and more. The following represent a sample of the kinds of case studies you might find:

Beschta, R., and W. Ripple. 2010. Recovering riparian plant communities with wolves in Northern Yellowstone, U.S.A. *Restoration Ecology* 18:380–389.

Bonilla-Moheno, M., and K. Holl. 2010. Direct seeding to restore tropical mature-forest species in areas of slash-and-burn agriculture. *Restoration Ecology* 18:438–445.

Liu, M., G. Jiang, S. Yu, Y. Li, and G. Li. 2009. The role of soils seed banks in natural restoration of the degraded Hunshandak Sandlands, Northern China. *Restoration Ecology* 17:127–136.

Vieira, D. L. M., K. D. Holl, and F. M. Paneireiro. 2009. Agro-successional restoration as a strategy to facilitate tropical forest recovery. *Restoration Ecology* 17:451–459.

Young, S., J. Barney, G. Kyser, T. Jones, and J. DiTomaso. 2009. Functionally similar species confer greater resistance to invasion: Implication for grassland restoration." *Restoration Ecology* 17:884–892.

2. For more on implementation planning, see Society for Ecological Restoration International Science & Policy Working Group. 2004. *SER International Primer on Ecological Restoration*. Version 2. http://www.ser.org/pdf/primer3.pdf

3. For an interesting discussion of how to deal with restoration implementation in the face of global climate change, check out:

Rice, K. J., and N.C. Emery. 2003. Managing microevolution: Restoration in the face of global change. *Frontiers in Ecology and Environment* 1:469–478.

The Monitoring Plan

One of the best ways to learn about restoration is through understanding the successes and failures of projects. In order to continue to improve restoration ecology, we need to know what works and what doesn't—and most importantly, to begin to understand the reasons that underlie a project's "success" or "failure." As discussed in Chapter 3, restorations ideally incorporate scientific experiments so that restoration theory, both substantive and procedural, can advance, and individual projects can achieve good results with minimal frustration. Experiments expand our understanding of restoration ecology by testing our ability to predict the outcome of a project. If our predictions are supported by the experimental results, we gain confidence in the theories that underlie them. If the predictions do not hold, we rethink the situation and, in many cases, come up with alternate hypotheses. More pragmatically, experiments established on a portion of a site at the beginning of a project to test, for example, a variety of implementation techniques enable us to select the approach best suited to the particular situation before proceeding.

Due to time and resource considerations, it is not always feasible to incorporate formal experiments into every project, but you can and should monitor the progress of all the restorations you undertake, at least to some extent. What exactly does it mean to monitor a restoration? In the broadest sense, monitoring is a systematic process by which you periodically check, describe, and evaluate the status of a project. It is essentially a site inventory and analysis, with the added dimension of time. The purpose is to watch for site changes and new developments on or off the site, whether anticipated or not, and compare the site at the time of monitoring with the restoration vision.

Monitoring lacks the formal predictive power of experiments, but it is nonetheless an integral component of the adaptive restoration process. For example, using monitoring results for a particular site, you can make midcourse corrections (e.g., replant, add different species, add a soil amendment). When a restoration contract requires your project to achieve particular benchmarks before authorizing your final payment, you can use monitoring data to demonstrate that the objectives and performance standards have been achieved. Once the project objectives are met, monitoring data can alert you when management action thresholds (conditions that call for management intervention) have been reached. By examining monitoring data for many sites and/or data collected on a few sites for many years, you can find patterns and form hypotheses to be tested by later experiments.

In this chapter, we will focus on one site component—vegetation—to illustrate the kinds of decisions you need to make in order to adapt the inventory tools introduced in Chapter 5 for use in monitoring. Plants and plant communities are often foundational components of restorations, and therefore are included in most monitoring plans.

▶ 9.1 FEATURES OF THE MONITORING PLAN

Similar to site inventory and analysis, monitoring occurs at several stages during the implementation of a restoration project, and later at regular intervals after the initial goals have been met and routine management begins (see Chapter 10). The monitoring plan is an explicit, interactive, and flexible guide to the monitoring activities. The plan is also project-specific; it concerns the objectives and/or implementation performance criteria of the restoration for which it is written.

A monitoring plan describes inventory, measurement, evaluation, and analysis procedures and the rationales behind using them. Presented with text, maps, and other graphics, it provides enough information so that you can not only follow the plan as written, but also adapt your activities without compromising the original vision, as new techniques are made available or as the site changes in unexpected ways.

Ideally, the monitoring plan is written at the same time as the master, site, and implementation plans (Figure 9.1). It is usually organized around the answers to a series of questions (Figure 9.2). The answers take into consideration the project goals and objectives, the dynamics and potential stressors anticipated by the project community/ecosystem models, and the time and resources available.

As we have discussed, almost all restorations include goals and objectives that describe the desired biological outcomes of a project in terms of composition (species or life forms), structure (vertical and horizontal patterns, age distributions), and dynamics and interactions (mutualisms, predator-prey relationships) of natural plant and animal communities. Many goals and objectives also describe the physical resources needed to sustain these

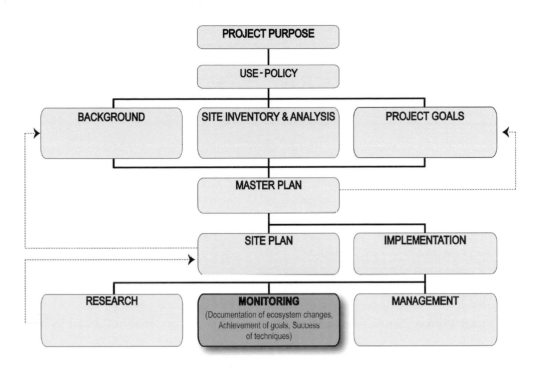

FIGURE 9.1. The Restoration Process: Monitoring.

- **WHAT is monitored?** What kinds of information are covered in the restoration goals and objectives? What kinds of data are useful in tracking anticipated dynamics or impacts? What kinds of information are needed to determine whether the action thresholds contained in the management plan have been reached?
- **HOW?** What techniques and approaches are to be used in collecting information, in storing and analyzing it, and in making it available to other researchers and/or the public?
- **WHERE?** Do different parts of the site have different objectives, and therefore different data needs?
- **WHEN?** During what season, stage of growth, weather conditions, etc., and with what frequency and over what time period should the monitoring be done?
- **WHO?** Do project staff members, volunteers, or paid subcontractors conduct the monitoring, i.e., collect and analyze the data?
- **WHY?** Who makes the decisions, using what assumptions and information?

FIGURE 9.2. Questions Addressed by the Monitoring Plan.

communities. Therefore, in order to evaluate how well a restoration is achieving its goals and objectives, monitoring plans collect information about vegetation and wildlife (Figure 9.3), and physical features such as hydrologic systems and soils. Monitoring plans for restorations that concern ecosystem processes—for example, carbon sequestration, nitrogen mineralization, groundwater recharge—contain monitoring protocols for collecting information that can be used to run tests of these functions.

All restorations are designed with people in mind, and their use-policies describe how people will or will not use the site. In some situations, usually in order to protect the restoration, plans call for keeping most people away from the site. Such prohibitions can either be temporary (e.g., when the soil is wet) or last for many years. In these situations, the aim is to monitor the site in order to determine how well the recommended strategies for keeping people out are working. Many restorations welcome people; they are encouraged

FIGURE 9.3. Monitoring Bat Populations. Researchers at the Wisconsin Department of Natural Resources are conducting a long-term monitoring study of bat populations. The monitoring station detects and records acoustic signals as bats fly by. The data are downloaded monthly and analyzed for bat activity patterns and species identification, based on unique echolocation calls. (Photo reprinted by permission from Steve Glass.)

to actively interact with the site—as volunteers involved with implementing and managing, as well as visitors coming for recreation and/or education. Goals and objectives describe desired visitation rates, knowledge acquisition, and user satisfaction outcomes. In these situations, monitoring plans include protocols for assessing human–environment interactions, such as documenting use patterns, measuring visitor and volunteer satisfaction, assessing rates of human impact, and evaluating the effectiveness of educational and other social programming activities.

You can find information about how to measure wildlife, physical features, and people in Chapter 5, and in the Resources sections at the end of this chapter and Chapter 5.

Many restorations are large enough to include several different zones, each with separate goals and objectives. It is often useful to develop separate monitoring protocols for each. You can create the boundaries of units to match vegetation, topographic, or cultural site features (such as roads and trails), and mark and record their locations using a GPS or other land survey technique. The monitoring plan includes a map of the units, together with careful descriptions of the boundaries. It is important to record and archive the original identity of the units, even as alterations in the restoration and/or the site goals cause changes in the size or shape of the units over time. This way, you can maintain site history.

The monitoring plan also includes at least two timelines that describe what should be happening in each monitoring unit over the course of a season or year. One timeline covers the period during which active restoration is occurring. The second is connected to management The timelines help keep everything organized and on task.

Last, but not least, the plan specifies the procedures you will use for archiving and managing the information collected as monitoring proceeds. And, as has been the case for all the different types of restoration plans we have discussed so far, you will need to include procedures for updating and modifying its policies and protocols.

▶ 9.2 MONITORING VEGETATION

You can describe the desired vegetation of a site in a number of ways, all of which can be used in setting restoration goals and objectives, and therefore the measurements used in the monitoring plan. Most often, restorations are concerned with two levels of vegetation organization—plant communities and individual species populations. Depending on the project, the objectives you specify in the site plan may pertain to one level or to both (see Chapter 7). For example, you may include objectives using measures that refer to community-level concepts, such as overall species richness or the presence of species with particular roles (keystone predator, spring bloomer; Figure 9.4). Or your objectives may relate to attributes of species populations, such as population sizes or rates of seed set (Figure 9.5), or to ecosystem functions in which vegetation plays a significant role (Figure 9.6). Because most of these vegetation measures concern species, it is important for the monitoring protocol to establish the taxonomic classification system to be used during the inventory. Generally this is done by using the nomenclature adopted by state, regional, or national herbaria, or by referring to local university experts or to the taxonomic publications the professional botanists in the region use.

Because every restoration project has a unique set of goals and objectives, every monitoring plan is also unique. It is possible, however, to recognize several questions about the vegetation of a site that most monitoring plans address.

1. **Composition**: The "parts" of the community listed and presented as the total recorded during a qualitative survey or quantitative sample.

   a. **Species**: Evolutionary, genetic, natural selection unit. A species list for an area describes its "floristics."

   b. **Life Form**: Growth form, including features such as size, life span (annual, perennial), degree of woodiness (herb, shrub), degree of independence (parasitic, saprophytic), general morphology (rosette form), leaf traits (evergreen), the location of perennating buds.

   c. **Guild (potential resource competitors)**: Communities with redundant species—several guild members—may withstand a disturbance better than less complex systems. Temporal, photosynthetic, reproductive.

   d. **Life History Stages**: Morphologically and often functionally distinct stages an organism goes through from birth to death—seed, seedling, sapling, tree, for example. Each stage may have different habitat tolerances.

2. **Diversity**: A concept generally including two components.

   a. **Richness**: The number of compositional units per unit area (generally the number of species per area).

   b. **Evenness**: The distribution of individuals among the units; equitability. Community A with 3 species, 10 individuals of species x, 10 of species y, 10 of species z is more even than Community B with 20 of individuals of species x, 8 of species y, 2 of species z.

   Sometimes the term "diversity" refers only to richness; usually it refers to both richness and evenness.

3. **Density**: A measure of abundance summarized from sampling data as the number of individuals in a sample. Density can be expressed in absolute terms (65 individuals in the sample) or as the number of individuals per unit area (stems per acre, plants per acre).

4. **Frequency**: A measure of dispersion or horizontal pattern; the chance of finding a unit in an area in any one trial. In sampling, frequency is the number of sampling points at which an element is found, expressed either in absolute terms (the frequency of a species found in 10 of 20 samples is 10) or as a percentage (the frequency of a species found in 5 of 20 quadrats is 25%).

5. **Cover**: The proportion of ground shaded or occupied by plants.

6. **Pattern (Dispersion)**: The spatial arrangement of units.

   a. **Horizontal**: The arrangement over ground surface.

      i. **Random**: The arrangement in which any spot has an equal chance of containing a plant.

      ii. **Regular**: The arrangement in which the presence of a plant reduces the chance of a similar plant being next to it.

      iii. **Aggregated**: The arrangement in which the presence of a plant increases the chance of a similar plant being next to it.

   b. **Vertical**: The arrangement between ground and sky; aerial layers (canopy, midstory, understory).

FIGURE 9.4. Vegetation Community Descriptors.

A. **Size**: The number of individuals in a population.

B. **Common/Rare**: Measures of relative abundance.

C. **Effective breeding population**: An estimate of the numbers of organisms within a population that are able to reproduce.

D. **Flowering, fruiting, seed-set, dispersal**: The number of flowers, fruit, and viable seed produced, measured per stem, or per individual; pollinator or disperser specificity and interactions.

E. **Pattern**: The spatial (generally horizontal) arrangement of individuals within a population.

   1. Random

   2. Regular

   3. Aggregated

F. **Metapopulation structure**: The interactions of individuals belonging to groups of populations more or less spatially separated; emigration and immigrations rates.

G. **Demographics/dynamics**: The statistical study of populations, especially age structure; birth and death rates.

H. **Genetic diversity**: The pool of genes contributed by all organisms within a population.

FIGURE 9.5. Vegetation Population Descriptors.

A. **Primary productivity**: The rate at which radiant energy is converted by photosynthetic and chemosynthetic activity of producer organisms (chiefly green plants) to organic substances.

    1. Gross primary productivity: The total rate of photosynthesis, including the organic matter used up in respiration during the measurement period.

    2. Net primary productivity: The rate of storage of organic matter in plant tissues; gross production minus respiration (often calculated per growing season or year).

C. **Biomass (primary production)**: The amount of material present at any given time.

    1. Plant biomass is often expressed as kg/ha or g/m^2; may be reported as aboveground and belowground portions; by biomass types (leaves, stems, roots, flowers, and fruits); by species or by life form.

    2. The biomass (and litter) of a plant community is often considered an important nutrient sink (a place where nutrients are stored rather than rapidly cycled). Especially important consideration in carbon and nitrogen cycles.

D. **Flux**: The rate of nutrient flow.

E. **Mineralization (ammonification)**: The rate at which nutrient elements are converted from organic to inorganic form through the process of decomposition of plant litter and soil organic matter.

F. **Cycling index**: The proportion of the total throughflow of a nutrient element that is recycled in an ecosystem (i.e., the proportion of incoming material that moves from one part of the system to another before exiting the system).

G. **Evapotranspiration**: Evaporation from soil plus water loss from vegetation.

H. **Index of water availability**: The ratio of rainfall to potential evapotranspiration (which depends on temperature).

FIGURE 9.6. **Vegetation Function Descriptors.**

At the community level:

    1. What are the native and exotic species richness and/or diversity of each unit, and how are they changing over time?

    2. To what extent does the community contain species or life forms of special concern (species considered to be rare or endangered), or species that, because they are most often found there, represent a particular type of community (e.g., community indicator species, modal species)?

    3. Are the desired species or life forms increasing in numbers and abundances relative to the undesired species or life forms?

    4. What is the age structure of the community dominants (life forms or species), and how is this changing over time?

    5. What are the vertical and the horizontal structures of the community, and how are these changing over time?

    6. What is the response of the community composition and structure to natural disturbances, such as windstorms and wildfires, and to human impact?

At the species or population level:

    1. What is the population size—the number of individuals in the population—and how has it changed through time? (Is it increasing or decreasing?)

    2. Where are members of the population located (in what parts of the site), and how has the distribution changed through time?

3. What is the reproductive capability of the population (number of flowers and/or fruit produced, seed viability), and how have these factors changed through time?

4. What is the age/size class structure (number of stems/individuals in each class), and how has this structure changed through time?

5. What is the spatial distribution of individuals in the population (clumped, random, regular)?

To answer these monitoring questions, you would use the qualitative and/or quantitative vegetation sampling techniques introduced in Chapter 5. There are many measurements to choose from, and each one provides a different insight about composition, structure, and function. There is no single "right" way to sample, or to evaluate the sampling data. For example, you could use the results from either a quantitative sample or a walk-through technique to determine whether the site has achieved a "minimum of 75 species from among those on the attached list." But you would need to use a quantitative measurement to determine if a tree canopy cover was "between 50% and 75%."

The case study on page 286 illustrates one way to use sampling data to monitor the composition and structure of a deciduous forest. See the Sidebar below to learn how you might use the data collected in this way to monitor a restoration, and to test your understanding of the strengths and weaknesses of the approach.

Sidebar | Evaluating Vegetation Monitoring Data

Tables 9.1, 9.2, and 9.3 present data collected in June 2006 from the restoration using the protocol described in the accompanying case study. The desired outcomes of this restoration include the following:

- Maintain at least 40 native species, including a minimum of 5 native tree species, 5 native shrub species, and 25 native herb and vine species.
- <5% of the species are non-native.
- <1% of the species are pest plants.
- Some herbaceous species are common, most are relatively rare.
- Pest species, if present, are rare.
- Total shrub cover is 15–40%.
- Tree density is 240–300 trees per hectare.
- The canopy is dominated by oak species.
- Oak reproduction is occurring.

FOOD FOR THOUGHT

1. Based on the monitoring results, determine whether or not each of the restoration outcomes had been achieved as of June 2006. Explain your reasoning, and indicate any assumptions you had to make in arriving at your decisions.

2. The site and monitoring plans for this restoration are due for a routine review this year. What changes, if any, would you make in way the objectives are written, and/or in the monitoring protocol? Explain. (Consider, for example, whether or not the current objectives [outcomes] are clear and measurable, and whether the current monitoring protocol as described in the case study, provides sufficient information to accurately assess the restoration in terms of the objectives.)

Table 9.1 • Forest Understory Sample: Presence/Absence Data

Sampling Points→	1	2	3	4	5	6	7	8	9	10	11	12	13	14	15	16	17	18	19	20	Freq
Tree Seedlings																					
Ostrya virginiana				1								1									
Prunus serotina						1									1						
Prunus virginiana	1				1	1									1			1		1	
Grasses, Herbs, Ferns, Vines																					
Agrimonia gryposepaala													1			1					
Arisaema triphyllum					1									1			1				
Athyrium filix-femina				1														1			
Chelidonium majus	1	1									1										
Circaea lutetiana	1	1	1	1		1	1	1							1		1	1			
Desmodium glutinosum											1		1						1		
Dryopteris sp.			1									1						1			
Galium aparine		1									1								1		
Geranium maculatum	1	1	1			1	1			1		1	1		1			1		1	
Geum canadense			1	1	1	1			1	1			1		1			1	1		
Oxalis sp.						1			1												
Parthenocissus sp.		1		1		1	1		1	1		1		1		1	1		1		
Potentila simplex									1												
Smilacina racemosa										1			1								
Smilacina stellata						1	1								1						
Viola pubescens		1		1		1				1						1			1		
Viola sp.				1																1	
Shrubs																					
Cornus racemosa															1						
Rubus allegheniensis	1	1	1	1	1	1	1			1	1		1		1	1	1	1	1		
Rubus idaeus					1								1								
Rubus occidentalis						1				1					1					1	
Viburnum lentago											1										
Grasses and Sedges																					
Sedge sp.			1			1			1												
Non-Native Grasses, Herbs, Ferns, Vines																					
Alliaria petiolata*				1													1				
Glechoma hederacea*				1											1						
Totals																					

*Pest species.

The data indicate the presence (1) or absence of each species in each of 20 lm² quadrats. The sample points are arranged in a grid pattern in the field. Quadrats 1–5 represent the first row and are numbered sequentially moving east to west. The second row (6–10) is located 20m south of the first row; the third (11–15) is 20m south of the second; the fourth row (16–20) is 20m south of the third.

Table 9.2 • Forest Tree Sample

Tree Sample (20 points sampled)

	# Pts.	# Trees	BA (cm^2)
Acer negundo	6	6	852
Ostrya virginiana	12	28	4,129
Prunus serotina	4	4	3,548
Quercus alba	18	36	91,057
Quercus rubra	2	2	403
Quercus velutina	2	2	142
Totals	**44**	**78**	100,131

Pts represents the number of quadrats within which a species is found.

Trees is the total number of trees found in the sample.

BA is the total basal area (the cross-sectional area of a trunk) of the trees sampled; it is one measure of dominance.

Table 9.3 • Forest Sapling and Shrub Sample

Sapling Sample

	# Pts.	# Saplings	BA (cm^2)
Carya ovata	2	4	168
Ostrya virginiana	2	2	1
Quercus ulbu	2	2	187
Totals	**6**	**8**	**356**

Pts, # saplings, BA defined as in Table 9.2

Shrub Intercept

20 transects x 15.2 m each = 30,400 cm total number of possible intercepts

Rubus allegheniensis–Average cover: 8230 cm/30,400 cm = 27% cover

9.2.1 Photography

Photography is a technique that is part of almost all most vegetation monitoring protocols. To understand how a site is progressing through time, a photographic time sequence is ideal. The first step is to establish a series of permanently marked ground-level photo points. The idea is to take photographs at each location at specified dates and times. By referencing the locations to a GPS, as well as marking them in the field, such as with a piece of iron rebar, you can return to the same locations even as vegetation landmarks change. It is important that you designate the date, time of day, and weather conditions each time a photo is taken. Also, be sure the direction in which the picture is taken, the height of the camera, and the lens focal length remain the same so that the views will be as consistent as

CASE STUDY	Using Sampling Data to Monitor Forest Composition and Structure

In order to monitor the progress of a 15-acre deciduous forest restoration located on private property near Madison, Wisconsin, the restoration team established a reference grid prior to implementing the site plan. The grid consists of twenty 20m × 20m cells. The measurements for monitoring and evaluating the site using the outcomes specified in the site plan are listed in Figure 9.7.

The monitoring plan uses 20 permanent sampling points, one of which is located within each grid cell. The team collects data according to the following protocol:

A. Understory

At each sampling point, construct a square 1 × 1 m quadrat, with the southeast corner of the quadrat touching the sample point. Record species presence data. Note herbaceous species (including ferns), ground-hugging vines, and tree seedlings. Use the data to compile a species list, and to determine one measure of abundance—how widespread the species is.

B. Shrub layer

Center a 15.2-m tape on each sample point at right angles to the direction of the grid line. Record the length of tape intercepted by each shrub species. Use the data to compile a species list and to measure shrub abundance by calculating the percent cover.

C. Trees and saplings

Construct a 100-m^2 circular quadrat at each sample point by standing on the point, holding one end of a 5.64-m rope. Using the rope as the radius of a circle, locate all trees within the quadrat that are 30.5 cm in circumference or larger, and record the species and circumference of each. To measure saplings, repeat the protocol, using a 1.79-m rope as the radius of a 10-m^2 circle and measuring tree species that are less than 30.5 cm in circumference and greater than 30.5 cm tall. Use the data to compile a species list, and to describe the density and abundance of trees and saplings.

Assuming the sampling points are distributed in such a way that they account for site variation, the data should provide a good idea of the status of a restoration. ∎

Composition: For each of three layers—ground layer, midstory shrubs, trees
- Number of native species present
- Number of target species present (found on reference list; see Chapter 8)
- Number of non-native species present
- Number of pest species present

Structure: For each of three layers—ground layer, midstory shrubs, trees
- Distribution throughout the forest

Structure: Midstory shrub layer
- Abundance

Structure: Tree layer
- Abundance
- Which species are dominant
- Which species are reproducing

FIGURE 9.7. **Vegetation Monitoring Criteria for Deciduous Woods.**

possible. It is helpful to take several pictures from each point, capturing wide and narrow views of the same landscape scenes.

You can use photo points both to track specific situations covered by the project objectives—for example, documenting the horizontal extent of a population of specific plants—and to represent the range of onsite variation. Trail intersections, peaks and depressions, and vegetation ecotones are good choices from which to document changes through time. The trails provide good site access and a scale reference; high and low points are good locations for wide views, and ecotones are places where change is readily apparent.

In the early years of a restoration, it is helpful to take photographs at least once each season. Changes often occur rapidly at this time, and young restorations are also most prone to impacts from a variety of disturbances. After the project has achieved the original restoration objectives, photos taken every 3–5 years will likely provide enough information to track site dynamics.

Most photographers now use digital imagery. Digital images have many advantages, but questions remain about how well the digital media last over time, and how retrievable the images will be if hardware and software continue to change as rapidly as they have over the past decade. Given this state of flux, it is helpful to include a combination of black-and-white, color, and digital images, in both electronic and print forms, if resources permit. Prints on archival paper are the most durable choice. Such prints are estimated to last at least 100 years, with proper storage (see Chapter 3).

Aerial Photography

Aerial photography is particularly useful for tracking the location of communities and species, and for noting structural changes through time. The level of detail you can interpret depends on the scale and the resolution of the imagery, as well as on the time of year during which you capture the images. For most restoration monitoring purposes, a scale of 1:5000 is acceptable, though 1:1000 would be ideal. Photos taken during the times of year when deciduous trees retain their leaves are especially good for determining upland community boundaries (e.g., between forests, savannas, and prairies). For some wetland systems in seasonal climates, it is helpful to have images taken with trees in leaf-off condition and at a time when water levels are high. Under these conditions, the reflectivity of the water can really stand out. Black-and-white images often show crisp details of vegetation structure. In some temperate zone systems, color or color infrared images taken in early spring can be used to map the location of cool-season pest species, which "green up" earlier than the natives

As discussed in Chapter 4, aerial photographs are available from a variety of sources. You can obtain regional aerial photographs for most of the agricultural areas of the United States from the U.S. Department of Agriculture's Farm Service Agency's Aerial Photography Field Office (APFO). Historically, new images were available at intervals of 5–10 years, making this source of information most useful for long-term monitoring. A recent reorganization, however, has resulted in an initiative, the National Agriculture Imagery Program (NAIP), to provide digital images on a basis of 3–5 years. The NAIP provides 1- or 2-m resolution images at scales of up to 1:5000, taken during the growing season (leaf-on condition).

Maps are increasingly available on the Internet. For example, Google Earth now allows you to view both contemporary and historic air photos online for many parts of the globe.

You can also contract with a local aerial photography service to customize the timing, image type (natural color, black and white, infrared, digital), scale, and resolution of the images. Some people are attaching small cameras to model airplanes, kites, balloons, and even long poles to create bird's-eye views of restorations.

9.2.2 Vegetation Sampling Design

Some restoration objectives concern particular species and/or certain site locations. In these situations, your monitoring protocols may have a narrow focus. For example, a restoration plan might specify that the site maintain a population of at least 15 blooming individuals of a particular orchid species. The protocol would likely be to locate the orchid plants wherever they are found onsite and count those that are in bloom. Or a plan might specify that a site remain free of particularly problematic pest species. In this case, the protocol could call for annual qualitative walk-through surveys with the explicit aim of looking for those species (see Chapter 5). In both instances, the data you collect will be limited. In many situations, however, restorations have plant community–based objectives in order to describe the vegetation as a whole, with regard to, for example, plant diversity, or spatial structure and organization. In these cases, especially on large sites, you will likely use a sampling system.

In designing a monitoring plan sampling protocol, in addition to specifying what information to collect and the particular data collection technique to use (quadrats, transects, etc.), it is important to decide where to place the sampling points, and how many to include. Also, because monitoring by its very nature extends for a number of years, monitoring plans using quantitative sampling techniques need to address the issue of whether to use permanent sampling locations or to relocate the points with every visit.

Permanent Sampling Locations

Permanent sampling locations are often the preferred choice because they have one very important advantage. By repeating the data collection at the same location through time, any changes that are noted are quite likely to be real. This makes action decisions based on change (or the lack thereof) relatively clear. The use of permanent sampling locations also has some disadvantages, however.

One potential problem is that, unless the points are well marked, it may be difficult to find them if the interval between sampling times is large, such as 5–10 years. Since the logic of the protocol relies on repeatedly collecting data from exactly the same spot, even slight variations in location can introduce errors in the data, and thus in their interpretation. To avoid loss, make sure that any markers you place in the field can withstand climatic factors, such as wind, frost heave, snow cover, flooding (and, in our experience, snowmobiles), and management techniques, such as mowing or fire. In many situations, the markers also need to be both conspicuous enough for you to find easily, and designed to withstand damage from people and/or wildlife. One widely used marker is a length of iron rebar pushed into the ground, either at the exact location of the sampling point, or at a known distance and direction away from it. You can fit each rebar with a labeled tag to make identification even more certain.

In addition to marking the points in the field, you must also record their locations in a database. In all but heavily forested situations, a GPS is a very useful tool for this purpose because it lets you record spatial geographic coordinates. Depending on the sophistication of the GPS system used and the climate and satellite conditions at the time of the sample, your readings can be accurate to within a few centimeters. Even less precise readings can direct you near enough to the sampling points to be able to find them quickly, using landmarks and, in the case of metal markers, a metal detector.

Some issues remain even if you can easily find the permanent sampling points each time you conduct a monitoring survey. One problem is that over time, locations can be compromised, for example, due to changes in land use or accidental destruction. A case in point occurred in Curtis Prairie at the University of Wisconsin Arboretum, where a number of permanent plots were covered over by dredge spoils from a retention pond built to control stormwater flow from offsite development. The problem is that the reliability of the data may depend on the number of sampling points used. If too many are lost over the years, the accuracy of the information can suffer.

Another issue with permanent sampling points is whether the locations accurately reflect the composition and structure of the community as a whole. If the locations of the points turn out to over- or underrepresent certain species or microhabitats, the decisions you make based on the sampling data may be in error. The factors involved in determining placement include site variability, the type of information specified in the restoration objectives, and the monitoring resources available.

Two approaches are commonly used. The first is to place sampling points in targeted locations, such as areas that you believe (1) are typical (representing the community as a whole), (2) contain species that are of special concern, and/or (3) are especially vulnerable to impacts from human use or cross-border influences. The second approach is to place permanent sampling points at regular intervals throughout the restoration, creating a grid of sampling points (as in the previous case study). In this way, all portions of the site are included, a situation that can prove useful if changing circumstances (e.g., an offsite land use change) bring previously remote areas into prominence.

Selecting targeted sampling locations is particularly useful when a site has areas that are relatively homogeneous, and/or you can indentify potential issues in advance. On the other hand, such points may be difficult to locate if you visit them infrequently. Spreading the samples in a regular pattern makes it relatively easy to establish and relocate the samples, making this approach attractive, particularly if your monitoring plan calls for long intervals of time between data collections, or if turnover among your project personnel is frequent. The ideal solution is to use both.

Repeated Random Samples

An alternative to using permanent sampling points is to randomize the sampling locations for each survey. This approach avoids the issue of the loss of permanent sampling locations over the years, and it is often faster to implement if relocating permanent sampling locations proves difficult.

One way to locate random points is to establish two perpendicular baselines at the borders of the site to represent x and y axes. Then, knowing the lengths of these lines, you use a random numbers generator to determine the x and y locations of each sample. This will result in a situation in which the location of each new sample is independent of the locations of the previous ones—the definition of a random pattern.

Sometimes a random sample will cluster the majority of the points in one part of a monitoring unit, leaving the rest of the vegetation unexamined. In order to be sure that sampling data represent the entire unit, you can use an approach that ecologists refer to as a stratified random sample. This process involves dividing the area into subunits, and randomly establishing a sampling point within each. For example, you could superimpose a virtual grid of 10- × 10-meter squares over the monitoring site. Then, using the sides of each grid cell as x and y axes, choose pairs of random numbers to locate one or more sampling points within each cell.

With this technique, the locations of the samples are randomized separately for each survey; therefore, you rarely collect data from the exact same place more than once. Due to the spatial variation that exists even in the most uniform sites, it is likely that the results of two random surveys would not be exactly the same, even if you conducted both on the same day. In order to make decisions using monitoring data, you need to be confident that any vegetation change you detect represents actual temporal change. One way to accomplish this is to establish thresholds; in other words, for each measurement, set the magnitude of change that is sufficient for you to accept as being real.

Determining the Number of Sampling Points

The number of sampling points to include in a monitoring protocol is another important factor that contributes to your ability to track "real" change. The more locations you sample, the more likely you will represent site variation, whether you are using permanent plots or randomized samples. Exactly how many sampling points you need depends on the variability of the vegetation, as well as on the desired level of detail specified in the restoration objectives. For example, you will probably need more sampling points if you have to take measurements about all the species present on a site than if you only need information regarding the common species. And the number of samples may differ if you have to collect cover measurements as opposed to presence-absence data. More samples, of course, means additional costs.

How can you determine the number of sampling points you need in order to make informed decisions without wasting resources by collecting too much information? One place to start is by searching the academic and professional literature for conventions—that is, the sample sizes others in the field are using. The reported successes of the sample sizes used in previous studies involving similar plant communities can provide you with good estimates of the number of samples to include. Another approach is to conduct an initial survey using a large set of sampling points—say, 60—and then summarize the results using a series of subsets—such as 10, then 20, then 30, and so on—until the results change little with the addition of more samples. A more sophisticated approach is to conduct a "power analysis," a statistical tool that allows you to, among other things, select an optimum sampling size for avoiding sampling error. In this case, the goal is to determine the sample size needed to detect biological rather than sampling variability. Power analysis software programs are readily available.

▶ 9.3 ESTABLISHING MONITORING TIMELINES

Once you establish all your data collection protocols, the next step is to develop one or more timelines that schedule these activities well into the future. The protocols specify what is to be monitored, where, and how; the timeline establishes when monitoring will be conducted. More specifically, you need a set (collection) of timelines, as the protocols for measuring different resources often differ. For example, if a restoration objective concerns the density of active songbird nest sites, you will need to monitor in the nesting season. If an objective on the same site concerns the flower production of a fall-blooming plant, the timing of your data collection will likely be different.

As mentioned at the start of the chapter, monitoring plans often include several sets of timelines—one to cover the implementation phase of the restoration, the other to cover the management phase, for instance. In either case, you can measure time in months, days, or years, or in terms of benchmarks, such as "achieving a tree canopy cover of >60%." Ideally, you design the timeline so that a site manager can tell at a glance what is expected at any given moment.

CASE STUDY · The George Washington Carver National Monument Monitoring Timeline

George Washington Carver National Monument, the boyhood home of the famous scientist and educator, is located in southwestern Missouri, and managed by the U.S. National Park Service. Part of the site is being restored to the natural vegetation of the region, a mixture of grasslands and forest communities. A master plan prepared in 1999 includes several vegetation units (A–F), each of which has a set of unique goals and objectives. The timeline proposed in the monitoring plan includes the following recommendations:

1. At least once every 3 years (preferably every year if resources permit) at designated times throughout the seasons, take a set of 6 photographs at each permanently marked photo station, using the specifications included in the site plan.

2. At least once each year, walk through all of the monitoring units and record the general state of the restoration, taking note of anything that might trigger management interventions—for example, the appearance of a new pest species. (In some situations, this reconnaissance should take place at least once a season.)

3. Conduct detailed walk-through inventories of a few units each year, listing the species that are present and describing the degree to which the restoration goals are being met. If necessary, in order to take into account different blooming periods, visit the units several times during the year to complete the inventory. Schedule these detailed surveys so that each unit is visited at least once every 5 years. Only a few units need to be inventoried each year (Table 9.4).

4. Conduct a quantitative sample in each unit at least once every 5 years in order to evaluate how well the restoration is achieving the measurable objectives set forth in the site plan. Use a sampling scheme appropriate to the restoration objectives. For example, conduct both an overall assessment of the communities within the unit, and targeted assessments to address a specific issue, such as the rate of mineralization. ∎

Table 9.4 · Vegetation Monitoring Timeline

The following vegetation monitoring activities should occur each year:
- Spring and fall walk-through reconnaissance of all vegetation units.
- Seasonal photographs taken at scheduled photo stations.
- Walk-through monitoring surveys of scheduled units.
- Quantitative sample of scheduled units.

Site	2009	2010	2011	2012	2013	2014	2015	2016	2017	2018
A	Q		W			Q		W		
B		Q		W			Q		W	
C			Q		W			Q		W
D	W			Q		W			Q	
E		W			Q		W			Q
F	Q		W			Q		W		

Site A: Southeast Prairie Unit. Site B: Northeast Prairie Unit. Site C: West Prairie Unit.

Site D: Southwest Woodlands. Site E: Streamside Forest. Site F: Northeast Woodland.

Q = Quantitative sample, following plan protocols.

W = Qualitative walk-through survey, following plan protocols.

▶ 9.4 ANALYSIS AND STORAGE OF MONITORING DATA

Determining how to analyze the collected data is a decision that goes hand in hand with determining which data collection strategy to use. In essence, the objectives drive both decisions, as the specified measurements guide the form of the analysis and thus the data collection method.

If one of the purposes of the restoration is to support a breeding population of cranes, for example, one of the objectives might be: "Maintain a minimum of 3 breeding pairs of cranes." The monitoring technique could be to walk the site at a specified point in the breeding season, checking for nesting activity, and the corresponding analysis would be a count of the number of observations made.

If a goal is to maintain a high level of tree species diversity, the corresponding objective could be: "Maintain a Shannon Index number >3.0." The Shannon Index measures diversity by taking into consideration both the number of species present and the number of individuals that belong to each (see Sidebar). Therefore, the data collection strategy would

---

Sidebar | Comparative and Composite Measurements of Vegetation

In addition to the quantitative descriptors of plant communities listed in Figure 9.4, a number of comparative and composite measurements can be useful in establishing restoration objectives, and therefore monitoring protocols. The concepts of being "rare" or "common" are two familiar examples of comparative measures. The terms have meaning only relative to each other; a rare species is low in abundance or found in fewer places than a common species. However, without more information, it is not possible to tell if a trees species that is present on a site with a density of 30 trees per hectare is rare or common at that site. Many restorations are designed to include both rare and common species; to be clear which is which, the objectives specify abundances for each species. Composite measurements, as the name implies, use two or more individual descriptors in combination. For example, diversity-rating indices, such as the Shannon Index, combine measures of species richness and evenness in calculating a single rating value:

$$H' = -\sum p_i \log_e p_i$$
where p_i = the proportion of the ith species

John Curtis used a number of comparative/composite measurements to develop community/ecosystem models of the composition and structure of the "natural" communities of Wisconsin (Curtis 1959). These are often used to create restoration objectives for projects in the midwestern United States (see Chapter 7). If data are available, they can be applied to any community worldwide:

Importance value (IV): The average of relative frequency, relative density, and relative dominance using a 0–100 scale. Importance values can also be calculated using only relative density and relative frequency, and in other ways.

a. **Relative density**: The percentage of all individuals in a sample belonging to a given species.

b. **Relative frequency:** The frequency of a species divided by the sum of all frequencies in a stand, expressed as a percent.

c. **Relative dominance**: Used for trees, the percentage of total basal area contributed by a species. Basal area is the cross-sectional area of a trunk at breast height (1.4 m).

Prevalent ("representative" species): The species most likely to be encountered in a community, based on being found in many different examples of a community type; high presence percentage.

Modal species: A species is modal to a community in which it achieves its highest presence percentage.

be to collect information on not only the number of different tree species present, but also the number of individuals of each species. One approach could be to use a quadrat sample.

No matter which analysis tools you use, it is essential to document exactly how to gather and analyze the data—that is, what sampling equipment and statistical software packages to use. This kind of information becomes particularly important as, with the passage of time, staff turnover occurs and new thinking or technology renders old statistical tests and computer programs obsolete.

You also need to allocate sufficient time and resources to allow for analyzing the data. All too often, raw data accumulate and are left unanalyzed and neglected in a file, thus failing to serve their purpose.

Similarly, it is essential that the plan include specifications on record keeping and data storage (see Chapter 3). The specifications should cover:

1. The kinds of information that need to be kept (original field notes, data summaries).
2. The format that will be used (digital or paper, pressed specimens or photographs).
3. How the information will be indexed.
4. Where the information will be stored.
   a. Location (one place; multiple copies in different locations).
   b. Type of facility (herbarium cabinets, temperature-controlled facility for photographic images, filing cabinet, specific computer, etc.).

Data storage approaches will, no doubt, change with time. It is important to develop a procedure whereby the old records are archived and/or converted to the new systems as they are introduced. This could mean using original raw data to perform new analyses, for example.

▶ 9.5 WHO DOES THE MONITORING?

Of course, it takes people to actually implement a monitoring program, and no plan is complete without a discussion of who will do the monitoring. There are several things to consider, including:

1. Who will be in charge of the monitoring program.
2. The kinds of skills that are required to implement the different sections of the plan, and how much of any required training can be provided by the project managers.
3. Whether the monitoring will be carried out by permanent employees, by people hired on a short-term basis, by volunteers, or by some combination.

It is helpful to include a section in the monitoring plan to address the answers to these questions. At minimum, your plan should specify the kinds of skills and training required to collect and manage the data according to the management protocol. It is best if you also include an organizational structure that describes who is in charge of what.

The biggest advantage of using full-time or part-time staff members to conduct a monitoring program is that they are likely to be very familiar with the site and with the purpose and goals of the restoration. In addition, unless the turnover rates are high, staff members will likely be able to participate in the program over a number of years. Experience with the site, the project, and the monitoring techniques leads to monitoring efficiency and accurate data collection and analysis.

Maintaining a staff can be expensive, especially if the monitoring plan calls for a wide range of expertise. Therefore, your organization may choose to hire firms that specialize in monitoring and management procedures to collect and/or analyze the data. In such cases, it is particularly important that you have a detailed monitoring protocol so that nothing is overlooked or misinterpreted.

Volunteers can be invaluable members of a monitoring team, as long as they are adequately prepared, trained, and integrated into the project (see Chapter 13). Volunteers are often enthusiastic and tireless workers. However, because they are not on the payroll, it is easy to overlook the fact that they are not employed without cost. Someone needs to coordinate their services, and provide any necessary training and ongoing supervision. Training is particularly important for volunteer monitors, as many of the protocols call for sophisticated skills. Sometimes an experienced volunteer can fulfill this role in fine fashion. This can be an excellent fit for energetic retirees. However, it is often necessary to devote staff time to developing a volunteer program that will be an asset, rather than a liability.

In restorations that are open to the public, observations made by site visitors can sometimes greatly aid monitoring efforts. For example, visitors can be asked to report sightings of birds or other wildlife, or acts of vandalism, or the locations of storm damage. Many visitors are very enthusiastic to participate in protecting a restoration in this way. Doing so makes them feel part of the project, and raises their awareness of conservation issues.

▶ 9.6 THE MONITORING BUDGET

No monitoring plan is complete without a budget—an itemization of the costs of the required resources for carrying out the monitoring protocol into the future. Ideally, a monitoring budget informs, and is shaped by, the restoration planning process in a mutually iterative way. A monitoring budget includes line items for administrative and supervisory time, labor, materials and supplies, and expenses, as well as data analysis, maintenance, and storage. The time and labor line items should include the time the regular staff spend on monitoring and the staff resources needed to train and supervise the work of volunteers. Depending on the organizational planning context, the typical monitoring budget will either describe and itemize the project's anticipated and needed expenses or, on the other hand, prescribe acceptable expenditure limits. In the latter case, the monitoring plan may need to be modified to accomplish work within these limits.

The length of a budget cycle will vary, but it typically runs for 12–24 months, necessitating a renegotiation of the budget (and perhaps the monitoring plan) with each new fiscal year. You can make adjustments to the monitoring budget because of unanticipated costs or savings during an annual budget review.

It is important for the monitoring budget to include a section that addresses planning for the long term. If all goes well, restorations will survive for many years and require periodic monitoring. You should project the budget 5–10 years into the future, following the monitoring timeline, to estimate and plan for future resource needs. Based on these projections, you should recommend procedures to follow in order to ensure adequate long-term funding.

KEY CONCEPTS

- Monitoring provides information necessary to determine how well a restoration is meeting project goals and objectives at any point in time.

- A monitoring protocol addresses what information is to be collected, when, where, and how often. The plan specifies the techniques to be used and who collects the data; it explains how to interpret, communicate, and store the information.

- The specifics of a monitoring plan are unique for each project, and are based on the objectives found in the site, implementation, and management plans.

- Vegetation monitoring protocols include qualitative and/or qualitative approaches and community and/or population measurements.

FOOD FOR THOUGHT

1. Site inventory and monitoring use similar techniques. In what ways do they differ?

2. You are in charge of creating a monitoring plan for a newly restored 40-acre marsh/sedge meadow complex surrounding a shallow pond. The area was protected in part because it contains a population of rare orchids, and because it is an important stopover area for migratory waterfowl and a nesting site for sandhill cranes. The restoration objectives are as follows:

   - Maintain a minimum of 50 native plant species.
   - Maintain <5% cover of exotic species and <20% cover woody species.
   - Maintain the orchid population such that it is at or above the minimum viable population size.
   - Maintain quality habitat sufficient to support a minimum of 25 species of waterfowl during spring and fall migrations.

   Propose a monitoring plan for this site. Explain the reasoning behind your decisions. Is there additional information that you need?

3. Given that ecological theory and data collection and storage technologies continue to advance, what steps can a restorationist take to create a monitoring plan that remains relevant and useful over the life of a project?

4. Contact managers of restorations in your area. What kinds of monitoring plans, if any, do they administer? What are the monitoring issues that are of most concern?

5. Discuss the pros and cons of (a) qualitative and quantitative monitoring approaches and (b) permanent and repeatedly randomized sampling locations.

Resources for Further Study

1. If you are interested in learning more about the importance of restoration monitoring, you can begin by checking out the following two resources:

   Busch, D. E., and J. C. Trexler. 2003. *Monitoring Ecosystems: Interdisciplinary Approaches for Evaluating Ecoregional Initiatives*. Washington, DC: Island Press.

   Holl, K. D., and J. Cairns Jr. 2008. "Monitoring and Appraisal." In *Handbook of Ecological Restoration*, vol. 1, edited by M. R. Perrow and A. J. Davy, 411–432. New York: Cambridge University Press.

2. In most instances, the techniques used in monitoring programs are the same as those used in site inventory and analysis. If you are interested in learning about the tools used to investigate the biological and physical resources of a site, be sure to look at the Resources at the end of Chapters 4 and 5. In addition, take a look at the following sources:

   Morrison, M. L 2009. *Restoring Wildlife: Ecological Concepts and Practical Applications*. Washington, DC: Island Press.

   Sutter. R. D. 1996. "Monitoring." In *Restoring Diversity: Strategies for Reintroduction of Endangered Species*, edited by D. A. Falk, C. I. Millar, and M. Olwell, 235–264. Washington, DC: Island Press.

3. A number of organizations maintain excellent websites that share information about restoration monitoring programs. It is always helpful to review such precedents before initiating a restoration monitoring plan. Here are two aquatic systems examples, the first from Sacramento River Conservation Area Forum (California), a nonprofit conservation organization, and the second from the U.S. Environmental Protection Agency:

   Sacramento River Conservation Area Forum. *Sacramento River Monitoring and Assessment Project*, Collaborative approach to determining ecosystem health. http://www.sacramentoriver.org/SRCAF/index.php?id=sacmon

   U.S. Environmental Protection Agency. *Voluntary Estuary Monitoring: A Methods Manual*. EPA-842-B-06-003. http://www.epa.gov/owow/estuaries/monitor/

10

The Management Plan

LEARNING OBJECTIVES

After reading this chapter, you will be able to:

- Identify the components of a restoration management plan.

- Explain the underlying logic of a restoration management plan, and its place in the restoration process.

- Evaluate situations to determine whether management interventions are required.

- Plan your management actions.

- Use some basic vegetation management tools.

What is restoration management? Simply stated, it is a set of strategies and techniques restorationists use so that sites continue to meet the restoration objectives, once they have achieved them. Management begins as soon as implementation is over, in response to information provided by monitoring (Figure 10.1). For example, if the purpose of your restoration is to provide habitat for a particular species of butterfly, and the population size or the abundance of its preferred food plants declines to a level below that specified in the restoration objectives, you would likely decide to take steps to reverse the decline. Sometimes management involves taking no action (the "hands-off" approach). You might decide that the declines in the butterflies or food plants are part of an anticipated cycle and will self-adjust, or monitoring may reveal that no changes have occurred.

FIGURE 10.1. The Restoration Process: Management.

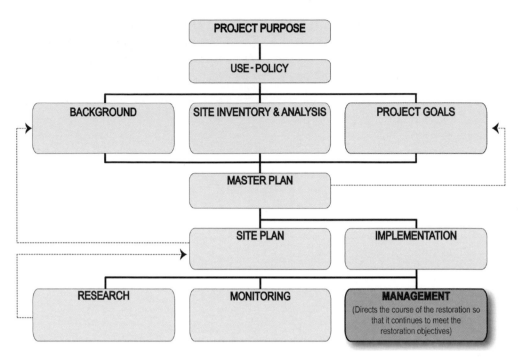

The management plan is your guide in making both kinds of decisions—to act or not to act. The idea is to maintain boundaries within which change can occur, not to prevent any change from happening.

▶ 10.1 THE NEED FOR RESTORATION MANAGEMENT

Why is management needed? Why not let nature take its course, once you have implemented the restoration? First, most restorations require management basically for the same reasons that restoration is needed in the first place—human impacts. Human activities can impact a site whether people visit it or not. Because of our past activities, restorations may face one or more of the following situations (see Chapters 1 and 12):

1. Sites consist of isolated patches that are constantly prone to direct and indirect human influence.

2. Landscape- or ecosystem-level processes may be missing or changed.

3. Plant and animal populations critical for restorations are missing, because they require more space than the size of a site, are not able to coexist with people, or have become extinct.

4. Pest species are widespread both inside and outside the restoration site.

5. Sites are subjected to the effects of global climate change that will vary with location.

Second, natural communities/ecosystems, with or without the presence of people, are dynamic and constantly changing (see Chapter 2). If these changes move a restoration away from the project objectives, intervention may be needed.

It should not be the goal of management to prevent or respond to every ecosystem change, but only to those changes that have the potential to cause a restoration to stray from the project objectives. How do you identify potential undesirable changes in your restoration? You can refer to several sources of information to help you decide:

- The community/ecosystem model that was used in establishing the goals and objectives for the site plan. You can use the model to identify the kinds of stressors the restoration may experience, such as the nature of the impacts stemming from the situations metioned above.

- The original site inventory and analysis, especially with regard to the site context and therefore the type and magnitude of cross-boundary influences.

- The degree of flexibility called for in the restoration goals and objectives.

- The nature of the site use-policy, and the potential for and nature of continuing onsite human impacts. These use impacts are also restoration stressors.

To understand how, over time, impacts to a restoration site can cause changes to ecosystem structure and function, consider the case study that follows.

CASE STUDY   Management of a Pine Forest Restoration

Impacts of water and resulting erosion are some of the most common restoration ailments. In one large pine forest restoration, a sparse herbaceous ground cover and a thin shrub layer, combined with steep slopes, have resulted in sheet erosion and gully formation during rain events. At this site, many trails and service roads run straight down the slopes and have become badly eroded, exposing tree roots and causing safety hazards. Erosion is especially bad on one popular trail segment that connects northern and southern portions of the forest separated by a highway.

The planting site is the 17-ha (40-ac) Leopold Memorial Forest, an attempt begun more than 60 years ago at the University of Wisconsin–Madison Arboretum to re-create a red pine (*Pinus resinosa*) and white pine (*Pinus strobus*) forest. Although the Madison site is several hundred kilometers south of the normal range of white pines in Wisconsin (with the exception of naturally occurring southern relict stands in cooler microclimates, such as gorges and ravines), the trees have thrived. However, typical northern pine forest wildflowers never established, despite repeated plantings.

Various small stormwater erosion channels run through the restored forest toward a low point in the landscape near a highway. At this point, the eroded sediment is carried under the highway via a culvert and deposited in the northern, or low-lying, portion of the pine forest. Here, the water has created a deep, braided erosion channel through the woods. If that were not bad enough, the storm water carries seeds of herbaceous pest species that thrive in the runoff's excess sediment and nutrients. The pest species have become well established, represent an obstacle to restoration, and serve as a continual source of reinvasion throughout the site.

Management goals for the troubled forest include:

- Reestablishing locally appropriate and sustainable native shrub and ground layer species with thick canopies and abundant basal leaves to protect the soil from rainfall.

- Remedial planting of species whose dense root mass can protect against soil erosion.

- Remedial planting of other species whose deep root systems can infiltrate water.

- Rerouting of the existing footpath away from steep slopes.

- Creating a bioinfiltration basin to capture sediment and water before it is carried to the receiving lake.

For more on this story and to read how the restoration team analyzed the situation, see the Sidebar. ∎

Sidebar | Pine Forest Management Planning

Faced with the complex of management problems in the case study of soil erosion in the pine forest restoration, where should a manager start, and how should she proceed? The managers in this case are taking the following approach. They have defined the situation as a combination research problem, a need for immediate action, and an opportunity to encourage involvement from volunteers and local university students.

To begin, the researchers determined what they knew, what they didn't know and would need to learn to solve the problem, and what were the constraints for getting the answers. Specific questions for this site included these:

- Are "tools" (meaning the appropriate plants) available for slowing surface-soil erosion but need to be deployed? Or do the tools need to be developed?

- Is it known that pest species are the cause of sparse ground cover or a symptom?

- Will removing the pest shrubs destabilize the soil even more?

- Is it known that increased ground cover would slow the erosion?

- What, if any, ground covers can the site support following shrub removal?

The restoration team identified some possible constraints that might indicate if action or no action right now is the best answer for the time being.

- Is the problem so large and the need so urgent that immediate action is required based on existing knowledge?

- Is the area large enough to support an experimental approach (i.e., room for replicates), or should various alternatives be tested, using the results to make gradual improvements?

- Given everything else the restoration team is working on, where does this project fit on the priority list?

FOOD FOR THOUGHT

1. Can you think of other "unknowns" that need to be answered? Are there opportunities here the team has not mentioned?

2. Can you think of other constraints? What would your restoration team need to know to address the constraints? What would you recommend in this situation?

▶ 10.2 FEATURES OF THE MANAGEMENT PLAN

Management begins when the objectives of a restoration have been achieved. However, management planning begins early in the restoration process—ideally, at the same time as you put the site and implementation plans together—well before management interventions are required. In this way, your site design can include provisions for potential management needs, neighbors and other interested parties will be prepared when actions are taken, equipment can be gathered, and the management team will be ready for action.

As is true for the site, master, and implementation plans, the management plan includes text and graphics. It contains a set of management goals that are based on the site plan objectives, a set of if/then strategies for continuing to meet those goals, and built-in strategies for review and making modifications. There are five basic steps involved in developing the management plan:

1. Anticipate and identify potential future problems—ways in which the site could begin to deviate from the restoration goals and objectives—due to potential impacts from inside or outside the restoration, or to natural dynamics (stressors). In so doing,

    determine action thresholds—site conditions indicating that a management activity may be needed.

2. For each process or problem identified, consider and evaluate several potential alternative strategies or actions. Always include a "hands-off" alternative, and use an adaptive approach to select the desired alternatives.

3. Develop a management prescription for each of the chosen actions.

4. Identify management units, and link management prescriptions to each one.

5. Develop a budget and procedures for reviewing and updating the plan.

10.2.1 Anticipating Future Problems

The key to developing a management plan is for you to anticipate the kinds of unwanted changes that will most likely occur, given the nature of a restoration, and to plan what you will do when they happen. A good way to proceed is by establishing a series of if/then scenarios: The potential problem is the "if," and the possible solution is the "then." You can start by developing and answering a set of questions. For example:

- How might the composition and structure of the onsite communities change through time in the presence or absence of such things as fire cycles; water cycles, including seasonal or periodic floods, saturated soils, and droughts; grazing activities; microscale soil or canopy disturbances; and regular immigration or emigration of individuals or species?

- What kinds of impacts have been noted as affecting area remnant communities of the types found in the restoration?

- What onsite or offsite activities affect or have the potential to affect the restoration site?

- What pest species are known to be in the region and are likely to establish in the restoration?

- How stable are the land use practices in the region?

The answers to all of these questions are important. The first question relates to what we discussed in Chapter 2—the fact that communities and ecosystems are dynamic and variable in space and through time. Events such as fires, floods, and windstorms are important drivers of change, as are species interactions such as competition, changes in microclimate brought on by the inhabitants of a site, or the movements of individuals from one site to another. In some cases, such events create disturbances that move a restoration away from its goals, as, for example, when a fire destroys the canopy of a forest restoration. In others, they lead to stability, when, for instance, a fire rejuvenates the grasses and prevents tree colonization in a grassland.

    Because modern human activities have substantially altered the occurrence and scale of these processes and interactions, one of the major roles of management is to reinstate them in the restoration or to mimic their effects. We will discuss two examples later in this chapter—prescribed fire and grazing. Your community/ecosystem model will help you decide whether to introduce them or prevent their occurrence.

    This is also the time for you to set management action thresholds—the conditions under which a management action may need to be taken. For example, in a grassland you could establish a litter depth that, when reached, would trigger a prescribed burn. Or using our earlier butterfly example, the threshold might be set in terms of a population size estimate. Keep in mind that as you restore a community/ecosystem, you will gain practical experience that can be applied to the process, and that other "future changes" might be increases in your knowledge about how the system functions.

10.2.2 Considering Alternative Strategies

For each of the if/then scenarios, it is important to consider the pros and cons of more than one strategy before taking action. In deciding which option to use, you should consider how successful it is likely to be; its potential positive or negative affects on nontarget species or community/ecosystem components; and the costs of the labor, equipment, and supplies involved. Here are some typical questions: Is enough known about how the system works for you to design an effective action? Are you certain that the planned cure will not be worse than the problem? Do you have the time, resources, and experience to take action now? Will the action affect neighbor relations, and, if so, what it would take to address peoples' concerns?

It is also important to examine each technique in the context of any policies or regulations that apply to the site, as well as potential user conflicts. For example, it is common for municipalities to place restrictions on conducting prescribed management or research burns or the use of herbicides. By thinking through these issues in advance, you can avoid many headaches (e.g., using fire in a fire-sensitive neighborhood).

After considering the pros and cons of various alternatives, you may find that there is no clear-cut best choice. In such cases, it is a good idea to use the adaptive restoration approach and establish a small-scale experiment or field trial to determine which technique performs best on that particular site (see Chapter 3).

One of the strategies you should always consider in approaching a management problem is the nonintervention or "hands-off" approach. Choosing to take no action is a different strategy from "letting nature take its course," unless that is the explicit goal of the restoration. Even in situations where human impacts are common or the restoration requires management-induced disturbance cycles, it is always important to consider what would happen using a hands-off approach. By comparing more active strategies against nonintervention, you clarify what you hope to achieve. In essence, you are asking the question: What is the likelihood that the problem will worsen, remain the same, or diminish without intervention?

To some extent, including a hands-off option is meant as an exercise to think through whether any successional change or pest species invasion, or even a management intervention, is likely to push a community off course in relation to the restoration goals. This type of consideration is similar to that required by Environmental Impact Assessments, in which you discuss what would happen to the site if the proposed project were not carried out.

You should choose the hands-off option only after careful deliberation. Situations for which this is the best choice might include communities where disturbance cycles have very long return times, or no action thresholds have been reached to date.

10.2.3 Developing Management Prescriptions

A management prescription specifies the conditions under which you implement a particular management strategy. It includes a description of the management threshold—the resource conditions that trigger the response—and the anticipated outcomes (goals). The prescription contains a detailed set of instructions for when (time of year, weather conditions, time of day), how, where, and by whom management is to be accomplished. The timing of management activities is particularly important for achieving the desired results, while avoiding collateral damage. For example, walking on saturated soil in pursuit of a pest species may compact the soil, trample desirable vegetation, and create erosion rills that turn into gullies. Therefore, a management prescription details not only when to initiate a management activity, but also when not to carry out a management activity.

10.2.4 Identifying Management Units

Especially on large sites—those comprising 4 or more hectares—it is helpful to divide the area into management units. These are sections of a restoration site that are managed differently from one another, either in terms of the tools used, or in the timing of conducting a management prescription. The units are often based on patterns of vegetation, topography, or cultural features, such as roads and trails. You might do this, for example, if the resources affecting a management prescription are unique in particular areas of the site.

The boundaries of the units may coincide with the community boundaries identified on the site plan. Often, however, management units are smaller or larger than this. For example, you may decide to divide a grassland into several different management units with the goal of conducting prescribed burns at different times in each. This can be important to avoid impacting all individuals of a single plant or animal species at once. Or you might create a management unit that includes a cross-section of a number of different community types in order to mimic natural disturbance patterns that do not stop abruptly at such boundaries.

Management unit boundaries can also be fluid and vary with the management treatment (e.g., fire, grazing, flooding, cutting) and the management target (e.g., pest species, fuels, or stem densities). In other words, you may create different sets of units, one for each of several strategies aimed at managing different goals. And for some activities, management units can also be combined. It is a good idea to establish the smallest divisions at the outset; it is generally quicker to combine small units than to divide a large one.

Regardless of the method used to identify the units, you must mark, record, and map the boundaries using land survey techniques. Include the maps, together with careful descriptions of the boundaries of the units, in the management plan document. You may decide to consolidate or break the management units apart over time as the conditions change. It is important to document these changes so that the management history of each part of the restoration site is preserved.

10.2.5 Developing a Budget and Management Plan Review and Updating Procedures

The restoration management plan also includes procedures and timetables for establishing and modifying the management budget, as well as for the periodic review of the plan itself.

The management plan that you write while developing the site plan is forward-looking. It is intended both to help you design the restoration with management in mind—such as by including firebreaks when restoring a fire-dependent community—and to anticipate the kinds of equipment and personnel that may be required for the long-term stewardship of the project. Of course, the actual expenses incurred from year to year will depend on which of the if/then scenarios that you explore actually occur. The budget focuses on which management prescriptions you decide need to be employed. Therefore, the management plan includes a process and framework for building and approving a budget on a regular basis—ideally, once a year.

The development of management prescriptions will inevitably generate adjustments to the budget as you discover that additional tools, personnel, and/or time are required to carry out the work. Because of this, it is particularly important that you develop a straightforward and transparent budget process to handle new budget requests.

It is essential to create a procedure for reviewing and modifying the management plan and updating procedures. Of all the restoration plans, this is arguably the one that will be most dynamic, as it is intended to be in active use for many years. It is helpful to conduct a formal review of the plan every 5 years or so, and to revisit it any time a stressor changes.

▶ 10.3 VEGETATION MANAGEMENT TOOLS

Tools are the instruments with which you carry out a management strategy. There are many different management tools, and most of them are very similar to those developed for other conservation practices, such as maintaining a sustainable timber harvest, stormwater control, the management of nature preserves, and implementing restorations (see Chapter 8). Information about management tools and strategies is readily available from natural resource agency websites, through networking at professional conferences, in books and journals, and by simply comparing notes with your professional colleagues.

In this section, we provide an overview of some of the most commonly used vegetation management tools. We have chosen to concentrate on vegetation because the management of almost all restorations involves the use of these tools. We will discuss another management situation common to most restorations—reducing the human impact by working with people—in Chapter 13. The goals of vegetation management may include the simulation or enhancement of disturbance cycles (e.g., buffalo wallows), the removal of pest species without harming desired plants, enrichment plantings in cases where species do not colonize successfully, remedial plantings to correct unexpected erosion problems, or dealing with the aftermath of storm events. Reducing the competitive abilities of dominant species, or increasing the vigor and reproduction of desired species, may also be vegetation management goals. Each of these calls for a different kind of tool. Vegetation management tools fall into four main categories: mechanical (pulling, cutting, mowing, etc.); chemical (herbicides); fire; and biological (grazing and other forms of herbivory). Before exploring these categories, we will discuss some of the general principles of using management tools.

10.3.1 General Principles of Tool Use

The effective and efficient use of any tool depends, of course, on choosing the right tool for the job and using it according to directions, and in the manner intended. In deciding which management tool to use, you have to consider (1) the management objectives, (2) the effectiveness of each tool for the purpose at hand, (3) the size of management area to cover, (4) the time available for implementation, (5) the difficulty of obtaining necessary use permits from local authorities, (6) the availability of resources and labor, and (7) the stage in the plant's life cycle (such as dormant, bud break, flowering). These variables combine to make a complicated management situation. The issues need to be addressed by using a planned management strategy, or prescription, rather than by scrambling to improvise a solution in real time.

Just because a tool, technique, or strategy works in one site at one time does not guarantee that it will achieve the same results again, at a different time at the same site or at a different site. This is because a whole host of variables can influence the effectiveness of a tool—and we cannot emphasize this point enough. For example, it is not uncommon for a restoration team to use the same tool, technique, and methods (say, brush cutting on the same target species), yet be puzzled when results (reduction in shrub cover) vary throughout the season. Upon further investigation, the team will discover that seasonal changes in plant physiology (growing season versus dormant season) produce different responses to cutting (resprouting vigor and depletion of root reserves). Trees and shrubs respond differently to being cut in May than they do in December. However, physiology alone may not explain all the various responses. Another source of variation is human error (inconsistent use of the tool), which can be corrected somewhat through training and supervision. Because working with people is a big part of the restoration equation, this is a topic we address in Chapter 13.

10.3.2 Mechanical Approaches

Mechanical approaches to vegetation management include pulling, cutting, girdling, mowing, brushing (simulated grazing), mulching, and digging (such as scraping away soil to remove pest species, sod, etc.). The tools range from hand trowels and weed wrenches to chain saws and larger machines, such as gas-powered brush cutters.

Pulling and Cutting

Restoration managers are often faced with the need to remove unwanted vegetation that colonizes the site from external sources, or that creates an overabundance (relative to the objectives) of species already on site. Pulling removes plants from a site, roots (at least in theory) and all. You can use hand pulling to control small herbaceous species and the seedlings of woody plants. Pulling is most effective on small groups of plants on small sites, and is only feasible on larger sites if you have enough workers and ample time. You can use a weed wrench, tractor, or bobcat and chain to pull larger material, such as saplings or shrubs. A weed wrench has a jaw that clamps around the base of a sapling or shrub with a diameter up to the maximum width of the jaw (about 6.5 cm). A handle acts as lifter arm that you manually pump until the roots of the plant break loose of the soil. Pulling will have a long-lasting effect only if you remove enough of the root system so the plant will not regrow. A drawback of pulling, in addition to being time consuming, is that it disturbs the soil surface, which may stimulate the germination of undesirable species present in the seed bank.

Cutting involves removing some or all of the aboveground portions of a plant. Cutting shrubs and trees can be efficient and effective for species that do not resprout, such as conifers. You can cut using a hand saw, power saw, and/or brush cutter. In addition to the time and labor involved in the actual cutting process, a major concern with cutting is disturbance to the surrounding environment and the expense of removing the cut material from the site, and disposing of it later. You will probably cause the least significant damage to a site if you cut when the ground is frozen or dry. In some instances, you can stack and burn the cut material onsite. In temperate climates, this is best done in the winter with snow on the ground, when the chance for an escaped fire is considerably less.

Girdling

Girdling is used in place of cutting to kill saplings or trees of species that have the tendency to aggressively sprout at their base, or sucker from their root system, when you cut and

FIGURE 10.2. Girdling. Bigtooth aspen (*Populus grandidenta*) were girdled to open the canopy during the initial steps to restore oak savanna that had occurred on this site prior to the suppression of fire. (Photo reprinted by permission from Steve Glass.)

remove the aboveground portion. The idea is that girdling prevents the hormonal trigger that tells a tree to send up sprouts after the trunk is removed. You accomplish this by removing the portion of the plant's vascular system, the phloem, found in the bark of a tree. Girdling removes a wide section of the bark around the circumference of the trunk (Figure 10.2) that provides food to the roots and transports hormones, without disrupting the water transport system (the xylem or "wood"). Girdling essentially starves the plant over time without triggering sprouting. The technique is done using a tool you can make yourself (see Sidebar).

Not all tree species appear to be susceptible to girdling, nor is girdling an effective strategy in all situations. It may be impractical, for example, in very large areas or dense stands. Proper timing is critical, as girdling is possible only for a few weeks early in the growing season when the bark is susceptible to the treatment. Time to death varies with stem size and, if the species is clonal, with the extent of the clone system. For example, to be effective, girdling of the clonal species trembling aspen (*Populus tremuloides*) requires that all trees and suckers in the clone are girdled or removed. Otherwise, the population will recover.

Mowing

Mowing is the act of removing the top portions of plants using a sickle, a scythe, or a machine with a rotating blade—a mower. Mowing is most often used in restoration management to prevent seed production and dispersal by herbaceous pest plants, or to simulate the effects of grazing by large herbivores when the restoration site cannot support the animals. (When used as a surrogate for grazing, the mown debris is removed from the site.) You can also use mowing to constrain the growth of young woody seedlings and small shrubs; in some cases, mowing is used in place of fire to remove litter in dormant grasslands.

The timing of mowing is a major factor in its effectiveness. To reduce resprouting and the regeneration of undesired plants, mowing is most effective when done at the time of leaf or flower emergence, particularly if it is repeated throughout the growing season and

Sidebar	How to Make a Girdling Tool

Restorationists need effective tools that are inexpensive, easy to make, and easy to use. One of the cheapest, simplest to make, and handiest is a girdling tool. You may want a variety of sizes to use on different trees, but one of the best girdling tools is one that slips in your back pocket—always there when you need it. Here are instructions on how to make your own.

Get a small but rigid piece of metal, 15.25 cm (6 in.) to 30.5 cm (12 in.) long. Knives will not work because they are too sharp and dangerous, and under pressure a pocket knife can fold in on your hand. Automobile spring leaves are popular because they come in a variety of sizes, are cheap, and readily available.

For girdling sapling-sized stems that are 2.5–10.1 cm (1–4 in.) diameter at breast height (dbh), a good tool can be made from an old butter or table knife. You can cheaply purchase butter knives from second-hand stores, and in an hour a craftsperson or handyman can fashion a dozen girdling tools, which look like small, flat chisels that fit in the palm of your gloved hand.

To cut off about one-third of the tip of the blade end of the butter knife, you can use metal shears or a grinding wheel. Use a grinding wheel to sharpen the cutting end of the remaining blade; it should be sharp enough to pierce tree bark. Next, you will want to put a small inward curve in the blade (think of the round stem) so that the tool slides more easily around the tree. Finally, to safely carry the girdling tool with you, fashion a scabbard out of a scrap piece of leather or an old glove. Or slip it in the scabbard of your pruning shears, and you are all set to go.

for several years following. The concept is that the continued removal of top growth will slowly deplete the resources of the roots and "starve" the rest of the plant. For many species, unless a mowing treatment is repeated or followed by a supplemental action of another sort (such as herbicide application), the density and distribution of the pest plant species may be greater after the treatment than if no action had been taken. (This is also true for most management that uses some type of cutting.)

Mowing does not "distinguish" between species, and desired species as well as undesired species are affected if their height is sufficient. Continual mowing can modify the growth form of some species as well. Instead of growing erect, the growth form becomes more prostrate, with the plant's height remaining below that of the mower blades.

10.3.3 Chemical Tools (Herbicides)

Herbicides are chemical toxins used to control or suppress unwanted plants either by killing them outright or by interrupting processes needed for their long-term regeneration and survival. If you use herbicides in management, it will most likely be to manage pest species (see Chapter 11). Their use in restoration management requires knowledge, skill, and sensitivity to the environmental and societal concerns that accompany their application. Herbicides can be highly effective and often require less time and lower labor costs than many other treatments. However, they also have great potential to damage the restoration and the surrounding environment. In addition, the long-term effects on people are unknown for many of the compounds. Therefore, it is critical that when you choose to use them, both the site manager and those who will be applying the chemicals have training in their safe use. Here are some basic good rules: (1) Apply the herbicide in the manner in which it has its greatest effectiveness, (2) understand its toxicity to humans and other nontarget organisms, (3) use the least amount of chemical to do the job, and (4) know and use the proper technique to avoid unintentionally spilling the chemical.

Herbicides are best used as a last resort—that is, on pest species that have few, if any, other controls—or when the effectiveness of other controls is limited or questionable, and when other controls require a significant time frame to be effective, during which damage to the restoration by the pest species would continue. Since the 1940s, hundreds of herbicides have been developed, mostly for agricultural applications. When selecting herbicides for use in restoration management, it is important to consider the chemical's mode of action, as well as its selectivity, volatility, movement through soils, toxicity levels, and duration or half-life.

Proper Use

It is essential that you read the labels on herbicide products for instructions on the proper use and handling. The herbicide label is a legally binding document, and it is your responsibility to adhere to the label's requirements. The label contains information about which species the product is intended to control (target species), application methods and rates, the environmental conditions under which the herbicide may be used, and safety precautions. One of the most important safety precautions is to wear proper clothing, which includes protective outerwear, gloves, a hat, and a respirator. This clothing is for your safety; it is designed to keep you from absorbing chemicals through your skin or breathing in the fumes.

Some government units require that you obtain a commercial applicator's license to purchase or use a specific herbicide. Many herbicides have been developed for agronomic use, and instructions for use in restorations may not necessarily be provided with these products. In such situations, you must contact the manufacturer and other area managers, or attend education programs on herbicides, to learn the proper way to use them.

Herbicides are available in concentrated form or in a ready-to-use (RTU) formulation. You will have to dilute concentrated herbicides. You can determine the most effective dilution for your purpose by reading the herbicide label, consulting with local experts, or conducting field experiments. Cut stump and basal bark applications generally require a smaller quantity of herbicide than foliar applications.

Public notification of an herbicide application is often mandatory, depending upon provincial, state, or federal regulations. Often this consists of notifying the owners of nearby properties in advance of the application, and posting signs on the borders of the site after the treatment is done. Before planning an herbicide application, it is important that you consult with the proper authorities for notification and posting requirements.

Choosing an Herbicide

The choice of herbicide depends on the species to be controlled, size of the area to be treated, whether the target species is a monoculture or grows interspersed with desirable vegetation, weather conditions, and the knowledge and skills of the people who will be applying it. Your choice also depends in part on the chemical formulation, whether the herbicide is selective or nonselective, and its mode of action.

Chemical formulation. Herbicides are sold under a trade (or brand) name. They contain an active ingredient (the actual control agent), a medium that facilitates its application (the carrier), and various additives. Surfactants, which enhance the herbicide's effectiveness, are one type of additive. Surfactants can help break water tension, in much the same way that soap does, increasing the spreading and sticking of the herbicide when in liquid form. Dyes are also added to herbicides to identify where the herbicide has been applied. This is particularly helpful where some individuals are cutting brush and saplings and another follows behind, applying the herbicide. Dyes can also be used to identify treated areas as a warning to later users.

Selectivity. Herbicides are classified as being either selective or nonselective. Selective herbicides are effective against only a limited group of plants, based on morphology (e.g., monocots and dicots), growth period (preemergent and postemergent), and physiology. Nonselective herbicides kill any plants with which they come into contact. Obviously, they must be controlled and applied with care in restoration situations.

Mode of chemical action. Systemic herbicides are carried throughout the plant, including the roots, by the vascular system. Preferred by most restoration managers, systemic herbicides are generally effective at suppressing or killing the target species. In contrast, contact herbicides are not carried throughout the plant and work only on the area over which they are applied.

Persistence in the Environment

Not all herbicides break down rapidly into inert substances, and some can persist in soils for years. Soil organic matter and clays, if present, will absorb herbicides, thus limiting leaching into groundwater. Sandy soils are often more susceptible to herbicide leaching than soils with high clay and organic matter content. In some instances, systemic herbicides can move out of the vascular system of one plant into the soil and be taken up into the roots of other plants.

Application Methods

Once you have worked your way through the options and selected an herbicide, the next step is to determine the application method. Several different methods have been developed for applying herbicides. Here are some of the more commonly used techniques:

- *Paint.* Apply a thin coat of a systemic herbicide, formulated as an ester in oil, onto the trunk or bark of a woody plant.
- *Cut and paint.* Cut and remove the trunk or stem. Apply a systemic herbicide, using a brush or spray nozzle set to a drip, on the cut surface of the stump. You need to apply water-based herbicides to stump cuts soon after the cut is made. Oil-based herbicides can be applied a week or more after the cut is made.
- *Wick.* A wick is a perforated tube surrounded by a cotton wick (or similar absorbent fiber). The wick draws herbicide from the tube and is brushed across the plant or plants to be treated. You can use a long ropelike wick pulled between two vehicles to cover large areas.
- *Injection.* A hatchet or lance penetrates the bark of a woody stem to the cambium, and a systemic herbicide is then injected into the cut.
- *Spray.* Using either a handheld or vehicle-mounted tank of herbicide and a spray wand or arm, you apply herbicide to foliage, bark, or incisions.

The various painting techniques minimize herbicide use and damage to nontarget species, but they are relatively time consuming and labor intensive. The spray application can cover a wide area efficiently, but it increases the risk to nontarget plants because it can release herbicide into the air.

Evaluating an Herbicide

If you face a management situation that you determine requires the use of herbicide, you have to select the appropriate chemical and decide whether the herbicide you chose, including method of use, is successful. But your ultimate responsibility is to prove to yourself, in the field, that the tool works.

By "work" we mean two things: performance and outcome. Number one: Does the herbicide perform the way you want it to by, say, killing the target species? Number two: Does killing the target species really achieve the overall desired restoration outcome, whatever that is in your case, and according to your pest species management plan? These are two different measurements.. Outcome monitoring protocols for pest species management also need to be included in the restoration monitoring plan (see Chapter 9).

10.3.4 Fire

Fire is one of the most widespread of natural phenomena, occurring almost everywhere, except possibly in alpine communities, northern Europe, some tropical systems (such as humid and wet forests), and mangrove swamps. Fire is an essential natural disturbance force that shapes the structure and function of the world's fire-tolerant or fire-dependent plant communities, such as grasslands and savannas, and various open pine-oak and oak woodlands (see Chapter 2). Fire is widely employed as a conservation tool—for example, in chaparral restorations in Mediterranean systems around the world, in pine-oak and pine forests in Mexico, in parts of the African savannas (which make up more than half of the African continent), and in North American grasslands and the ponderosa pine forests of the southwestern United States. The main uses of fire in restoration vegetation management are to restore fire cycles to communities whose composition and structure are maintained by periodic fire, and to control pest species.

Fire damages plants selectively, and by removing or reducing litter, and in some cases canopy cover, fire changes the microclimate. As a result, fires affect the composition and structure of plant communities by favoring some species over others, sometimes leading to the start of a new successional sequence, and in others to community stability (see Chapter 2). These effects also can be used to control some pest plants (see Chapter 11). For example, cool season grasses (species that grow actively in spring and fall) from Europe (*Poa pratensis*, among others) that have entered into warm season grass (species that grow actively in summer) dominated communities of North America. Periodic burns are used to constrain or eliminate the establishment of these grasses (that are not adapted to fire), as well as to eliminate the spread of woody plant species. The post-burn environment often differs from pre-burn conditions in terms of greater light penetration at the soil surface, a reduction in soil moisture and air humidity, and nutrient availability. For some wildlife species, food and cover are lost. Many of these changes may be short-lived, as vegetation returns, but for some species, even one to two weeks of a more arid exposed environment can be a disadvantage. Other species that are adapted to such environments are then benefited.

But fire is not appropriate in all situations. Fire can be destructive or catastrophic in certain communities, such as temperate and tropical rainforests. And, depending on the timing of a burn with regard to life cycle events, it can even cause the death of desired species in fire-adapted communities. For example, the introduction of fire can convert forests to brushland.

Fire suppression can also change community structure dramatically. In these situations, prescribed fires are necessary to maintain restorations with goals related to composition and structure. A good example of the large-scale application of prescribed fire to reverse ecosystem degradation and to achieve ecosystem restoration goals is demonstrated by the Malpai Borderlands Group in the southwestern United States.

CASE STUDY The Malpai Borderlands Group

The Malpai Borderlands Group (MBG) is a grass-roots organization perhaps best known for its habitat restoration and land protection efforts. The MBG is organized and led by ranchers to preserve a way of life and a wild landscape, and to implement ecosystem management over an area of about 323,749 ha (800,000 acres) in a 3,237-km² (1250-m²) triangle of land along the borders of Mexico, southeastern Arizona, and southwestern New Mexico.

The MBG came together in 1994 after ranchers began to see the conversion of fire-dependent grasslands and savannas to shrubs due to fire suppression. They viewed this loss of grasslands as a threat to their way of life and livelihoods. In the Malpai Borderlands, fire-dependent ecosystems historically burned every 5–10 years with low-intensity ground fires, but with the advent of cattle grazing in the late 1880s, grassy fuels were reduced and shrubs began to increase. Along with cattle, roads and trails further interrupted burning, and by the twentieth century, government agencies were actively suppressing all fires.

As of 2010, the MBG had 27.923 ha (69,000 ac) under prescribed fire management. Burn units run into the thousands of acres; burns may last several days, and sometimes require many years of planning. The Thomas Tank fire in June 2008, for example, blackened 1,112 ha (2,747 acres) within a total burn unit perimeter of 1,740 ha (4,300 ac.) In the bottom photo of Figure 10.3, the foreground is part of the Malpai Borderlands and is in the prescribed burn program. In the distance, where junipers and other shrubs are establishing upon the land, is Mexico.

The Malpai prescribed fire program was initiated after a wildfire opened up land to more grasses,

FIGURE 10.3. **Fire as a Restoration Management Tool.** The Malpai Borderlands Group is a collection of ranches along the southeast Arizona, southwest New Mexico, and Mexico borders. The top image shows the borderlands that are being managed with fire in addition to grazing. The bottom image shows the managed borderlands in the foreground, with Mexico in the background. Notice the increased density of junipers south of the border where fire management has not been applied. (Photos reprinted by permission from John Harrington.)

greater biodiversity, and reduced shrubs, thus demonstrating the potential to use fire to recover the native grasslands and savannas. Area ranchers then acted to bring together scientists, public agencies like the U.S. Forest Service, and conservationists in an effort to return fire to the land through prescribed burns. The management plan that the MBG developed calls for enhanced biodiversity, while at the same time allowing for continued ranching. The plan includes prescribed fire management goals with a reduction of shrub cover and density and the rejuvenation of the grasslands, not only for cattle but also for deer, pronghorn antelope, and bighorn sheep.

The MBG is also well known for its land protection initiatives and has protected 30,351 ha (75,000 ac)

through conservation easements. The MBG engages in "innovative cooperative land management" and invented the concept of "grassbanking," whereby a rancher could rest land (e.g., stressed from drought) by grazing his herd on a neighboring rancher's property in exchange for the first rancher granting land use easements that prevent future "development."

The Malpai Borderlands Group takes what it learns and disseminates this knowledge through an active community outreach program. The organization achieves its desired restoration outcomes not only through well-planned and well-implemented restorations, but also because it is skilled at working with people and has cultivated an informed and engaged community (see Chapter 13). ■

Fire Effects and Behavior

Fire effects and fire behavior are complex topics, and a thorough discussion is beyond the scope and purpose of this book. (You can find more information on wildland and prescribed fire in the Resources at the end of the chapter.) However, in order to use fire as a management tool in a safe, efficient, and effective manner, you will need a basic understanding of the factors that influence the character of a fire and its ecological impacts. As Stephen J. Pyne puts it: "The biological response to a fire can vary widely. It will depend, first, on the physical properties of the fire—its intensity, size, frequency and time of occurrence—all of which influence the chemical potential for combustion and determine the nature of the chemicals liberated by combustion. It will depend, too, on the genetic potential stored within biota, which may also be released by a fire, and on the mechanisms or relationships for exploiting a fire that may exist with the biota (1982, 38).

Fire behavior describes the manner in which fuels ignite and flames develop, the fire's rate of spread, and the fire intensity measured by flame length and British thermal units (BTUs). Fire behavior is an important determinant of how a burn event will influence the vegetation. The management prescription specifies the desired fire environment and behavior to match the management goals for a particular site. Understanding fire behavior boils down to being familiar with the fuel, weather, and topographic conditions that collectively influence the start and spread of prescribed fire.

Fuel Conditions That Affect the Start and Spread of Fire

The moisture content, size and shape of fuels, and the site's fuel loading determine fuel conditions (Figure 10.4). Fuel moisture content is a product of recent precipitation and the relative humidity of the air, as well as fuel size and shape. The fuel absorbs moisture from precipitation and water vapor, and loses moisture through evaporation. The time it takes for fuel to come to equilibrium with the moisture in the surrounding environment is referred to as time-lag. Finer fuels, such as grasses, respond quickly to moisture changes and are labeled 1-hour time-lag fuels (fuels <0.6 cm in diameter). As fuel size increase, the time it takes to come to equilibrium increases. Small-diameter fuels can lose moisture

A. Fuel types
   1. Grass, a light fuel
   2. Shrub, a light fuel
   3. Timber litter, a heavy fuel
   4. Logging slash, a heavy fuel
B. Fuel characteristics
   1. Fuel moisture, the amount of water in a fuel, expressed as a percentage of the oven-dry weight of that fuel.
   2. Size and shape, the physical characteristics of the fuel.
   3. Fuel loading or quantity of the fuel.
   4. Horizontal continuity and vertical arrangement: fuels can be uniformly or patchily distributed and consist of ground, surface, and aerial fuels.

FIGURE 10.4. **Fuel Factors Influencing Fire Start and Spread.**

quickly when exposed to air on all sides, as would be the case in standing grass litter in late fall in midwestern prairies. Flat fuels, such as tree leaves and bark, when laid horizontally and compacted on the ground, retain moisture as long as the ground is moist. Surface fires in woodlands often only burn the upper litter layer, as the lower litter remains too moist to ignite.

The horizontal continuity of fuels influences how far and fast a fire might spread. Fire will continue to spread until it meets material that will not ignite, such as bare soil, an area in which much of the fuel has been removed by grazing, or water. If a fire reaches into the canopy of a tree, the density of trees nearby will be a factor in whether it continues its horizontal spread through the canopy. Fuel loading (weight of fuel per hectare) and the flammability of the chemical properties of the fuel are two additional factors that influence how fuels ignite and with what intensity a fire may burn.

Weather and Topographic Factors That Affect the Start and Spread of Fire

Fires require oxygen, fuel, and heat. Weather conditions that influence these factors and therefore how easily a fire will ignite and its behavior once it has gotten started include temperature, wind, relative humidity, and atmospheric instability. In general, the higher the temperature, the lower the relative humidity, and the stronger the wind, the more intense the fire will be. Temperature, winds, and relative humidity work together to influence burning conditions. Winds flowing through and over fuels bring oxygen, thereby increasing the fire's intensity. If the winds are dry, they will wick off moisture and dry out fuels, which increases fire intensity and the rate of spread. Wind causes both radiation and convection currents to heat fuels in the direction it is blowing. The rule of thumb is: As long as fuel type and topography remain constant, as the wind speed doubles, the rate of fire spread will quadruple.

Atmospheric instability is the rate at which a parcel of warm air rises. As a dry or unsaturated warm air parcel rises, it cools at approximately 9.8°C per 1000 m (5.38°F per 1000 ft). Eventually the parcel will cool to the same temperature as the surrounding air and will stop rising. You can judge the atmospheric instability by looking at the clouds overhead. Horizontally spreading stratus clouds are an indication of a stable atmosphere, whereas rising cumulus clouds are produced by an unstable atmosphere. Atmospheric instability influences how high smoke columns will rise, and it is an indicator of potential wind shifts and cold or warm fronts that may pass through, causing erratic fire behavior. As a result, highly unstable atmospheres make it difficult to control fires. On the other hand, a highly stable atmosphere will not allow smoke to dissipate. A moderately unstable atmosphere is desirable for the dispersal of smoke.

Several topographic features are important to the start and spread of fires, largely because topography influences microclimate and wind and weather patterns. Features of importance include (1) aspect, the direction in which a slope faces or exposure to the sun; (2) slope, the amount or degree of incline of a hillside; (3) landforms, the overall shape of the country and the presence of box canyons, narrow canyons, and other topographic features; (4) elevation, the height of the land above mean sea level; and (5) barriers, roads, rivers, lakes, or anything that impedes the spread of fire.

Slope aspects that face the afternoon sun tend to be hotter and drier than aspects that have little exposure to direct sun. Such sun-exposed habitats are more prone to fire. Slope steepness influences fire as well. When fire burns, it releases heat in three ways: by radiation, which travels in all directions; by convection, which is the upward movement of warm air; and by conduction, or the direct transfer of heat from one object that is touching another. These processes influence the microclimate close to the ground. Radiation ahead of the fire heats fuels before the actual flame reaches them, resulting in a quicker ignition. Fires travel more quickly upslope than downslope because the flames and radiation reach the fuels more quickly than when traveling downslope. The rule of thumb is: All other conditions being equal, as percent slope doubles, the rate of fire spread quadruples.

The general distribution of landforms across a landscape will generate local wind and precipitation patterns and thus influence fire behavior. Similarly, elevation influences microclimate, and the effects are more pronounced in regions with mountainous terrain and high elevations—in general, the higher the elevation, the cooler and moister the microclimate and the shorter the fire season window. Topographic features such as streams or bare rocks serve as firebreaks that can stop or at least impede the progress of a fire.

Prescribed Fire

You cannot control fire, but you can predict its behavior by understanding the issues discussed above. Thus, through careful planning, you can use fire as a management tool by conducting a prescribed burn to achieve management or research purposes. To do this, you write a fire management prescription. The prescription will specify the conditions under which you can safely conduct a burn.

Fire is potentially very dangerous; therefore, in order to use this tool, you will need to complete training courses before participating in prescribed burns.

Prescription fire plans. Because of the nature of fire, it is particularly important to consider, in detail, all aspects of its application. This means that you will need to spend considerable office and planning time to create a burn implementation plan. The plan specifies the management goals you hope to achieve by using this tool and, in the case of fires that you ignite, describes in detail how you will conduct the burn. You will need to locate and map important features of the management unit you intend to burn. These include the topography of the site, fire breaks, hazards, research plots, escape routes, safe zones, and the location of downwind smoke-sensitive areas, as well as any additional points of interest to the burn crew.

The plan also lists all the pieces of equipment you will use. In the case of a grassland fire, your list might include drip torches to ignite the fire, protective fire-retardant clothing, backpack water pump sprayers or water tanks mounted on trucks to contain the fire, and radios or other communication devices. You also specify the needed size and skills of the field crew, and assign roles to each crew member. You should also include contact numbers for emergency assistance.

Timing and frequency. As is true of all restoration management activities, you must determine when and how often to implement prescribed burns. The timing and frequency of a burn is based on maximizing the effects you want the fire to accomplish, as well as minimizing unwanted impacts on other parts of the restoration. Here are the main factors you will need to consider:

1. The frequency and seasonal timing of the natural disturbance cycle.
2. The growth stage of a plant species that is to be encouraged or discouraged.
3. Wildlife activity, in order to avoid unintended harm.
4. Weather conditions.

In situations where undesired saplings and woody seedlings are of concern, early growing season and hot burns are desired. For litter removal in grasslands, early to mid-spring burns are effective. Try to write the management prescription to avoid too much regularity. Any management tool, including fire, that is applied at the same time at a consistent frequency (annually, biannually, etc.) will favor some species and disfavor others. By varying the time, frequency, and weather conditions under which restorations are burned, you can increase or maintain diversity.

Techniques for implementing a prescribed fire. In writing a burn plan, you can choose from a variety of techniques for igniting a fire, controlling the direction and speed with which it moves, and containing and extinguishing the flames. Firebreaks—places that act as barriers to the progress of a fire—can be permanent site features, such as a wide stream or river; or you can create a temporary break just before the fire is ignited. A plowed field or a previously mowed or burned strip of grassland can be an effective barrier. In addition to writing the fire prescription to specify the temperature, wind speed, and relative humidity, you can influence the fire's behavior by igniting it with reference to the direction from which the wind is blowing. This will create different burn patterns:

- *Back fires*: Fires that burn into the wind. This type of prescribed fire is accomplished by creating a firebreak along the border that is farthest from the wind. The fire is ignited on the inside of the firebreak so that it can only burn into the wind. Back burns are slow-cool fires with relatively low flame lengths. These are fires that allow you to operate relatively close to the flames. Their greatest disadvantages are the time it takes for a burn to be completed and the fact that the smoke produced does not move quickly off the site. Also, because they move slowly, the flames have time to transmit a lot of heat to the soil and surrounding air, thereby thoroughly consuming most of the fuel in the area. This can be desirable for reducing large litter, but is undesirable when you wish to maintain unburned patches of vegetation as refugia for seeds and animals.
- *Head fires*: Fires that burn in the direction the wind is moving. These are rapidly moving fires with long flame lengths. Head fires can cover considerable ground quickly;

although they burn hot, because they travel quickly, they do not necessarily burn all fuels in their path. The major drawbacks are the heat produced and the difficulty of controlling the fire. Typically, you would use head fires only after all firebreaks are secure.

- *Flank fires*: Fires that are pulled into the wind in the direction of the fire's tail. The fire then spreads sideways into the wind.
- *Strip fires*: Lines of fires that you ignite perpendicular to the wind behind a firebreak. You could use these on large sites where control may be an issue. Strip fires move only a short distance before running into the black area (burned-out fuel) of the strip ahead of it.
- *Ring fire*: One of many fire sequences for burning an area. Figure 10.5 outlines a simplified ring fire technique.

The sequence of a prescribed burn in longleaf pine (*Pinus lacustris*)–wiregrass community of South Carolina is presented in Figure 10.6.

Constraints on the use of prescribed fire. Fire managers usually have to obtain a variety of permits from municipalities and/or other regulatory agencies, such as fire departments and natural resource management agencies. The permitting process is a means of ensuring that fires are conducted according to guidelines that support the health and safety of those doing the burn, and those living or working in the area. To obtain a permit, you usually will have to submit your fire prescription to the permitting authority.

Two issues that have become important in recent years are concerns over smoke management and the potential impacts of the fire on air quality. You will want to check with your municipal fire department or natural resource management agency for current regulations and procedures before you begin to write a fire management prescription. Air quality and smoke management concerns should be addressed in the planning process. Sometimes,

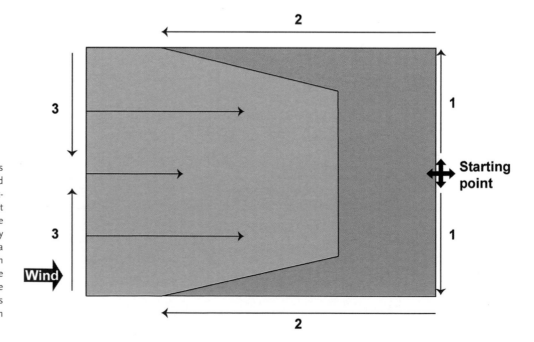

FIGURE 10.5. A Ring Fire. (1) The fire is ignited at this point. Two fire lines proceed in opposite directions, setting off a back-burn situation. (2) When both lines are at the corners, they will proceed along the lines labeled 2, in parallel fashion. As they burn these lines, the black area, or the area devoid of fuel, will increase. (3) Once both lines turn the corner onto edge 3, the fire will become a head fire and burn with the wind into the black. Modifications to this sequence will be required with changes in wind direction.

FIGURE 10.6. **The Sequence of a Prescribed Burn in a Longleaf Pine-Wiregrass Community.** The top left image shows the dense sapling and litter layer that has built up with fire suppression. The top right image shows a strip fire burn pattern. The middle left image shows a head fire that has been released after surrounding borders were secured. Notice the influence of convection currents. In the middle right image, smoke is concentrating and impacting visibility. The bottom left image shows the postfire environment, and the bottom right shows scorched limbs from the impact of intensely heated convection currents. Although the heat from the convection currents will not kill the trees, they will be stressed for several years, allowing opportunities for other species to colonize. (Photos reprinted by permission from John Harrington.)

however, the location of the burn unit and its surroundings make the use of prescribed fire impossible. In this case, it is best to choose an alternative management strategy.

Neighbors will be interested in all of your management activities, but the drama and effects of a prescribed burn generally draw the most attention. Some people will become involved out of curiosity and a desire to learn more; others will be concerned about safety issues and possible property damage. As always, it is best to deal with potential problems early on and in a straightforward manner (see Chapter 13).

10.3.5 Biological Controls

The biological control of vegetation involves using herbivores or disease-causing microorganisms to influence the growth of plants and community composition and structure. These control agents can be native, but often they are not. The two most common restoration management situations in which biological control is used are mimicking the grazing pressure of native mammals, and limiting the growth of non-native pest plants. In both cases, managers need to be sure that the prescription used minimizes damage to desired species.

Mimicking Grazing

Grazing and browsing by mammals, such as cattle or goats, is used in many regions of the world to manage restorations and provide a livelihood for people. In the midwestern United States, grazing is being used to manage both natural and restored prairie and savanna. In these natural grassland systems, grazers such as bison, antelope, or elk act as keystone species to maintain high plant species diversity (see Chapter 2). Grazing is thought to be as important as natural fire cycles in maintaining community composition and structure. For example, in 25 years of study at Konza Prairie, Kansas, researchers have found that carefully applied grazing tends to increase species diversity more than the management tool of using fire (Figure 10.7). Grazing by mammals also helps prevent the establishment of woody species.

A grazing technique that shows promise for maintaining restorations and in raising domestic animals for profit is intensive rotational grazing. Grazing/browsing animals are continuously rotated through a series of fenced pasture units or paddocks. Grazers are kept in one paddock for a brief time and then moved to the next paddock. The first paddock then "rests" or recovers before grazers are returned to it once again. Several grazing cycles are used to address the management goals. If the period of grazing is timed correctly with an appropriate stocking ratio, the negative effects of grazing on the landscape will be minimized, and plant species diversity and community structure will benefit.

One of the concerns raised about the practice of managing with the use of domestic animals is that their grazing strategies are different from those of the native grazers. For example, the studies at Konza Prairie indicate that bison (the native grazer) preferably eat more grasses, whereas domestic cattle eat the forbs. Bison will walk in loose packs through an area, whereas cattle tend to walk in single file, creating paths. Consequently, the results of grazing will be different, depending on which species are used. There is some evidence that although species differ in their grazing strategies, the more important factor in using grazing as a management tool is how the herbivores are managed, particularly with regard to the stocking density and the length of time the animals are left in any single location.

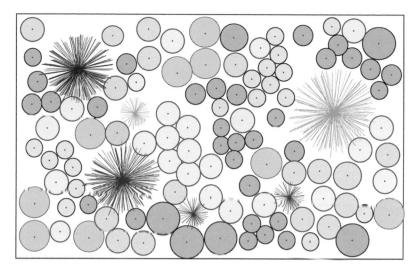

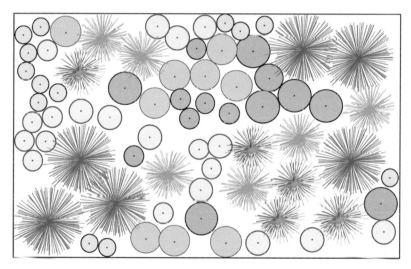

FIGURE 10.7. The Influence of Fire and Grazing Regimes on Species Composition. Top: Grasslands under a fire regime provide a greater opportunity for grass establishment. Bottom: Grasslands under a grazing regime often provide an opportunity for greater forb diversity. The effect of fire depends on several variables including timing. Spring burns, in particular, favor grass growth. Star shapes represent grass; all other shapes and colors are indicators of forbs.

Controlling Pest Plants with Insects

Insects can be used to control invasions of pest plants in restorations (see Chapter 11). However, because of the potential risks to nontarget species, this practice is usually tried only after other techniques have failed. The most common situation in which biological controls are used in restoration management involves introduced plant species that have few known native "enemies." In such cases, you would look to the plant's native habitat to find insects that specialize in disrupting the growth of the plant. The idea is to import those insects, after screening tests to ensure that the insects will not affect any native plants, and release them in the field to control the pest. Within the United States, biocontrol insects are extensively tested under the supervision of the Animal and Plant Health Inspection Service (APHIS) to ensure they will not "escape" and naturalize, or become parasites and predators on other species.

The introduction of a non-native insect to manage a non-native plant raises several concerns. For one thing, the numbers of insects available for release are limited, and they often take a long time to disperse across a site and develop adequate populations. A second concern is that, in spite of the screening tests designed to limit the risk, the insect will eventually adapt to its new environment and begin to affect nontarget organisms, thus becoming a pest species itself.

One example of insect pest plant control agents being used in natural areas and restorations in North America involves the exotic herbaceous perennial, purple loosestrife (*Lythrum salicaria*), which has invaded many wetland systems. The insect controls include *Galerucella pusilla*, a leaf-feeding beetle; *Galerucella calmariensis*, also a leaf-feeding beetle; and *Hylobius transversovittatus*, a root-mining weevil. The insects are able to survive in the field and show promise in reducing purple loosestrife populations. Although the insects were tested in advance of their release and found not to affect related native loosestrife species, there are now a few reports that crossover feeding on the native plant, winged loosestrife (*Lythrum alatum*), may occur. Although biological controls will not likely eradicate pest plants, they can slow or halt their spread.

▶ 10.4 REDUCING MANAGEMENT IMPACTS

Sometimes management activities can damage a site. This reality underscores the importance of setting up a responsible management plan so that you can anticipate and respond by establishing policies, procedures, and practices that avoid or mitigate undesirable impacts.

Potential management impacts, in addition to those mentioned above, include herbicide damage to nontarget species, accidental cutting or pulling of native plants during a pest plant control work party, vegetation trampling, creation of unofficial trails during the course of a work party, soil compaction when staff walk on soil that is too wet, and rutting of soil when heavy equipment is used inappropriately.

You can help prevent site damage and injury to personnel through education, and training in the safe, efficient, and proper use of equipment. It is important to take the time to set clear expectations and establish standards. Here are some examples:

- To avoid soil compaction and rutting, do not drive or walk on soil that has water at or near the surface. Chronic wet areas should be designated as off-limits until after it rains and water has soaked into the soil.
- Hold plant identification walks and talks so that staff and volunteers know what plants to keep and which to pull, and why.
- Establish seed and plant collection policies.
- Establish safety policies and procedures that specify personal protective equipment to be used during risky activities, such as burning, herbicide application, animal control, or the use of motorized equipment. The field crew must wear protective equipment, including hard hats, leather leg and arm protectors, and steel-reinforced work boots.
- Use only those pieces of motorized equipment that are suited to the site conditions, and ensure that they do not compact or gouge the soil during operation.

Some restoration management activities require vehicles, such as tractors, that are equipped with a front end loader for moving bulk materials like soil and wood chip mulch, and have a rear power take-off drive to operate brush cutters and mowers. A skilled operator can perform wonders, but if not used with care, such equipment can damage a site. Proper training and supervision can minimize mistakes.

A well-thought out management plan, if followed, implemented, and updated as conditions change, will go a long way in ensuring the long-term success of a restoration.

KEY CONCEPTS

- The management phase of a restoration project begins once the initial restoration goals and objectives have been achieved. Management includes planning, implementation, monitoring, and revising; it directs, but does not prevent, change.

- Potential management strategies include taking no action (hands-off); the reintroduction of natural disturbance processes or their facsimiles; and the use of various tools chosen to match the biological, physical, and human use restoration goals. For each strategy, a prescription specifies when, how, where, and by whom it is to be applied.

- The success of any management strategy is influenced by the timing, frequency, duration, and intensity of its application.

- Vegetation management tools include mechanical approaches, and the use of herbicides, fire, and biological controls.

FOOD FOR THOUGHT

1. Describe the differences and similarities between the focus of the initial restoration and the focus of subsequent management of a restoration.

2. Choose a restored site in your area to visit. Look for, or inquire of the site's managers about, any natural processes that have over the years been impeded by human activities. As a result, what management issues might need to be addressed? Are these issues that can be addressed right now with available tools? Will new "tools" have to be developed?

3. Consider another restoration site you are familiar with or have read about. What are a few of its management concerns? Could you write a management prescription for it? What would you need to know? What management tools would you recommend?

4. Would it be possible, based upon your site inventory and analysis of the area surrounding your restoration, to anticipate and/or prevent the kinds of impacts that human activities might have on your restoration?

5. What site parameters would you consider prior to conducting a prescribed burn or initiating a grazing program to achieve management goals?

6. Under what circumstances might a management action cause more harm than good?

Resources for Further Study

1. Journals devoted to natural areas management include:

   Natural Areas Journal, published quarterly by the Natural Areas Association, an organization of public and private natural areas managers from across North America.

   Ecological Applications, a publication of the Ecological Society of America, containing articles on management and policy. It does not focus on the management of particular species or communities, but rather on broad issues of policy, theory, and perspective.

2. The following textbook provides a good introduction to practical conservation management. It presents real-world problems involving multiple stakeholders and guides you in making management decisions.

   Meffe, G. K., L. A. Nielsen, R. L. Knight, and D. A. Schenborn. 2002. *Ecosystem Management: Adaptive, Community-Based Conservation*. Washington, DC: Island Press.

3. There is a vast literature on the use of prescribed fire in community/ecosystem management. The following are a few resources that we have found to be particularly useful.

   A. This is a must-have book for anyone interested in understanding fire behavior and learning how to manage fire to accomplish management goals:

      National Wildfire Coordinating Group. 1994. *Introduction to Wildland Fire Behavior, S-190: Student Workbook*. Boise, ID: National Interagency Fire Center.

   B. The Natural Areas Association offers a CD with a collection of articles devoted to this topic:

      Fire in Natural Areas Compendium. CD. http://www.naturalarea.org

   C. This text provides a good overview of the topic of wildfires:

      Whelan, R. J. 1995. *The Ecology of Fire*. New York: Cambridge University Press.

   D. This is a helpful resource on fire prescriptions:

      Weir, J. R. 2009. *Conducting Prescribed Fire, A Comprehensive Manual*. College Station: Texas A & M University Press.

4. For more on the Malpai Borderlands Group, visit its website, where you can find more about its history, land protection, land management work, and outreach efforts. Peruse a list of books and other publications.

   http://www.malpaiborderlandsgroup.org/index.asp

11

The Role and Impact of Pest Species in Restoration

Throughout much of the world, pest species pose a major challenge to implementing high-quality restorations. They create the greatest ongoing need for labor, money, and planning, once a restoration is completed. Restoration practice is seriously confounded and complicated by the presence of pest species, because they constrain site preparation and restoration implementation, outcompete desired species, degrade wildlife habitat, transmit diseases, and alter ecosystem functions and processes.

▶ 11.1 CHARACTERISTICS OF PEST SPECIES

Pests, or pest species, are also referred to as invasive species, alien species, or exotic invaders. For the purpose of our discussion, we will define pest species as those native or exotic (non-native) plants and animals that interfere with restoration and management goals. This is a "situational" or "operational" definition, in that a species that behaves as a pest may, under different circumstances, at different times, or in the same or different places, be a beneficial component of the ecosystem being restored.

An example of a native species that can be a pest in some situations is trembling aspen (*Populus tremuloides*), an ecologically and economically important species of the Northern Hemisphere. Trembling aspen is clonal, and under some types of disturbance and environmental conditions, it can expand rapidly, spreading at high densities over large areas. When this rapid expansion occurs, ecosystem and community structure may be altered to the point that it no longer matches the restoration goals.

Some non-native species that are considered pests in most situations can, under different conditions, perform essential ecosystem functions. For example, researchers in Kings Bay, Florida, found evidence that "four notoriously invasive exotic macrophytes in Kings Bay ecosystem—water hyacinth, water lettuce, hydrilla and Eurasian milfoil—perform functions that potentially could be utilized within a holistic and adaptive program of ecological restoration" (Evans et al. 2007, 207). The idea is to control these plants, which are already present in the bay, rather than remove them, because of their abilities to remove nutrients and heavy metals from the water, outcompete blue-green algae that release toxins into the water, and provide food for the rare Florida manatee (*Trichechus manatus latirostrus*).

Sometimes pest species have become so established in an area that their control or eradication has unintended consequences. For example, in New Zealand when the introduced and invasive watercress (*Rorippa nastustrium-aquaticum*) was removed from the Walkoropupu Springs to save an endemic moss species (*Hypnobartlettia fontana*), a much worse species, the introduced rush (*Juncus nicrocephalus*), invaded the remaining native vegetation areas (Wittenberg and Cook 2001).

▶ 11.2 THE ROLE OF HUMANS IN THE PEST SPECIES PROBLEM

The success of pest species depends upon (1) their ability to disperse to a site, (2) their adaptability to the physical environment of the new site, and (3) their interactions with species already present on the site. Many factors contribute to species behaving as pests in a particular restoration. It is often hard, in any given situation, to identify all the contributing factors, much less be able to tease apart cause and effect. Nevertheless, three types of human impacts seem to be especially influential in creating pest species worldwide: the rapid human transport of species throughout the world, global climate change, and human land use practices.

11.2.1 Human Transport of Pest Species

The movement of plants and animals from continent to continent in the global market-place has resulted in the introduction of species to areas they did not previously inhabit. While the rate of introductions was relatively slow in the era before steamships and air travel, it now can occur in a matter of hours. This mixing of species is only likely to accelerate with expanding global trade.

Introductions of species by humans have been both purposeful—for agriculture, medicine, horticulture, and landscape design uses—and accidental—as hitchhikers on food, fiber, timber, and ships, and in ballast water. Not all introduced species behave as pests. Many are able to thrive only if people actively care for them, providing nutrients and water, removing potential competitors, breeding them, or otherwise practicing active husbandry. Those exotic species that do become naturalized—able to grow and reproduce on their own—usually are able to do so because they can outcompete native competitors and/or avoid predation.

Relatively isolated island ecosystems have proven to be particularly vulnerable to introduced species behaving as pests. Historically, the tropical forests of Puerto Rico and Hawaii have long been battling pest species. Scientists estimate that about twenty new alien invertebrates establish in the Hawaiian Islands each year, some of which may turn out to be pests.

11.2.2 Global Climate Change

The predicted biological changes resulting from global climate change are expected to alter where individual species live, and how and with what other species they interact. The effects of climate change may be felt first at the species level and then later at the ecosystem level, as species displacements and rearrangements influence community structure, composition, and function. As these changes occur, ecosystems and communities become more vulnerable to invasions by pest species.

Some ecologists expect that over time, more exotic species will take on the status of pests, as introduced species react in a variety of ways to a range of factors, including elevated carbon dioxide levels, increased nitrogen deposition, warmer temperatures, increased precipitation, rising sea levels, extremes of flood and drought, altered hydrological regimes, new species interactions, and new community assemblages.

Elevated atmospheric CO_2 levels, for instance, are known to increase the growth rate, size, and biomass of common pest species and to change their chemical composition. This might make them more invasive and hard to manage, thereby possibly contributing to the alteration of natural disturbance events, such as fire and grazing. It is changes such as these that Lewis Ziska of the U.S. Department of Agriculture's Agricultural Research Service believes are to blame for the increased flammability of cheat grass (*Bromus tectorum*), an exotic species in the western United States grasslands. Cheatgrass is less nutritious than the native grasses, and has limited wildlife habitat value, yet in many areas it has become a dominant grass. Cheat grass is extremely flammable and fire tolerant to start with, and where it has invaded the western U.S., natural fire regimes have been disrupted with increased fire frequency—now every 3–5 years instead of the historic return interval of once every 60–100 years.

In the prairies of Saskatchewan, crested wheatgrass (*Agropyrum cristatum*), a cool-season, non-native grass, has long been blamed for causing the decline of native flora. Recent research has shown that earlier spring warmth and more consistent summer rainfall favor this species over warm-season native grasses. Increased nitrogen deposition resulting from the burning of fossil fuels, and from fertilizers in agriculture, is also suspected of favoring a whole range of species. Elevated atmospheric CO_2 levels since the Industrial Revolution are also credited with producing healthier and more vigorous poison ivy (*Toxicodendron radicans*) plants, which produce a stronger form of *urushiol*, a skin irritant.

Global climate change impacts are expected to include increases in fire frequency and vegetation flammability in some parts of the world. This increase may make the native communities more vulnerable to occupancy by pest species. South Africa and New Zealand, for example, are expected to be disproportionately impacted by introduced pest species.

11.2.3 Human Land Use Practices

Human activities not only move species around the globe, but they also transform the local landscape and the atmosphere and are responsible for disturbances that create new environments capable of being colonized by pest species. Herbaceous, perennial, biannual, and annual plant species that are adapted to disturbance-prone pioneer habitats spread along roads, walking tracks, and field edges because of constant disturbance from traffic. Additional land uses that create disturbed habitats include timber harvests, farming, flood periodicity changes, the drainage of wetlands, and various types of urban and suburban development. All the habitats created by these activities become sources of plant species that readily invade the early (and even later) stages of restoration implementation and behave as pests, outcompeting the desired natives.

Land use changes also lead to habitat range changes for wildlife species, which can further distribute the seeds of pest plants. Birds and mammals, in particular, are quite effective at distributing seeds from many plant species either on feet, in hair, or passed through digestive systems.

▶ 11.3 IMPACTS: WHY RESTORATION ECOLOGISTS ARE CONCERNED ABOUT PEST SPECIES

One of the reasons we initiate restoration actions is that pest species have come to dominate a site, resulting in the decline and/or elimination of established native species populations. Also, during the process of restoring a site, implementation activities often create disturbances that provide many opportunities for additional pest species to colonize.

The presence of pest species in native ecosystems has long been associated with local reduction of native species populations, even though direct cause-and-effect relationships have rarely been established by research. Around the world, many extinctions or reductions in numbers of species of fish, plants, birds, and other animals have been anecdotally attributed to pest species. Although pest species have played a role in biodiversity reduction in some situations, the ultimate causes of species loss are ambiguous and hard to tease apart from the impacts of human activities that disrupt ecosystem structure, function, or processes.

A historical analysis of global species invasions and extinctions (Sax and Gaines 2008) shows that (1) the extinction of many native wildlife species on islands are linked to predation by nonindigenous species (including humans), but that (2) competition by introduced plants has caused few plants to go extinct, and (3) many islands demonstrate an ability to absorb species invasions without displacing native species. Although global species loss is difficult to attribute to pest species alone, many studies have found the loss of biodiversity within single sites after pest species invade them. And even though attributing species extinctions to pest species has not been proven, changes in species abundance and population size appear to be highly corelated to a combination of pest species and habitat change.

In Guam, for example, the brown tree snake (*Bolga irregularis*), introduced from Australia and Papua New Guinea, has had severe repercussions for the island's bird and other small animal populations. This species of snake was introduced to Guam shortly after World War II and by the 1970s had spread throughout the island, with up to 13,000 individuals per square mile. Without natural predators and an abundance of vulnerable prey animals, the brown tree snake is implicated in the loss of a large percentage of native bird, lizard, and bat species. A dozen endemic bird species are thought to be extinct, and nine of twelve native lizards are expected to go extinct—all attributed, directly or indirectly, to the brown tree snake.

Insect outbreaks, diseases, and viruses can also serve as pests, and these have impacts on individuals of a species, as well as species populations and community/ecosystem structure. Dutch elm disease (*Ophiostoma ulmi, syn. Ceratocystis ulmi*) has all but eliminated the once-dominant American elm (*Ulmus Americana*) from northeastern forests, and the emerald ash borer (*Agrilus planipennis Fairmaire*) is expected to eliminate ash species (*Fraxinus* spp.) from eastern North American woodlands if it continues to spread as projected.

11.3.1 The Impact of Specific Pest Species

Pest species have numerous impacts on ecosystem functions and processes. These impacts, in turn, significantly influence a restoration's final and future composition and structure. For example, the cinnamon fungus (*Phytophora cinnamomi*) has had a dramatic effect on forest and scrublands in Australia because it can attack 50–75% of the plant species present in a forest. This has a considerable impact not only on plants but also on the animals that depend on them for food and habitat. A fungal chestnut blight, *Cryphonectrica parasitica* (formerly *Endothia parasitica*), introduced from China in the early 1900s, decimated the American chestnut (*Castanea dentata*) and, in the process, altered the forests of the northeastern U.S. in ways still not fully understood. Prior to the introduction of the blight, the chestnut was the most numerous and major nut-producing tree in the northeastern deciduous forest; now it is functionally extinct. The chestnut supported a vast number of animal and plant species, and its loss left large gaps in the forest that have since been filled by other species.

Woody pest species in South African savannas deplete groundwater supplies; in North America, climbing and strangling vines, such as oriental bittersweet (*Celastrus orbiculatus*) and kudzu (*Pueraria Montana var. lobata*), alter plant community composition and structure when they kill canopy trees. As these canopy trees die, the overall height of the forest is lowered. The loss of canopy trees reduces competition for water and nutrients and leaves significant gaps in the canopy, which allow increased light levels to reach the forest floor. These changes favor the introduction and establishment of a different set of plant species—many of them non-native pest species—from what grew there previously.

Where reed canary grass (*Phalaris arundinacea*) dominates former remnant wet prairie at Faville Prairie near Lake Mills, Wisconsin, the aboveground productivity is higher than where it does not dominate, but belowground productivity is lower. This is a potentially permanent change (Jelinski and Anderson 2007).

Buckthorn (*Rhamnus cathartica*), a small forest and edge tree, is thought to alter soil characteristics, such as fertility, or impact soil microflora and microfauna. In areas where large dense stands of buckthorn have established, few ground layer species persist, and altered soil conditions make restoration difficult because they are not easily reversed, even after removal of buckthorn stands (Figure 11.1).

FIGURE 11.1. A Buckthorn Thicket. These forest scenes show before removal (top) and after removal (bottom) of *Rhamnus cathartica*, a common woodland edge invader in urban areas throughout the northeastern United States. Buckthorn can reach midstory densities that are typical of a closed forest. Buckthorn alters soil chemistry and microfauna and limits the available light to ground flora. Where buckthorn occurs in large colonies, the ground beneath is often devoid of plants. (Photos reprinted by permission from John Harrington, top, and Steve Glass, bottom.)

In riparian areas of the southwestern United States, the Eurasian tamarisk or salt cedar (*Tamarix* sp.) absorbs vast quantities of groundwater. Introduced for erosion control, this species has spread into wetlands and nutrient-poor uplands. Tamarisk can transpire up to 300 gallons of water per day, severely limiting the availability of water for other plants. Soil salinity builds from salt secretions that the plant emits from its leaves. Over time, tamarisk creates an environment where it is highly competitive at the expense of native species (Figure 11.2). Similarly, in the Cape Region of South Africa, woody pest species are monopolizing and drying up local catchments and reducing stream flow. Australian *Acacia*, *Eucalyptus*, and *Hakea* species and European and American *Pinus* and *Prospis* species account for about 90% of the problem in the invaded area, which covers about 1.7 million ha (Wittenberg and Cook 2001).

FIGURE 11.2. **The Invasion of a Riparian Basin by an Exotic Species.** Salt cedar or tamarisk (*Tamarix ramosissima*) is an exotic tree or shrub that invaded the riparian basins of the southwestern U.S. prior to the 1940s. Its mechanisms for colonizing and establishing in new environments is similar to that of many other introduced pest species—high reproductive potential and the ability to modify its environment to one that is beneficial to it but that is disadvantageous to native species. These photos show the establishment of tamarisk (and efforts to control it) in Canyon de Chelly, Arizona. (Photos reprinted by permission from John Harrington.)

Some plant species, such as yellow star thistle (*Centaura solstitalis*), alter grazing lands and grasslands in California. In North American forests, exotic earthworms drive ecosystem change by consuming litter and decreasing soil fertility. The ecosystem changes wrought by earthworm species set the stage for invasion by other non-native species, including buckthorn and garlic mustard.

Habitat loss, fragmentation, and degradation are often accompanied by a decrease in native species and an increase in pest species, thus affecting the quality of wildlife habitat. This correlation is well established. Less well known, in any particular situation, however, are the causal agents and interacting forces responsible for the loss of native species. For example, land use changes might facilitate the local or regional expansion and establishment of pest species either (1) because habitat degradation provides more suitable and available space for pest species or (2) because land use changes remove an agent that inhibits the spread and establishment of pest species, or (3) introduces an agent that promotes the spread of pest species. In some cases, all of these pathways may be involved.

In the initial establishment stages of a restoration, fast-growing plant species appear to interfere with the establishment of many desired species. Competition for nutrients, water, and light often favors species that establish rapidly, are effective at utilizing resources, and persist for several or more years. When shade-intolerant species that are slow to germinate emerge under dense canopies of fast-growing species, they are often at a disadvantage. In new plantings where pioneering native or introduced species establish early, a decline in the establishment of native species is often observed. The causes are speculative. Competition for light is one, but alleopathy and resource acquisition of nutrients and water are possible others. In the Curtis Prairie restoration in Wisconsin, continued patches of bare ground enabled the pest species white sweet clover (*Melilotus alba*) and wild parsnip (*Pastinaca sativa*) to establish and persist while their native counterparts, old-field thistle (*Cirsium discolor*) and wild lettuce (*Lactuca canadensis* and *L. ludoviciana*), did not. Speculation about the causes of the success of non-natives over natives (Kline and Howell, 1987) led to an experiment that not only illuminated a reason for success of the pest species but also developed a management technique to control sweet clover. This experiment is discussed below.

CASE STUDY | The Development of a Protocol to Control White Sweet Clover

White sweet clover (*Melilotus alba*) is an introduced legume that has become a serious pest species in both remnant and restored prairies of the midwestern United States. It has often been planted for its forage value and was brought into the University of Wisconsin–Madison Arboretum in the early 1930s as a cover crop for grassland birds. By the mid-1970s, sweet clover was running rampant and had achieved a 17% frequency in Curtis Prairie, a planted restoration, compared with a 10% frequency for its native counterparts, wild lettuce (*Lactuca canadensis* and *L. ludoviciana*) and old-field thistle (*Cirsium discolor*).

The prairie had been burned at 2- or 3-year intervals every spring since 1950. The burns were timed to occur when most of the native prairie species were still dormant. Following the burn in 1977 (after a period of three growing seasons with no burns), Arboretum Ecologist Dr. Virginia Kline noticed a stand of first-year white sweet clover (*Melilotus alba*) plants that was even more vigorous and dense than usual. Many of these plants survived the winter, and in 1978 there was a "nearly continuous canopy often over 2 meters high" (Kline 1984, 149) in a large central section of Curtis Prairie. Dr. Kline

decided it was time to institute some serious control measures.

Her first step, following best management practices, was to find as much information as possible about the ecology of the species. She discovered that white sweet clover is an obligate biennial legume. In the first year after germination from seed, the plant grows vegetatively and remains small; the plants overwinter and in the second year put on growth spurts, reaching 5 feet in height, bloom, set seeds, and die. Its seeds are long-lived in the soil, and early spring prescribed burns trigger synchronous germination of its soil seed bank. Thus, if a prescribed burn regime has a return interval of 2 years or greater, a white sweet clover population explosion is just about guaranteed. This was the exact situation that Dr. Kline observed in Curtis Prairie in 1977. Dr. Kline also found that this weedy response phenomenon had been noted in Curtis Prairie in 1948 by Professor John Curtis and Arboretum Manager Max Partch during their pioneering work that showed the importance of periodic fire in promoting prairie growth and discouraging weeds. After one March burn, they noticed in one plot that the sweet clover population "resulted in the nearly complete exclusion of all other plant species" (Kline 1984, 149).

Based on this information, Dr. Kline reasoned that the key to the long-term reduction of sweet clover populations would be the prevention of seed production. Individual sweet clover plants may be killed before seeds set by pulling the plant out by the roots. At a larger scale, a more efficient and effective treatment is mowing and/or burning. Second-year sweet clover plants are not vulnerable to mowing or burning until after shoots have emerged from dormant buds at the crown of the root. In Wisconsin, sweet clover plants reach this stage of development in early to mid-May. Because of its early spring burning regime, Arboretum managers were actually encouraging the spread of sweet clover. Not only were the spring burns promoting seed germination; they were not affecting the seed set of second-year plants by being conducted too early in the year.

This situation provided Dr. Kline with an excellent opportunity to test a variety of mowing and burning treatments as potential control methods. She set up a trial using seven combinations of growing season mowing, dormant season and growing season burns, and intervals of no burning. The treatments were followed for 6 years with each sequence of treatments repeated twice. "A combination of an April (dormant season) burn in year one followed by a May (growing season) burn the next year was more successful at reducing *Melilotus alba* than any of the other treatments tested" (Kline 1984, 152). The study also found that mowing when the sweet clover is in bloom reduced its growth substantially, and could be a useful substitute where burning is not possible, or as a follow-up to control second-year sweet clover not killed by the fire.

Arboretum managers used the recommendations of this adaptive management research to inform the restructuring of the Arboretum's prescribed fire schedule and other management activities. As a result, by the mid-1990s, white sweet clover was virtually eliminated from the prairies. Seeds still come in on storm water, but new plants are quickly evicted by timely spot hand pulling. ∎

▶ 11.4 FACTORS THAT EXACERBATE THE IMPACT OF PEST SPECIES

The dynamics and patterns of pest species invasions of a restoration are—like everything else about restorations—situational in nature. The composition of a suite of pest species and their onsite distribution reflect site-specific environmental conditions, visitation and use patterns on the site, weather and climate, land use types in the surrounding built landscape, sociocultural features of the community, and the characteristics of the nearby pest species. It is difficult therefore to make general, sweeping predictions about the potential of a restoration being bothered by pest species. However, certain factors can exacerbate the various impacts of pest species on restorations, and you can anticipate them when developing your implementation and management strategies (see Chapters 8 and 10). These factors include (1) the proximity of the restoration site to a source of pest species colonists, (2) continuing human impacts that create conditions more favorable to pest species than to desired natives, and (3) situations in which circumstances (human disturbance plus available colonists) combine to create conditions that favor pests. A few examples follow to illustrate these points.

One factor that influences the impact of pest species is, of course, the likelihood that the species will colonize a site. And one thing that influences pest species colonization is the distance of a site from a source of colonists. Restoration is often done on a site that is surrounded by urban lands or old-fields filled with pest species. You will naturally be concerned that there will be repeat invasions—or new invasions—from across the administrative boundary, and that all your past pest species management efforts will go for naught. The good news is that at least when talking about pest plants in urban areas, not all natural areas surrounded by urbanization are heavily invaded, and that predictions of high invasion rates in urban areas in general may not always come true. However, it is important to be aware of this risk factor and to be vigilant, so that, if an invasion does occur, you can take steps to manage it (see Chapter 10).

Pest plants can take advantage of previous ecosystem changes, thereby facilitating further change. Reed canary grass (*Phalaris arundinacea*) and hybrid cattail (*Typha x glauca*) thrive in areas where the hydrological regime has been altered. Reed canary grass quickly moves into ecosystems disturbed by stormwater runoff, thriving on the higher water temperatures and sediments and nutrients therein (Kercher and Zedler 2004). Artificially high and prolonged water levels in lakes and wetlands, caused by damming, lead to colonization or expansion of cattails (*Typha angustifolia* and *Typha x glauca*) (Figure 11.3). Fluctuating water levels in wetlands result in periodic drawdowns, allowing phosphorus (a limiting nutrient) to bond with iron in the soil and become unavailable. However, when the wetland remains flooded for long periods, the resulting anaerobic conditions and chemical activity promote the release of phosphorus, and excess phosphorus encourages the growth of cattails.

Combinations of circumstances can lead to interactions that together create more problems for a restoration than any one of them would acting separately. The case study below is an example of one pest species (the native bark beetle) exacerbating the impact of another (an introduced pine blister rust), and human impacts acting locally and globally, adding to the problematic situation.

FIGURE 11.3. **The Impact of Pest Species.** This photograph shows the invasion of a sedge (*Carex* species) meadow by cattails (*Typha* species) in Gardner Marsh, Madison, Wisconsin. Notice the sedge meadow (light green) on the left and the advancing front of cattails (beige color) on the right. Cattail invasion is facilitated by relatively stable water levels in the marsh maintained by a dam and by nutrients and sediments in stormwater runoff. (Photo reprinted by permission from Steve Glass.)

CASE STUDY The Threat of Local Extinction to Whitebark Pine

If you travel anywhere in the mountain forests of western North America today, you will probably see the characteristic orange-brown color of dead needles on individual trees, small forest patches, or entire forest mountainsides. Periodic outbreaks of native bark beetles (*Dendroctonus, Ips, Dryocoetes,* and *Scolytus* species) are a natural disturbance feature of forested ecosystems in western North America. Although there are hundreds of species of bark beetles native to this region, only a few, the "aggressive" species capable of killing trees, are responsible for large areas of dead trees scattered across millions of acres of the West. Since about 1995, infestations of mountain pine beetle (*Dendroctonus ponderosae*), a type of bark beetle, have killed more than 570,000 acres of whitebark pine (*Pinus albicalus*) and limber pine (*Pinus flexilis*) in the northern Rocky Mountains of the United States (Figure 11.4). Bark beetles attack older trees, and trees that are weakened by stress and disease, thereby making room for younger trees to renew the forest. Bark beetles also feed on dead and downed trees, thus speeding up the process of decay and nutrient recycling. But the outbreaks since about 1990 are occurring to a degree not previously recorded. They are extraordinary because of their intensity, extensive geographic range, and the fact that they are happening all at once in many different ecosystems across the continent.

There are so many interacting variables that it is hard to untangle and attribute cause and effect, but certain patterns appear to occur at both regional and site-specific scales. On a continent-wide basis, scientists believe that contributing factors include:

- A changing climate across the entire outbreak area, thus both stressing the trees and favoring the beetles.
- A history of management practices (timber harvesting and wildfire suppression) in some forest types and geographic areas.
- An altered history of natural disturbances (wildfire suppression).
- Air and water pollution of forests, leading to increased stress and vulnerability to the beetles.

FIGURE 11.4. **A Tree Killed by Bark Beetles in Rocky Mountain National Park.** (Photo reprinted by permission from Steve Glass.)

- The introduction of an exotic pathogen, a blister rust, that both kills and stresses the trees.

Of course, over such a large geographic region, the patterns of infestation will vary also with the tree species and species of bark beetle involved. In a specific forest, contributing factors may include the species of trees, how densely they are packed, and local soil and weather conditions. In addition, wildfire suppression and lack of use of prescribed fire as a management tool may contribute to intensifying bark beetle infestations.

The whitebark pine (*Pinus albicaulis*) is a tree of high elevations found in the upper reaches of subalpine systems (near the timberline) in the Canadian Rockies, Washington, Oregon, the high Sierra of California, Idaho, Wyoming, and Montana. In addition to the bark beetle, the whitebark pine is also under attack from white pine blister rust (*Cronartium ribicola*), an introduced species that has infected more than 65% of the whitebark pine stands in the same area. Blister rust either weakens a tree, making it more susceptible to the beetle and reducing cone and seed production, or kills the tree outright.

Survival of the whitebark pine is critical because its seeds help support a number of wildlife species, such as red squirrels (*Tamiasciurus hudsonicus* and *T. douglasi*), black and brown bears (*Ursus americanus* and *U. arctos*), and Clark's nutcracker (*Nucifraga columbiana*), the tree's primary seed disperser.

The Clark's nutcracker not only disperses seeds of the whitebark pine; it depends upon the tree for survival. Clark's nutcrackers cache and eat the seeds of the whitebark pine. In turn, the whitebark pine relies on the nutcracker for its survival. Seeds of the whtebark pine are released only by the nutcracker and dispersed only by it because seeds do not drop from cones, nor are they dispersed in the wind. The whitebark pine depends on the Clark's nutcracker to disperse the seeds to suitable germination sites (open, pattern-rich landscapes created by fire with sparse ground cover and little snowpack in winter). Whitebark pine trees also serve as nurse trees for other high-altitude species; by stabilizing the soil and slowing the snowmelt in the timberline zone, whitebark pine trees, as they mature, create conditions favorable for the germination and establishment of Engelmann spruce (*Picea engelmannii*) and other timberline species.

Doubly troubling is the fact that the mountain pine beetle is killing whitebark pines that have developed genetic resistance to the blister rust—together, bark beetles and blister rust threaten whitebark pines with local and regional extinction. While restoration managers cannot protect all trees over such a large area, they are taking steps on a local scale to protect healthy trees. One tactic is the use of pheromone packets to repel mountain beetles on selected trees (Figure 11.5). Seeds of trees that survive by such measures may also be used in propagation and reforestation efforts. (For more exploration, see Sidebar.) ∎

FIGURE 11.5. Protecting Healthy Trees from Pests. Like others in the white pine group, the limber pine (*Pinus flexilis*) is susceptible to attacks from mountain pine beetles. On this tree, like many others in Rocky Mountain National Park, researchers have attached a packet of the pheromone *verbenone* in an attempt to deter attacks from the mountain pine beetle. (Photo reprinted by permission from Steve Glass.)

| Sidebar | Implications of Bark Beetle Outbreaks for Restoration |

Given the scale and results of bark beetle infestations and the apparent role of human activities in the outbreaks try to imagine the range of consequences—long term and short term, opportunities and constraints—and their implications for restoration and management. Think about how you would address them. Use the information in the text, as well as other sources. For more background, you can visit the website of the Rocky Mountain National Park and search for information on bark beetles and management activities.

FOOD FOR THOUGHT _____

1. What role(s) does restoration have in dealing with a complex pest species management problem on such a large scale? What about the role for restoration on a more local scale?

2. What does it mean to "manage" a pest species in this situation?

3. Does the fact that a native species is involved change your view on what can or should be done?

4. What are the goals, objectives, and targets of pest management in the case of the bark beetle infestations in the mountain west?

5. What can be done on a local (watershed) or site-specific scale?

6. What are some restoration strategies that are, or could be, used?

7. What would you do if novel communities developed in the wake of a bark beetle tree-killing episode?

8. What factors (biotic and abiotic, social and cultural) would you consider in making judgments about the above questions? Is this strictly an environmental problem? Does the problem have only environmental-based solutions?

11.4.1 Pest Plant Traits That Favor Invasion

Pest plant species have several life history traits that prove advantageous for colonizing disturbed settings. These include:

- High population growth rates.

- Well-developed mechanisms for pollination and seed dispersal.

- Rapid responses to resource availability (e.g., to eliminate or reduce resources, interfere with resource assimilation).

- Relatively low levels of natural controls in the environment into which they are introduced.

- Adaptations to disturbed habitat (broad range of tolerance, good competitive ability).

Restorationists and natural area managers are not always able to predict which species will become pests and cause problems for the restoration. Efforts to develop lists of probable invaders based on life history traits have been difficult because many additional variables play a role. Many species have one or several of the above characteristics, but having them does not necessarily mean a species will be invasive.

The ecological condition of the restoration site is another predictor of its potential for invasion by pest species. Because many pest species invade following ecosystem change, as in an altered disturbance regime that then drives ecosystem change even further, the more disturbed the site, the more likely it will be troubled by pest species. For many managers, simply finding what has been invasive under similar climatic and geographic conditions, and in similar community types or restoration settings, can be a great asset to developing management plans.

▶ 11.5 DEVELOPING A PEST SPECIES MANAGEMENT PLAN

The restoration management plan will often include a separate plan for managing pest species. The pest species management plan follows the logic of the restoration plan itself, and its development includes the same general steps as the development of the restoration plan.

The Nature Conservancy's pest species planning process relies on adaptive management principles. An adaptive pest species management plan includes key components of the restoration process, and follows a strategy that uses the lessons from previous seasons of work to mold future efforts. Figure 11.6 lists the various phases of an effective pest species management strategy.

11.5.1 Plant Pest Species Management Principles and Strategies

Restoration of areas colonized by pest species face the interrelated challenges of eradicating, containing, and/or controlling the pests, as well as replacing them with native species appropriate to the site. But attempting to eradicate, contain, or control all the pest species present on a site is rarely logistically feasible—or, in some situations, necessary. Therefore, your plan should include *priorities* for each pest species' control or elimination.

Early detection and prevention of possible new invasions is the gold standard of pest species management. (Examples of local, state, and regional early detection programs are listed in the Resources at the end of the chapter.) To minimize the potential for a long-term workload, you should act to prevent new pest species from entering the site. A good prevention tool is to identify those pest species that have proven invasive elsewhere in similar ecological and geographic situations, and that also have infestations nearby and outside the restoration. The team should focus on nearby pest species that are the fastest growing, most disruptive, and most likely to affect the most highly valued area(s) of the site. But early detection and most prevention are not always possible, and in that case your options are eradication, containment, and control.

---

1. Establish the restoration/management goals for the site.

2. Identify the pests, describe how they interfere with management goals, and then assign control priorities based on their impacts.

3. Determine what aspects of the site (ecosystem structure, function, process) may be facilitating pest species problems, and develop plans for correcting these underlying site problems.

4. Decide on effective control options. Assess the likely effects of the control efforts on the target (and nontarget) species. If necessary, adjust the control methods.

5. Develop and implement the management plan.

6. Monitor and assess the impact of management actions.

7. Evaluate the effectiveness of methods (measured against the site goals), and use this information to refine the control priorities, methods, and goals.

FIGURE 11.6. **Phases of an Effective Pest Species Management Strategy.**

Eradicating Pests

Eradication is the elimination from the site of an entire population of a pest species. When prevention has failed to stop the introduction of a pest species, an eradication program is one preferred method of action. Globally, there have been a few successful eradication programs. The Giant African snail (*Achatina fulica*) in south Florida, Siam weed (*Chromolaena odorata*) in Australia, white-spotted tussock moth (*Orgyia thyellina*) in New Zealand, and rabbits on Phillip Island in the South Pacific are examples of successful eradication efforts. There are examples of failed attempts as well. A notoriously failed eradication program was the effort directed at the introduced fire ant (*Solenopsis invicta*) in the southern United States. An insecticide reduced the populations of not only the introduced ant species but most other ant species as well, and the introduced ant species was able to repopulate the area more quickly than the native ant species. To accomplish a successful eradication program, you need to detect invaders early and act decisively to remove them swiftly before a sustaining population is established.

Eradication efforts are most effective on "islands" with isolated populations, or places where the chances of reinvasion are limited. Eradicating charismatic species, such as deer, can raise public opposition. Eradication is scale-dependent, meaning that you can undertake eradication efforts in one area of concern in conjunction with containment in another area of concern. For example, in Hawaii, *Miconia calvescens* is a serious pest plant. There, an eradication program aimed at the local or island level is combined with containment elsewhere.

Eradication is costly. But if the desired outcome is achieved, it is a much less expensive option than a long-term control or containment program. However, failure may mean you have wasted those resources and time.

Containing Pests

Containment of nonindigenous invasive species is a special form of control. The premise is that resources, knowledge, and/or the capability to eradicate the species from the site are not available. Instead, the aim is to restrict the spread of a species and to contain the population to a defined geographic area. To be successful, it is important that you clearly define containment goals. For example, you could illustrate the goals on a map to show the areas or "line in the sand" beyond which the pest species cannot go.

Controlling Pests

The control of pest species involves setting goals of species density and abundance to below a preset acceptable threshold. The idea is that you determine a population level threshold low enough that the harm to the community/ecosystem caused by low numbers of the species is acceptable. The control strategy is used in situations where eradication is not feasible, but it is possible to keep the species numbers low. The premise, in part, is that the costs and available time for eliminating a species entirely from a site, even if it were possible, are extremely high. The ability to have resources and time to devote to other needs overrides the need to completely eradicate a species, particularly one that can be controlled at levels where its interference with project goals is relatively insignificant.

Pest plant species control strategies are often based on life history characteristics. Based on some of the life history traits listed above, here are some control guidelines:

- For plant species with high population growth rates, it is best to initiate control efforts when populations are small and new.

- For plant species with well-developed mechanisms for pollination and seed dispersal, concentrate control efforts on preventing flowering and seed maturation.
- For plant species that grow rapidly when resources become available, eliminate or reduce the resources.
- For plant species with few natural controls in the environment, introduce physical or biological controls.
- For plant species with a broad tolerance to disturbance, control the occurrence and timing of disturbances; introduce disturbance-tolerant natives.
- For plant species with good competitive ability, weaken the species, such as by removing some of the biomass through mowing.

11.5.2 Identifying Pest Species to Target

In deciding which of the management strategies to initiate, you must first identify which pest species to manage immediately and which ones to manage later, if at all. One way to guide decision making about potentially undesirable species and the appropriate management strategy is to ask questions related to the goals of the community/ecosystem model. Rarely will an answer be simply yes or no. The answers will help determine, from a restoration perspective, the level of management priority a species should be given.

1. *How and to what extent does the species alter community composition, structure, or function across space and through time?* Pest species that you suspect are initiating ecosystem change by altering ecosystem structure, function, or process make restoring native species difficult or impossible, and they will probably receive a higher ranking in the risk analysis than species that colonize an already altered system. In Australian tropical savannas, for example, the invasive Gamba grass (*Andropogon gayanus*) alters the fire regime by increasing fuel loads and fire intensity, thus damaging native species that only tolerate the lower-intensity fires of the native grasses.

   Pest species also interfere with species interactions directly and indirectly—food chains, nutrient cycles, and so on. The leaf litter of buckthorn (*Rhamnus cathartica*), an introduced woody pest species in the northeastern United States, has a low carbon-to-nitrogen ratio. Non-native earthworms speed the decomposition of the buckthorn leaf litter; the rapid nutrient recycling favors both the earthworms and the non-native shrubs, creating an invasion complex.

2. *Does the species replace or diminish key species or life forms that play pivotal or essential roles in the survival of a community? If so, what is the mechanism by which it exerts influence?* An example is Australian pine (*Casuarina equisetifolia*), now widely established in the Hawaiian islands, on Florida's coasts, and in Puerto Rico, the Bahamas, and other Caribbean islands. Its dense, thick needle mulch shades out native dune and beach plants; ultimately, stands of Australian pine alter the microclimate of the dune environment.

3. *Is the pest species infestation likely due to an underlying site condition? Does the site condition need to be corrected before the pest species can be managed?* In the U.S., invasions of reed canary grass (*Phalaris arundinacea*), facilitated by urban stormwater runoff, have formed extensive monocultures that have displaced native prairie and wetland vegetation. In situations where the species is not intermixed with desirable species, it can be removed, but preventing its return without control over storm water is difficult.

11.5.3 Setting Priorities for Action

Your priorities for management actions should reflect the current onsite status of each pest species, the likely severity of present or future harmful impacts, the value of the habitats the pest could invade, and the feasibility of control. There are several standard approaches for evaluating the risks posed by pest species. A priority-setting process takes a risk analysis approach and may use a number of categories or filters to screen out the worst pests. The approach discussed here, and summarized in Figure 11.7, is based on the pest species priority ranking system developed by Hiebert and Stubbendieck (1993).

You can assign weighted criteria to each of the categories in Figure 11.7 that refine the rankings; the lower the score, the more serious the threat. Hypothetically, the highest priority for control possible for a restoration site would be an easily controlled, ecosystem-changing pest species, of which a few individuals have been found in a single remnant community that is likely to recover quickly (such a species would obtain a score of 4). Conversely, the lowest-priority pest species for control (a score of 15) would be one that is at or near saturation levels, or one that follows ecosystem change, has invaded a buffer or degraded area, and is difficult or currently impossible to control.

You assign ranking scores using the best available information culled from a variety of sources, including research reports, area restoration experts, land management agencies, onsite experience, and local experts. For best results, you should revisit the priority ranking regularly to incorporate new knowledge, to drop from the list species that have been controlled, and to add new species of concern.

You should also factor a cost/benefit analysis into the decision-making process. For example, in the South African Working for Water Program, the cost to clear the 1.7-million-ha infestation of woody shrubs because of their use of scarce groundwater was estimated to be about $1.2 billion in 1998. However, by strategizing for solutions that have

---

1. Current extent of the species on or near the restoration, or likelihood of the species establishing onsite.

   - A few individuals or small isolated patches, at low levels, but likely to increase in size and aerial extent.

   - Many scattered populations that, in disturbed areas, are expanding in size and aerial extent.

   - Small but persistent populations that are contained or unlikely to expand.

   - At saturation levels, stable and not expanding.

2. Current and potential significance of the impacts of the species in relation to the restoration goals.

   - Those species that alter ecosystem structure, function, or processes. In this case, the pest species triggers ecosystem change.

   - Those species that outcompete natives and dominate otherwise undisturbed areas. Colonization occurs simultaneously with ecosystem change.

   - Those species that overtake natives following disturbance. Ecosystem change triggers colonization.

3. Value of the habitats/areas that the species infests or may infest.

   - Remnants.

   - Legacy plantings or long-established restorations.

   - Recent and successful restorations.

   - Buffers and severely degraded areas.

4. Ease or difficulty of control.

   - Easy to control; native species will recover on their own.

   - Easy to control but may need to replant native species.

   - Difficult but possible to control, and controls may have undesirable side effects.

   - Hard or impossible to control, or control methods unknown, or little of the native system may remain.

FIGURE 11.7. **A Pest Species Risk Analysis and Priority Ranking System.**

multiple benefits, in this case a more efficient water delivery system, the estimated cost was reduced to $400 million.

As discussed in Chapter 4, it is important to note species that are not yet on the restoration site, but are found nearby and could be problems if they spread to the preserve. You can detect any new infestations by regularly monitoring the site and the larger watershed or regional landscape. Your risk analysis should also factor in the population density, distribution, and frequency of pest species onsite. Pest populations may range from species that occur as just a few individuals or isolated patches that are not likely to spread, as scattered populations expanding in size, to species that are at saturation levels and are unlikely to increase in density, distribution, or frequency. Each situation requires different kinds and degrees of attention, from casual observation to active management.

KEY CONCEPTS

- Pest species are plant, animal, insect, and disease species that interfere with site-specific restoration goals. Preventing the introduction and establishment of various pest species is a major goal of restoration management.

- Pest species may be native to the site or introduced from other areas. Many introduced species do not cause restoration challenges, but those that do can be difficult to control.

- Three factors in particular make ecosystems more vulnerable to pest species invasions and control more difficult: human transport, climate change, and human land use patterns.

- A pest species management plan includes criteria for identifying species that are behaving as pests on a particular site, sets management priorities, and sets specific strategies for eradication, containment, or control.

FOOD FOR THOUGHT

1. Consider a few common pest species that occur in your area. What impacts have they had on a restoration with which you are familiar?

2. Team up with a partner. With each of you taking opposite sides, discuss whether the pest species from question 1 are drivers or followers of ecosystem change.

3. Conduct a risk analysis of the pest species in your area, and assign a priority ranking for each. (Alternatively, you could do this together with your partner.)

4. What steps would you take to manage these pest species? Are they amenable to control or eradication, or is there an underlying ecosystem disruption that must be corrected before they can be controlled?

5. What factors would you consider in making a choice between a pest species eradication, containment, or control program? What would be the key elements of your management approach? Consider your response if the pest were an animal instead of a plant species.

Resources for Further Study

1. Pest species invasions of natural areas are a global problem. It is not surprising that a number of groups use the Internet to share resources and information. Here are links to several of these organizations:

   The University of Georgia's Center for Invasive Species and Ecosystem Health focuses on invasive species of the forested systems of the southeastern United States, but also covers other invasive species of the U.S., with good links to and coverage of European, Mexican, and Pacific Island invasive species issues. http://www.invasive.org/

   The North European and Baltic European Network on Invasive Alien Species (NOBANIS) provides information on invasive species in North and Central Europe. http://www.nobanis.org/default.asp

   The European and Mediterranean Plant Protection Organization is an intergovernmental organization focused on plants. http://www.eppo.org/

   Weeds Australia is the portal to all things weeds in Australia. http://www.weeds.org.au/index.html

2. The following publications discuss some of the theory behind pest species control:

   D'Antonio, C. M., and J. C. Chambers. 2006. "Using Ecological Theory to Manage or Restore Ecosystems Affected by Invasive Plant Species." In *Foundations of Restoration Ecology*, edited by D. A. Falk, M. A. Palmer, and J. B. Zedler, 260–279. Washington, DC: Island Press.

   Mooney, H. A., R. N. Mack, J. A. McNeely, L. E. Neville, P. J. Schei, and J. K. Waage, eds. 2005. *Invasive Alien Species: A New Synthesis*. Washington, DC: Island Press.

3. For more information about the whitebark pine case study, see the website of the Whitebark Pine Ecosystem Foundation. http://www.whitebarkfound.org

Managing User Impacts and Cross-Boundary Influences

People interact with plant and animal communities/ecosystems in a variety of ways, ranging from preservation to exploitation, and at a variety of scales, from local to global. Many human interactions with the biotic world result in changes in ecosystem composition, structure, and function at the local scale, which may or may not be desirable from a restoration point of view. In such cases, people can be said to be acting as disturbance agents (see Chapter 2). In other cases, the impacts have global effects that match or even exceed the magnitude of those resulting from naturally occurring fires, floods, or windstorms. Climate change is an example of a global impact resulting from human activities. So, too, are increases in the rates of nitrogen fixation through the industrial production of fertilizers; the introduction of novel chemicals to the Earth's ecosystems in the form of plastics or pesticides such as DDT; and the rapid transportation of organisms across continents, some of which become serious pests (see Chapter 11). It is important for you to make restoration decisions that minimize the potential for global environmental impacts—for example, limiting the use of inorganic fertilizers or pesticides—and to be aware of the predicted effects of global change on the ecosystems in which you work. However, because the day-to-day impacts at local and regional scales are often the ones of most immediate concern, and the most amenable to corrective action, we focus this chapter on those human impacts you can plan for and mitigate in the context of individual restoration projects.

Restorations are influenced by the sociocultural setting in which they are conducted. People are major disturbance agents both within, and outside of, restorations. Thus, a restoration is just as likely to be as affected by activities occurring in the community and landscape beyond its administrative boundary—that is, cross-boundary influences—as by onsite user impacts, such as the activities of tourists, local visitors, contractors, utility agencies, and restoration operations.

User impacts and cross-boundary influences are management concerns for restorationists in both urban and rural areas throughout the world. Restoration activities are expanding globally and related visitor impacts are increasing, especially in the tropics where ecotourism and restoration have become important components of the economy of such countries as Costa Rica and Belize. And yet the typical attraction features—rock outcrops and cave complexes, geyser basins, waterfalls and coastlines, flowering plants, animals, and scenic overlooks—are all highly susceptible to disturbances from human activities.

Experience has shown that if visitor impacts and cross-boundary influences are ignored or not addressed early on—before the problems become expensive to fix or before use patterns become established—restoration goals will be more difficult to achieve. Here are the primary reasons:

- Impacts happen quickly at either the initial use or very low levels of use.
- Impacts are cumulative, building up gradually to degrade the resource over time.
- Restoration goals and objectives are compromised; managers may be unable to fulfill their resource protection mandates.
- Impacts create other user problems, such as visitor dissatisfaction.
- Overuse can make it difficult to achieve the social and cultural goals the restoration was designed to support. This may lead to reduced community participation and financial support.

Because restoration ecologists and other land managers seek to protect the natural resources of the sites for which they are responsible, a restoration plan will include a section on managing people's activities onsite, as well as methods for counteracting external influences. This chapter suggests a range of planning approaches and strategies to mitigate potential human impacts.

▶ 12.1 USER IMPACTS

User activities can have a negative effect on the resource (vegetation, trails, water, or wild-life) or on the quality of a user's experience of the site (e.g., overcrowding, user conflicts, visitor dissatisfaction). Impacts result from one-time actions or long-standing practices and behaviors. Some impacts are largely unavoidable, such as vegetation trampling and soil compaction, but many potential impacts due to inappropriate visitor behavior—litter, tree damage, plant harvest, disruptive visitor behavior—can be avoided or minimized.

Damage to natural resources is a concern where visitor use is high. In places ranging from Central American ecotourist centers to high-use wilderness recreation destinations in the Pacific Northwest, user impacts have resulted in trail proliferation and increased soil erosion resulting from bare soil, exposed roots, and water running on deeply worn trails. Similar user impacts are felt in remote spots like the Daintree Tropical Rainforest World Heritage Site in Australia's northern wet tropics, as well as in world capitals like Stockholm's heavily used National Urban Park, which contains the restored Isbladskarret Fen.

Direct user impacts at Oregon Caves National Monument include broken cave formations, rock darkened and polished by human hands, and graffiti, along with less obvious impacts, such as sediment compaction and the rearrangement of animal bones and fossils. Managers at Oregon Caves have also observed indirect user impacts resulting from environmental changes caused by altered airflow due to the creation of large entrances and passageways, or the introduction of light into what would otherwise be a dark environment. These changes have caused the growth of algae on the fragile rock formations within the caves, which can hasten their decay.

Other examples of impact include damaged coral reefs, compacted soil around tourist viewing benches at Volcan Poas in Costa Rica, the trampling of sensitive vegetation, and the breaking of thermal crusts and chemical pollution of geyser channels at Great Fountain Geyser in Yellowstone National Park. In Norway, years of visitation to the popular King's View overlooking Tyrifjorden, a large inland fjord, has rutted the tourist trail, exposed tree roots, and erased the forest undergrowth surrounding the overlook (Figure 12.1).

If left unchecked, user pressure can damage the natural features that draw visitors to a site. In one restoration, managers observed that it took only a few people wandering off trail for a closer look at an endangered orchid to kill vegetation and expose the soil to rainfall. Over time, managers noted that one visitor-created trail led to another, and another, eventually developing areas of bare soil. As the exposed soil eroded, nutrients were lost and pest species invaded. Soon the changed habitat and pest species reduced the orchid population to a few individuals that were below the viable population level. In this case managers failed to take action, and when the orchid ultimately vanished, visitation to, and public support for, the site declined. Thus, not repairing an eroded trail, failing to protect an endangered plant from poachers, or not picking up litter can be as much of an impairment as the original impact.

As visitor numbers increase and the activities allowed on a site expand, the greater the need for devising accommodations through design, supervision, or education to avoid harmful impacts generated by undesirable user behavior. Managers of the Crater Lake Trail in the Snowmass Wilderness near Aspen, Colorado, took early and decisive management action to close a section of switchback and volunteer trail by covering it with deadfall trees and logs.

FIGURE 12.1. User Impacts. Tyrifjorden is a large inland fjord accessible from the Kings's View in Norway. A popular and age-old tourist walk, the undergrowth and ground layer plants have been erased by heavy human foot traffic. (Photo reprinted by permission from Steve Glass.)

▶ 12.2 CROSS-BOUNDARY INFLUENCES

Some human impacts to the restoration site are external; the activity or event happens or originates outside the restoration, and the results cross the administrative boundary. For example, stormwater runoff and groundwater depletion originating from offsite can bring sediments and pollutants into a wetland and at the same time reduce its base flow. Ecologist Dan Janzen recognized that "a preserve boundary is a sponge rather than a wall" (1986, 288), acknowledging the porous nature of artificial boundaries and their susceptibility to biological influences from outside. In fact, as Bader and Egan (1999) suggest, the flow goes both ways because restorations influence, and are influenced by, the world around them.

The U.S. National Park Service points to air pollution, aesthetic degradation, noise pollution, and pest species encroachment as critical cross-boundary influences. Poor air quality days in a region may preclude the use of prescribed management fires for a day or two; conversely, when limits are lifted, smoke from prescribed fires can adversely affect sensitive populations downwind. Likewise, herbicide drift and noisy power equipment can have negative cross-boundary influences on public health and safety.

12.2.1 Shared Resources, Conflicting Goals

Utility lines, transportation corridors, and communications towers are a special and dramatic case of cross-boundary influences. For example, cell phone and radio towers in a restoration area restrict public access and management options, such as prescribed fire. Utility lines that physically bisect a restoration or protected natural area can be a management problem by fragmenting the restoration and opening it to human trespass and invasion by pest species. Routine management and repair of these facilities can be detrimental or beneficial, depending upon the restoration goals. Annual brush mowing of a utility corridor may be contrary to the restoration goal of developing mature woodland, but it could be ideal preparation for grassland restorations and prescribed fires.

CASE STUDY   The WHA Radio Tower

An example of unintended consequences and how ecosystems can be damaged, perhaps irreparably, when there are unmediated and unresolved conflicting goals over shared resources occurred in Madison, Wisconsin. WHA radio, which is part of the University of Wisconsin Extension and a local affiliate of National Public Radio (NPR), operates a transmission tower in a remnant and restored wetland complex (the Southeast Marsh), managed by the University of Wisconsin Arboretum.

The UW system administration authorized building the tower (over objections from the Arboretum) in the early 1970s, when wetlands were viewed as expendable, waste places. The wetland was once part of the vast wetland complex that surrounded Lake Wingra. Southeast Marsh has suffered the usual insults of the urban environment but still contains the largest and highest-quality sedge meadow in the watershed. The presence of the tower, along with its support wires and 6 miles of underground copper wire antenna, have since prevented the use of prescribed fire as a research and management tool to maintain the marsh.

Management concerns extended beyond the prohibition on prescribed fire to the 6 miles of buried copper wire, which could cause the release of copper and chromium from corroded wire into the groundwater. In 2008, the tower required reconstruction because of metal fatigue. WHA administration turned down suggestions to relocate the tower outside the wetland, citing a tight timeline, plus federal regulatory hurdles and red tape. Pending construction in a wetland raised additional concerns, which included the likelihood of:

- Stripped vegetation and mineral soil exposed to invasion by pest species.
- Release of sequestered contaminants.
- Compacted soil.
- Altered local hydrology.
- Altered soil horizons.
- Introduction of non-native materials.
- Spread of herbaceous pest species.

These concerns were raised during the initial Environmental Impact Assessment (EIA) but were ultimately determined in the final EIA to be of no environmental consequence (see Sidebar). Construction began in the fall of 2008. You can see the surface impacts during construction in Figure 12.2 and the results in 2010, two growing seasons later, in Figure 12.3 ∎.

FIGURE 12.2. **The Construction of a Radio Tower in a Restored Wetland in 2008.** The surface impacts of construction equipment are obvious. Marsh vegetation was stripped, and the wet soil was highly disturbed because contractors failed to follow the precautions of working on frozen soil and using swamp mats to cushion the marsh from heavy equipment. Below the surface, a 6-mile wire antenna could leach copper into the groundwater. (Photo reprinted by permission from Steve Glass.)

FIGURE 12.3. **Results of the Radio Tower Construction in 2010, Two Growing Seasons Later.** Notice the patch of cattails (*Typha* species) that have invaded the ponded area shown in Figure 12.2. (Photo reprinted by permission from Steve Glass.)

| Sidebar | Impacts of the WHA Radio Tower Project |

Graduate students in the University of Wisconsin–Madison 2007 Water Resources Management Practicum studied the Arboretum's Southeast and Gardner marshes prior to construction, and were thus well qualified to speak to the potential impacts of the tower project on the sensitive wetlands. In testimony presented to the Environmental Impact Assessment (EIA) review panel, the students warned of the potential impacts to the wetland if work was done when the soil was not frozen or if precautions were not taken. In the practicum report, the students warned, "If the ground is not deeply frozen, heavy construction equipment will compact the organic soil, cause ponding, and increase habitat for invasive cattails" (WRM 2008, 56). This prediction came true.

Despite recommendations and language in the final EIA that mandated the project to take steps to minimize construction impacts on the Southeast Marsh, few of these simple precautions found their way into the final contract specifications, and none of them were actually followed in the field during construction (see Figure 12.2). A chief recommendation was to use swamp mats. Swamp mats are like portable ramps or bridges made of wood-lagging material that spreads the weight of construction equipment over a large area, thus resulting in a reduced distributed pressure. Why weren't swamp mats used in this project? The project manager explained that he did not deem their use necessary and were an added expense to the contractor. The result of this oversight was that cattails invaded the ponded area (see Figure 12.3).

12.2.2 Impacts from Neighbors and Local Residents

Area residents and neighbors of restoration projects who see the site as a resource to use for their livelihood or for recreation are a type of cross-boundary influence. Their activities can contribute to trail proliferation, water pollution, and litter. Neighbors and local residents can also be harvesters and/or consumers of natural resources when they engage in these activities:

- Traditional use patterns
- Subsistence agriculture
- Hunting and fishing
- Water diversions

These uses represent a form of competition for shared natural resources that can strain the relationships between local residents and protected area managers, unless the two groups work together from the outset.

An important tool for achieving a balance between site ecology and use is the visitor use-policy of the master plan, which spells out acceptable types and levels of impacts and the necessary steps for keeping those impacts within tolerable levels. The use-policy addresses the primary questions the restoration team and the manager will have: what types, extents, and locations of various impacts are allowed or tolerated without compromising public support, the visitor experience, and the quality of the resource.

▶ 12.3 POTENTIAL PLANNING APPROACHES

There are three major planning approaches restorationists can use to minimize human impacts and maximize resource protection:

1. Set visitor use-policy, visitor experience, and restoration protection objectives in collaboration with stakeholders.

2. Manage people through the site design; establish infrastructure (signs, trails, roads, etc.) to minimize potential use impacts.

3. Invite people to participate in managing the restoration.

These planning approaches will help to establish a balance between site use and site protection.

12.3.1 Setting Use-Policy, Visitor Experience, and Resource Protection Objectives

The restoration use-policy establishes guidelines for the ways in which people are encouraged to interact with the site and the kinds of uses that are not allowed (see Chapter 6). Use-policies cover two broad activity areas—those of visitors and those governing the restoration's implementation and management activities. You derive both sets of policies from the restoration purpose and the results of the site inventory and analysis. Over time, you can refine the policies based on your onsite experience and adapt them to meet the needs of changing conditions both inside and outside the restoration.

The restoration visitor use-policy describes the types and levels of impacts that are acceptable and the steps needed to keep those impacts within tolerable levels. Use-policies establish who can visit or work on the site, what kinds of activities are allowed, when and where they occur, and under what circumstances they can take place (i.e., time of year, weather conditions). For example, restorations established to provide habitats for threatened or endangered species are designed to limit human use. In such cases, the focus may be on finding ways to keep visitors away, at least from the most critical portions of the habitat. On the other hand, a large number or restoration projects are designed to encourage people to visit and interact with nature. In these situations, your goal is to find ways to facilitate use while limiting damage. Activities that restoration use-policies often explicitly prohibit include motorized vehicles, horseback riders, dogs, skateboarding, and collecting or removing seeds, plants, animals, and rocks. You may also have rules for closure during inclement weather, breeding seasons, and so on.

Although a common tendency is to define the problem of impacts only in terms of sheer numbers of visitors and to simply set strict limits on the number of people allowed on the site at any one time and place, situations are usually not that simple. A more constructive and helpful approach includes a careful consideration of the desired visitor experience in terms of the restoration goals and resource conditions the visitor is intended to view, balanced against the resources the restoration team has available to repair the user impacts.

The U.S. National Park Service (NPS) has a long and successful history of balancing visitor experience and resource protection, by upholding the following guidelines:

- Modifying the character of visitor use by controlling who goes where, when, for what purpose, and for how long.
- Modifying the resource base by making it more resistant to impacts.
- Reducing use in all or part of the property.
- Influencing visitor attitudes and expectations through displays and nature walks.
- Using site design (fencing, facility or trail locations) to minimize impacts.

The NPS often sets different use-policies for various parts or zones of each park, and establishes the management resources needed to keep the site in good condition.

Several approaches are available to balance access, visitor experience, and impact with regard for the particular sensitivities of a specific project, including managing people through site design, increasing management resources, and engaging the community to help with stewardship and management (Figure 12.4).

12.3.2 Managing People Through the Site Design

Whether or not a use-policy welcomes visitors and volunteer workers to the site, it is helpful to be able to set limits for the numbers of people allowed on the site, as well as to limit activities to certain portions of the property. Three design features are particularly helpful in controlling access: perimeters and entrances, trails, and parking lots. If they are designed appropriately, these features can facilitate both visitor experience and management activities. For instance, trails can be designed according to universal design principles to make them easily accessible, and also to ensure they are functional firebreaks during prescribed burns.

Perimeters and Entrances

Factors that determine whether a restoration site is more or less protected from external influences include the shape and size of the preserve, the ratio of edge or perimeter to area, and the nature of land use outside the preserve. Minimizing edge to interior, for example, reduces border access points (see Chapter 2).

One way to influence human access is to create edges that are either difficult or easy to cross. Protecting a sensitive restoration might require fencing the entire perimeter of a site. However, the drawbacks of fencing often outweigh its advantages. Fences are expensive to install and maintain, especially if determined visitors make their own openings. Fences can be barriers to the free movement of desired animals, and they create support for unwanted vegetation (such as vines). To the extent that they are visible from within a restoration, fences can have a significant impact on the user's aesthetic experience; they can be either an enticement or a challenge to overcome.

FIGURE 12.4. **Site Design Improvements for the Upper Geyser Basin (Old Faithful) in Yellowstone National Park.** One of the largest, most accessible, and most frequently visited features in the basin, this area was starting to feel the effects of visitors who left the small, existing boardwalk, walking on fragile vegetation and thermal crusts and putting themselves at risk of exposure to the thermals. In addition, the road edge was deteriorating and falling into the runoff channels that feed the geyser basin. If the problems were allowed to continue, visitors might get hurt, and permanent damage to the basin might result. The design solutions included lining the road surrounding the basin with weather-treated logs to contain vehicles, creating easier handicapped accessibility, and expanding viewing opportunities. (Photo reprinted by permission from John Harrington.)

In some situations, you can protect the edges of a restoration by encouraging the growth of dense woody vegetation along the perimeter. Such "living fences" can be very effective barriers to human trespass. However, such shrub masses would not work in many grassland situations; in fact, they might cause more of an impact on the restoration than the human impact they were designed to reduce. This is because many shrub species can colonize grasslands and often support birds that deposit the seeds of additional potential invaders.

Regardless of the accessibility of the perimeter, an effective way to manage human impact is to create one or more entrances—easy and visible access points designed to welcome and orient visitors. By locating entrances at trailheads and including kiosks or displays with an area map, you can help visitors find their way around a site and increase the chances that they will remain on trails. You can also use the kiosks to post information about seasonal site attractions, volunteer opportunities, and other upcoming community activities—all ways to continue to build a community. Kiosks and signs are also excellent means of informing visitors of use-policies, and explaining why minimizing use impact is important (Figure 12.5). If a project is large enough, an entrance might include a visitor center and/or a checkpoint with staff available to answer questions or otherwise assist visitors.

Trail Design

In addition to determining an appropriate level of access, another decision you will have to make in designing a restoration is whether or not to include trails. Trails are, obviously, convenient for visitors because they help people find their way into and out of an area and can provide a direct and controlled route to points of interest. They also help contain visitor impact by channeling the foot traffic. By careful placement, trails can direct visitors to specific areas of the site and keep them out of other more sensitive environments, or at some distance from rare species that are easily damaged (Figure 12.6).

FIGURE 12.5. **Communicating Visitor Use-Policies.** Placed by National Park Service managers near a spur trail along the Yellowstone River Canyon, this sign indicates the level of concern about the extent of user impacts. Informational and educational signs encourage desired behavior, in this case explaining the purpose for the restoration and urging photographers to keep to the footpath to avoid trampling trailside vegetation. (Photo reprinted by permission from John Harrington.)

FIGURE 12.6. **Effective Trail Design.** This raised boardwalk with railings meanders through a wetlands restoration. It controls access and limits damage to the wetlands. It also provides opportunities for birdwatching and small group discussions, without interrupting circulation, by including widened waysides or pullouts along the path. (Photo reprinted by permission from Cate Harrington.)

By channeling the use, trails concentrate the human impact—in effect, they sacrifice some portions of the restoration to protect others. On the other hand, the absence of any trails means that visitors may wander anywhere on the site. This dispersed usage, if visitors are few, may result in minimal impact. If there are lots of visitors, however, much of the site and/or sensitive areas can be damaged.

For educational guided tours, you can create trails that are wide enough for groups of three to five people to walk abreast. You can include periodic waysides, or pullouts, for individuals to rest or groups to assemble and talk. The waysides can be small and intimate or large and social, depending on the use-policy. To meet the needs of a diversity of visitors, you can design trails with a number of loops to allow visitors to walk for short and/or long distances and over level or rolling topography.

Surface quality can influence use as well. You can create trails that are paved with gravel, macadam, or concrete, or constructed of wood, thus welcoming the more casual visitor. Or surfaces can be rough, perhaps unpaved, thus providing a more challenging experience. However, wide and paved trails can impact the restoration by fragmenting the species habitat because the trail corridors act as barriers to movement. Developed trails that are highly visible to visitors can diminish the enjoyment of a site for those visitors who desire to immerse themselves in nature. On the other hand, most visitors are likely to remain on well-groomed established trails; therefore, visitor impact can be contained and minimized.

The placement of trails with regard to a site's topography, as well as the choice of surface materials, can influence the impact of visitors. Trail corridors running up and down a slope often become water and soil erosion channels, ultimately leading to the creation of gullies and deposition piles if the situation remains unchecked. There are many design options available for managing trail erosion, including switchback designs, and the use of check dams or porous pavers.

The Maroon Bells near Aspen, Colorado, is a busy tourist destination and jumping-off point for backcountry hiking. The popular Maroon Creek Trail and the more rugged trail to Crater Lake receive a lot of wear and tear from hiking boots, but the steep terrain does not provide managers with many options when the trails become overused and eroded. One solution was to temporarily close the segments of the rutted trails so that they can be revegetated and to open new trail segments paralleling the old ones.

Parking Lots

The size and location of parking areas contribute to making site access easy or difficult. This factor obviously has an effect on the numbers of people who will visit a restoration. If a parking area is full, people will often choose to postpone their visit. Similarly, if people have to park remotely from a site, casual visitors may decide not to come. To restrict the numbers of people on a site at any one time, restoration planners can limit the amount of available parking and place the parking lots at a distance. You need to make this type of decision in the context of the restoration use-policy.

12.3.3 Inviting People to Participate in Managing the Restoration

Most restorations reach out to the community to recruit citizens to join the team and train them to take on part of the job of maintaining the site. Frequently, restoration managers engage the community by recruiting volunteers to serve in several roles. These include:

1. Citizen scientists and students to monitor impacts on the land, sample vegetation, or conduct rare species inventories.
2. Volunteers to join workdays to perform some part of the restoration job.
3. Volunteer stewards and/or rangers to inform and educate public.
4. An engaged and attentive public who will look out for the restoration.

Here are a few examples of these approaches in action around the world.

Citizen Scientists

At the Oregon Caves National Monument, resource managers have monitored visitor impacts since 2005 through a combination of inventories, assessments, surveys, and digital photographs. Now, they are seeking funding for a project to train volunteers to map visitor impacts using a geographic information system (GIS). Similarly, at the University of Wisconsin, a botany/zoology class has participated in a 5-year pest plant control research project by annually sampling vegetation to measure the effectiveness of alternative treatments. These are just two examples of citizen scientists assisting in restorations.

Volunteer Workdays

Volunteer workdays are a popular way for you to attract individuals and groups to help with your restoration. Workdays, or work parties, can be one-time seasonal events or scheduled on a regular basis year-round—say, once or twice a month on Saturday. In many restorations, volunteers are recruited to serve as leaders of other restoration volunteers. (You can probably think of local examples close to your home.) These prospective team leaders receive extensive training in restoration techniques, leadership and communications skills, and general ecology. Many university students in natural resource management fields find this kind of volunteering to be a valuable supplement to their classroom work and a way to gain real-life restoration experience.

In the Atherton Tablelands in Queensland, Australia, the reintroduction of tropical forest trees is aided by the traditional ecological knowledge of indigenous peoples who identify native species and determine the proper time to collect fruits and seeds. Additional volunteers then help clean and sow tropical forest tree seeds, care for seedlings in the nursery, and ultimately plant them in stream corridor restoration projects.

Volunteer Stewards and Rangers

Volunteers sometimes serve as trailside interpretive rangers and land stewards, performing an important role as ambassadors of the restoration to the visiting public. This volunteer service is helpful, because, despite your best efforts to increase visitor satisfaction and minimize the impact of human use, by developing community support and designing restorations with people in mind, in some situations it will be necessary to actively encourage desired behavior. If possible, it is beneficial to have site rangers—staff members or trained volunteers whose role is to walk the site and interact with visitors. The idea is that visible rangers will act as a deterrent for visitors who knowingly violate a site's use-policy. If someone does not know the policies, the ranger can intervene and provide the necessary information.

The above planning ideas are suggested approaches, a list of menu items from which you can choose to customize a solution that fits your situation. Few restorations can afford to include all of these approaches in their management prescriptions, and you need not feel that you must use them all. Because the typical restoration project rarely has enough time, money, and staff to accomplish all the goals and objectives, you will have to be resourceful and creative and prioritize actions to carry out your work efficiently. Consider what is most important, and choose the combination of approaches that addresses your site's management needs. For example, if trailside education and outreach are needs, and you cannot afford a paid ranger, then concentrate on recruiting and training volunteer naturalists and stewards to fill these roles.

CASE STUDY | The Presidio in San Francisco

An example of an organization that cultivates and engages its public well is the Presidio in San Francisco, California. A former military base, the Presidio is one unit of the Golden Gate National Parks Conservancy, which stretches 70 miles north and south of the Golden Gate Bridge and includes more than 80,000 acres of federally protected parklands, including Alcatraz, Muir Woods, the Marin Headlands, Mount Tamalpais, and Point Reyes National Seashore. Protected natural features include ancient redwoods, shorelines, and rare and endangered species. At the Presidio itself, more than 1,400 acres of land have been saved as national parkland. Of that, more than 100 acres of shoreline park have been created at Crissy Field.

The Presidio has the dual mission of protecting natural resources and providing for public enjoyment. It has creatively found ways to do both. For example, a tidal marsh restoration in the shadow of the Golden Gate Bridge at the Presidio relies on attractive sturdy fencing, gateway portals, and informational signs to protect the site, keep beach access open, and inform visitors of onsite activities (Figure 12.7).

The Presidio has a well-developed volunteer program that provides a multitude of opportunities for individuals and volunteer groups. The volunteer activities include habitat restoration and monitoring, developing native plant nurseries, landscape maintenance, historic site restoration, trail maintenance and restoration, and beach cleanups and stewardship—all done under the auspices of the Golden Gate National Parks Stewardship Program. Begun in 1993, this program was created to bring people together to protect and restore ecologically sensitive areas. It is run with the assistance of its nonprofit fundraising partner, the Golden Gate National Parks Conservancy. Volunteers find opportunities for restoring four high-priority park sites, all of which support endangered species, at Milagra Ridge, Mori Point, Oakwood Valley, and Wolfback Ridge.

The Presidio engages large numbers of volunteers, some of whom are students such as yourself (see Sidebar). Volunteers participate in a wide range of activities, such as rare plant conservation, mapping of pest species, conducting baseline inventories, and salt marsh/tidal basin restoration at Crissy Field. Volunteers have written and illustrated a new educational curriculum as well as interpretive pamphlets, flyers, and signs.

In addition to restoration projects, the Presidio's Crissy Field Visitor Center hosts an assortment of events year-round for family and children. The Urban Ecology Educator program gives kids the rare chance to wade through the marsh, collect samples, and examine them in the lab (Figure 12.8). ■

FIGURE 12.7. **A Salt Marsh Restoration at the San Francisco Presidio.** A sturdy fence protects the restoration, in a very public and heavily used area. The educational sign keeps visitors informed about the project's progress. (Photo reprinted by permission from Steve Glass.)

FIGURE 12.8. **Sampling for Aquatic Organisms.** During an educational program at the Crissy Field Visitor Center at the Presidio in San Francisco, visitors have the opportunity to explore the organisms that live in the bay. (Photograph reprinted by permission from Sara Steele for the Golden Gate National Parks Conservancy.)

Sidebar | The Golden Gate National Parks Conservancy

As a restoration ecologist, you will most likely supervise, or at least work with, volunteers at some point in your career. Certain talents and attitudes are required to develop good supervisory skills. One of the best ways to get a feel for what works well is to put yourself in the shoes of those for whom you are responsible—in other words, volunteers. Below are some questions that you should approach from the point of view of a hypothetical, typical volunteer.

FOOD FOR THOUGHT

1. Go to the Golden Gate National Parks Conservancy website and read about the Park Stewardship Program. What types of community groups are targeted for participation? What kinds of activities do you think would have the most appeal for volunteers from different age groups?

2. What skills are required of someone looking to volunteer with the Golden Gate National Parks Conservancy, according to information on its website? Do agree with the skill set described, or do you think the requirement should be different?

3. According to the website, "Trails Forever interns work four days a week, ten hours a day with the same four- to six-person National Park Service trail crew." What do you think would be the highlights of this experience? What might be the challenges?

4. Figure 12.7 shows the fence erected at the Presidio. However, restoration ecologists often use only a simple rope fence with hand-lettered informational signs to convince visitors to respect the restoration. How might you determine what type of fence to use for a given project?

KEY CONCEPTS

- A restoration will be impacted by user activities that occur onsite, as well as by activities that happen beyond its administrative boundaries (cross-boundary influences).

- The users of a restored site include visitors, neighbors, local and regional residents, the restoration team (staff and volunteers), and contractors working on the site. Cross-boundary influences include such effects as air and water quality pollution, the spread of pest species, restrictions of management activities, population declines in migratory birds, and conflicts with neighboring land uses.

- Restoration success is enhanced by establishing support from citizens and public institutions from the outset of the restoration.

- Approaches to avoid or mitigate human impacts include establishing visitor use policies in collaboration with local residents and potential user groups, managing impact through site design, establishing trail systems, managing the size of parking lots, and engaging the community to help with stewardship and outreach.

FOOD FOR THOUGHT _____

1. Visit a local restoration or nature preserve and do the following: (a) Observe any signs of human impact that you see, (b) decide whether these are internal or cross-boundary impacts, (c) talk with the managers about the kinds of human impact they are dealing with and how they are addressing the issues, and (d) describe what you like about the managers' approaches and what you would change.

2. Assume you are in charge of managing a restored community in your area that is open to visitors for the purpose of nature study. Discuss what criteria you would use to determine whether or not hiking trails should be included in the design of the restoration. What are the benefits of including trails? What are the drawbacks? If you decide trails should be included, specify the design criteria in terms of trail length, grade, width, and surface materials.

3. Discuss the pros and cons of using each of the following techniques to minimize human impact on a restoration: (a) Limiting access through controlled entryways, (b) providing educational signage, (c) using volunteer rangers to monitor activities, (d) allowing access only with guided tours.

4. At the restoration site you visited in question 1, can you foresee how patterns and levels of internal visitation will change over the next 20 years? Are there likely to be unanticipated kinds of recreational and other kinds of uses? What do you foresee as possible changes in the nature and extent of cross-boundary influences in the next 20 years? What actions (planning, management, other) would you recommend the managers to take to prepare for these changes?

5. Determining the attitudes and expectations of visitor groups to natural areas and restorations is important but can be fraught with difficulties. For example, what are the characteristics of the average visitor? Is there an average visitor? What do you imagine are some of the variables that determine visitor attitudes and expectations? How would you go about answering such questions?

6. We discussed three major planning approaches:

   - Setting visitor use-policy, visitor experience, and restoration protection objectives in collaboration with stakeholders.

   - Managing people through the site design; establishing infrastructure (signs, trails, roads, etc.) to minimize potential use impacts.

   - Inviting people to participate in managing the restoration.

   We also illustrated a variety of ways restoration projects use these approaches. Can you think of ways that new technologies (say, social media, GPS, smart phones, etc.) could change the ways people interact with the natural world? Would these changes be beneficial or detrimental to restorations? How could managers use these new technologies to their advantage?

Resources for Further Study

1. There are many publications that document the impacts people have had on the environment. The two books below are of interest for different reasons. The first is of historical interest because of the interdisciplinary nature of the contributors, as well as indicating the thinking of leading scholars of almost 60 years ago. The second disputes the widely held idea (at least among members of the public) that the pre-Columbian population of the Americas was small and had little environmental impact.

Thomas, W. L. Jr., ed. 1956. *Man's Role in Changing the Face of the Earth*. Chicago: University of Chicago Press.

Mann, C. C. 2005. *1491: New Revelations of the Americas Before Columbus*. New York: Knopf.

2. For examples of identifying and managing human impact on natural systems, see the following:

Farrell, T. A., and J. L. Marion. 2001. Identifying and assessing ecotourism visitor impacts at eight protected areas in Costa Rica and Belize. *Environmental Conservation* 28:215–225.

Farrell, T. A., and J. L. Marion. 2002. The protected area visitor impact management (PAVIM) framework: a simplified process for making management decisions. *Journal of Sustainable Tourism* 10:31–51.

Marion, J. L., and S. E. Reid. 2007. Minimizing visitor impacts to protected areas: the efficacy of low impact education programs. *Journal of Sustainable Tourism* 15:5–27.

Schonewald-Cox, C. M., and J. W. Bayless. 1986. The boundary model: a geographical analysis of design and conservation of nature reserves. *Biological Conservation* 38:305–322.

3. The websites below contain examples of planning and design approaches to mitigating visitor impacts on natural areas. The first is a brief, concise case study description of management issues and opportunities and constraints in dealing with visitor impacts in the backcountry. The second is a case study of visitor impacts and management strategy and tactics.

Aldo Leopold Wilderness Research Institute. 2003. "High-Use Destinations in Wilderness: Visitor Impacts and Response." http://leopold.wilderness.net/research/nutshell/N015.pdf

National Parks Service, U.S. Department of the Interior. 2009. "Oregon Caves National Monument: Visitor Impact Mapping." http://www.nps.gov/orca/parkmgmt/visitor-impact-mapping.htm

Working with People

Restorations get off to a good start and are more successful in the long run if you plan, implement, and manage them using a collaborative process. You should invite all interested parties—restoration staff members, neighbors, colleagues, friends and their support organizations, citizens groups, public and private agencies, and any other interested members of the community—to work together in partnership (see Chapter 6). The need for collaboration is probably most evident for restorations on publicly owned property, but it is also true for projects run by nongovernmental conservation organizations as well as for those on private property. Working with people builds enthusiasm, interest, and understanding, along with political, financial, and physical support. Such public participation also generates valuable insights and new ideas.

Community participation is time consuming, and not without drawbacks. If the voices and opinions within the community are diverse, it may be difficult to achieve consensus. For successful community engagement, it is important that the views of all participants be heard and respected, and that the process for arriving at a decision be clear and well supported. In this way, most members of the community will respect the outcome, even if it is not the decision they would have preferred.

Over the life of a restoration, the membership of the community will change; therefore, community-building activities need to be ongoing. There are a variety of ways to engage community support for restorations during all phases of a project, from the initial planning meetings to the coordination of long-term management plans, including:

- Providing opportunities for volunteers to participate in ongoing management and monitoring activities.
- Reaching out to the community using multiple forms of communication to maintain a presence as neighbors come and go.
- Creating educational experiences and skill and leadership training opportunities.
- Organizing community festivals or other social events.

By creating a situation in which a restoration is considered to be an essential part of the fabric of society, the long-term survival of the ecosystem is enhanced. (For an example of how this civic engagement can be done successfully, see the Sidebar in Chapter 12 on the Presidio in San Francisco.) In this chapter, we continue our discussion of some of the social and cultural dimensions of ecological restorations. We begin with an overview of some of the types of concerns people may have about restoration projects and continue with a focus on a very special group of people, without whom many restorations cannot succeed—the volunteers. In a restoration project, collaboration, cooperation, and communication are crucial. Ecological understanding and technical restoration proficiency are necessary to overcome ecological challenges, but they are not sufficient to implement and manage a successful restoration in the modern world. Other variables are the economic realities and the social setting. These factors will ultimately determine what is feasible and acceptable.

▶ 13.1 THE SOCIAL DIMENSION OF RESTORATION

At the outset of a project, it is critical that you identify potential stakeholders—the individuals or groups who have an opinion about, or might make contributions to, the restoration. It is important to be as inclusive as possible—cast a wide net—and to not make assumptions about whether or not someone may want to be involved. It is helpful to invite people who have different perspectives, in order to identify common interests and to uncover and address conflicts.

13.1.1 The Restoration Team

The restoration team is made up of two or more people who come together to achieve a common goal—teamwork (see Sidebar). The team should include people who can fill one or more of the following key roles:

- Restoration planner
- Overall project manager

Sidebar	Teamwork

A restoration project is like a team sport—an endeavor that requires teamwork. Think about a team you have been a member of—a sports team you played on, a debate team, a theatrical play you have acted in, a planning group you participated in, or perhaps a class project you worked on.

FOOD FOR THOUGHT_____

1. How do you define teamwork?
2. Do you think teamwork is necessary for all projects? Why?
3. What are the qualities of a good teammate?
4. Over the course of the project, did peoples' roles change?
5. When your team was at its best, what did it do well?
6. When your team was not performing well, what was it like for you? What would you have changed?

- Onsite supervisor
- Field biologist (plant ecologist, botanist)
- Plant propagator
- Wildlife ecologist
- Researcher
- GIS specialist
- Database administrator

Depending on the project goals and the nature of the site, you may periodically also need to call on the expertise of a civil and/or environmental engineer, soil scientist, hydrologist, forest ecologist, arborist, marketing or public relations specialist, and/or volunteer coordinator.

Because people may fill multiple roles (e.g., onsite supervisor and field biologist), you might not need one individual for each role; only the most well-funded restoration projects can afford one person for every team role. You must remember to include volunteers on your restoration team, because volunteers bring a wealth of life skills and technical expertise to the project. Besides, volunteers help include the community in the planning and management process.

13.1.2 Stakeholders

Stakeholders in a restoration project may worry about and disagree with one another over a number of issues. Here are some of the most common topics of discussion:

- *Aesthetic perceptions.* Some people think restorations look weedy and out of control, at least in the early stages, while others understand and appreciate an untamed appearance.
- *Problem framing.* People view situations differently. Storm water that empties from a culvert into a floodplain forest on its way to a lake is considered a problem by civil engineers and city utility managers because it is uncontrolled and unpredictable.

Restoration ecologists see the unpredictability as a research challenge for the best bioengineering bank stabilization techniques and an opportunity to create habitat and recharge groundwater.

- *Expectations for the restoration.* Some people think a restoration should serve as a nature sanctuary, whereas others think it should be available for most forms of outdoor recreation.

- *Specific restoration practices.* Some people may be opposed to the use of herbicides or fire; others will resist any changes to a landscape they are familiar with.

- *Fear that certain species will have negative cross-boundary effects.* Project neighbors may worry that predators, such as wolves or coyotes, might prey on pets or livestock, or that native plants escaping into crops or gardens can act as weeds. However, the community at large may value the presence of large wildlife, and may be troubled that some cultivated garden plants can escape to become weeds in native plantings.

- *Concern that historic privileges will be prohibited.* Some people will dislike the fact that they will no longer be able to hunt or gather plant materials or hike whenever they want. As compensation, in some projects, hunters and gatherers can be recruited to target different things, like collecting seeds and cultivating plants for use in restoration plantings, or for research purposes, such as trapping live animals. In this way, they will trade activities, and trade for a different kind of access.

At the heart of some of these stakeholder discussion points are different definitions of nature and of restoration. The now-famous controversy surrounding early efforts by the Chicago-area Forest Preserve District to restore its forests to prairie and savanna (see Chapter 1) is worth reviewing again to illustrate how people, most of whom consider themselves conservationists, can disagree about restoration. The restoration controversy centered on differences over values, practices, and processes.

CASE STUDY | The Chicago Forest Preserve District

The controversy over restoration practices in the Chicago, Illinois, area erupted in DuPage County Forest Preserve District in spring of 1996, based on citizen objections to new (at the time and to the area) and poorly publicized restoration practices, such as cutting trees and setting prescribed fires. Traditional management of the Forest Preserve District had consisted of planting trees and preventing fires. Therefore, when managers began doing just the reverse, some citizens rose up in opposition. Protests were lead by a group called ATLANTIC (Alliance To Let Nature Take Its Course), and when the uproar hit

the local newspapers, the Forest Preserve District of DuPage County responded by declaring a countywide temporary moratorium on restoration activity. Soon, citizen protests spread to similar restoration work in Cook County (Chicago), and by fall the Cook County Forest Preserve District board also declared a moratorium on restoration work. The moratoriums lasted several years, but have since been lifted.

While this was all going on, a new restoration initiative was taking shape. Chicago Wilderness, a coalition of land management and conservation organizations, involving some of the key restoration leaders who were unfairly tarred by restoration controversy, was being formed. Chicago Wilderness includes the metro area plus portions of northwestern Indiana and southeastern Wisconsin. The Chicago Restoration Controversy, as it became known, spawned hearings, conference sessions, and discussion and coverage in every media forum available at the time. The restoration community did not remain silent and fought back with messages from its own restoration perspective. The Chicago controversy has become a touchstone in the restoration community and is well chronicled in the restoration literature (Friederici 2006; Stevens 1995).

The information presented below is drawn from the research conducted by Paul Gobster, a social scientist with the U.S. Forest Service in Chicago. In the early stages of the controversy, Gobster (2000) conducted a content analysis of the views of opponents and proponents, as expressed in newspaper articles, letters to the editor, radio interviews, transcripts of hearings, and fact sheets. Gobster's findings are relevant to our discussion—not because we want to belabor the Chicago controversy, but because we want you to learn from it and recognize the great service that the people involved have provided to future restorationists and the global restoration community.

As the result of his analysis of the controversy, Gobster identified the kinds of values that opponents expressed in relation to why restoration should not occur:

- *Functional*: The loss of privacy and solitude, and shade and cooling through prescribed burning and tree and brush removal.
- *Economic*: The fear that restoration would increase taxes and reduce property values of homes near where trees were removed.
- *Recreation and wildlife*: The loss of shaded recreation sites and habitat for some species.

- *Aesthetic*: Restoration would harm the wooded character of the forest preserves and impose a radically different idea of what is beautiful.
- *Symbolic*: Some felt that restoration was an attempt to control nature and impose an arbitrary point in time to which nature should be "turned back."

The critics also had objections to specific practices, including:

- *Removal of trees and brush*: The killing of healthy trees and large as well as small "brush"; defining too many tree species as "alien" or exotic.
- *Herbicide use*: Types being used; how, when, and at what strength they were being applied; the qualifications of those who applied them.
- *Prescribed fire*: Effects on air quality; safety of nearby homes; danger to wildlife.
- *Removal of deer*: Justification for control; concern over methods used.

Opponents of the restoration also had concerns about how and where the restoration efforts were being carried out. Some citizens also felt that information was withheld and activities were being concealed from the public. Critics felt they were not involved in decision making and wanted a greater voice in that process. Many felt out of the loop. Some felt there was "insufficient planning for restoration" and that the "restoration was being conducted without good plans" (Gobster 2000, 7). Critics also questioned the use of volunteers and were concerned about "whether or not the volunteers were receiving sufficient training and supervision." It is particularly interesting to note that many of the concerns on the second list might have been avoided if the planners had communicated with potential stakeholders early and often. ∎

13.1.3 When to Involve People

Public input and participation are critical at several stages in the restoration process:

1. At the very beginning, when information is first being gathered.
2. At the stage when alternative solutions are being discussed.
3. At the point when the final plans are ready for adoption.
4. At critical stages in implementation and management.
5. Subsequently when the plan is reviewed and revised.

Involving the public does not mean that you must always meet the desires of all stakeholders; indeed, this is rarely possible. It is important, however, for people to feel that their concerns have been considered, and that they are informed of the reasoning behind the decisions that are made. There are some rare circumstances in which the need for security outweighs the need for transparency (e.g., discussing the exact onsite location of a rare species). In such cases, it is helpful to explain why the information is being withheld.

13.1.4 Tools and Techniques for Increasing Public Involvement

As discussed in Chapter 6, common approaches for obtaining public input include (1) initial brainstorming sessions, in which problems are articulated and various solutions proposed; (2) public hearings, for discussing potential alternative solutions; (3) field trips, to view the project site and talk with project proponents; (4) providing opportunities to submit written comments on plan drafts; and (5) conducting Environmental Impact Assessments. In the United States, projects carried out on public lands usually require a procedure using one or more of the above approaches to involve the public. The same procedures are highly desirable on privately held properties as well, perhaps with a more limited number of participants.

Public Workshops and Meetings

Workshops in which members of the public are asked to voice their concerns and state their opinions are a vital component of the restoration planning process. Such meetings are important at the start of each of the planning phases of a restoration—design (master and site plans), implementation, monitoring, and management. By using a variety of participatory techniques, you can gauge the extent of community support and clarify concerns, which, if left unaddressed, could undermine the long-term survival of a project.

In Superior, Wisconsin, the planners of the Hog Island and Newton Creek Ecological Restoration Master Plan used a flexible, adaptive stakeholder participation process (see Chapter 6). The restoration team conducted three public workshops. In addition, two websites hosted meeting minutes, workshop materials, and posters to provide easy access for all. Developing a healthy and robust public participation process was not only a method but also a desired outcome to serve as a model for other restoration projects in the Great Lakes basin. Working with the U.S. Environmental Protection Agency, Biohabitats, Inc. noted that the "primary intention of this project is to define a process by which other AOC (Areas of Concern) in the Great Lakes basin can be restored according to the principles of ecological sustainability and stakeholder input" (2007, 78).

It is important to hold such meetings in advance of any anticipated change—either a change in restoration or management activities, or a change in programs or policies—or immediately after the occurrence of any unanticipated events (e.g., wildfires or floods).

Hog Island's first public workshop was held at least 9 months before the anticipated completion date of the restoration plan. Also, because people invariably move in and out of neighborhoods, it can be a good idea to meet with members of the public, at least briefly, once every 1–2 years. The updates could be part of the agenda of a regular neighborhood meeting, or they could be part of a field day, during which the public is invited to explore the restoration site.

Participatory Planning Techniques

A number of community participation techniques have been used by planners over the years in the development of projects ranging from building public parks and housing to constructing transportation networks, to designing conservation greenway systems. Recently in the United States, one of the most popular participatory planning tools has been the nominal group method. The purpose behind this tool is to allow all participants to have an equal voice, to feel that each of their ideas has been considered, and to attempt to uncover a wide range of perspectives from which to arrive at a consensus.

There are many variations of this technique, and here is an example of how it might be used in the brainstorming phase of creating a restoration master plan:

1. Invite as many stakeholders as possible to one or more public brainstorming sessions. (Be sure there is a comfortable room with food and drink available.)

2. Open the meeting, and after thanking everyone for coming, explain the purpose of the project about which you are asking for comments. (It is most effective if the facilitator of the meeting is a neutral party.)

3. Give the participants a stack of sticky notes and a marker, and ask them to write down anything that they want to say about the project, one comment per note.

4. Ask for a few volunteers (the number needed depends on the size of the group), representing different constituencies (neighbors, hunters, conservationists, public officials), and have them group the comments into themes. Then use some large sticky sheets, each labeled by one of the themes, and attach them to the walls of the room. Have another set of volunteers sort and attach the original notes to the appropriate theme sheet. If, in the process, one of the new volunteers thinks of a new theme, add it to the wall.

5. Give everyone time to review the themes and accompanying notes. Then ask everyone to rank the themes from most to least important.

Through this process, you create a dialogue. For example, after you have created alternative master plan solutions (see Chapter 6), the set of solutions can become the subject of another nominal group session, and you can ask the participants about the pros and cons of each. The result is often a new, even better master plan. The dialogue can and should be continued through all phases of the project.

Your important challenge is to balance and respect the wisdom of local knowledge, long-term cultural use patterns, and modern conservation principles. The nominal group process helps, but this insider/outsider divide can persist unless you address it directly. As is true for many aspects of a restoration, the steps you will need to take to gain mutual trust will vary. Fortunately, conservation groups have been working on creating partnerships for many years. You can find valuable insights by communicating with such groups as the WWF, The Nature Conservancy, the International Crane Foundation and Conservation International, and the Society for Ecological Restoration.

▶ 13.2 THE REGULATORY ARENA

Local, state, and federal units of government and regulatory agencies also play a role in restoration, not only by performing such work on public lands, but also by facilitating projects on privately held property through sponsoring, requiring, or funding restoration. State and national parks, wildlife management agencies, and bureaus of land management routinely practice restoration on at least a portion of their lands. In publicly funded projects, the government supervises spending and the wise use of these public funds, administers the many contracts involved, and ensures that the multitude of laws and regulations are followed. These laws and regulations promote public safety, and their existence can both faciltate and slow projects down.

Your restoration project's administrative boundary may intersect with the jurisdictional boundary of one or more units of government, such as village, town, shire, parish, county, state, or province. In this case, it is important to determine what laws or regulations are in effect, and what permits may be required (see Chapter 4). For example, ordinary restoration activities, such as prescribed burns, may be regulated by local ordinance and state air quality regulations; in some parts of the U.S., prescribed burns, for example, are prohibited if day-of conditions do not meet the Environmental Protection Agency's minimum daily ozone and particulate matter counts. One or more agencies may oversee activities in wetlands. The rules for safe and legal herbicide application may be set by a state agriculture department or department of natural resources. The management of threatened and endangered species in the United States is regulated by the appropriate state agency and the U.S. Fish and Wildlife Service (see Chapter 5).

▶ 13.3 WORKING WITH VOLUNTEERS

Once a project is well along, you can choose from several techniques that have been used successfully around the globe to continue drawing support for and interest in the restoration. Here is a sample:

- Connecting human and ecological history through participation in restoration activities is an effective way to build volunteer support.
- Providing opportunities for volunteers to assist in the planning and goal establishment phases can create a feeling of ownership or civic responsibility that can result in long-term involvement and support.
- Restoration work parties designed for volunteer participation encourage engagement and provide opportunities for learning about the natural world in general and restoration techniques in particular.

13.3.1 Engaging Staff and Volunteers

Restoration ecologists and land managers use many public engagement strategies that draw on the wide variety of human interests and talents. The types of civic involvements range from relatively passive support (e.g., being a dues-paying member of a local organization) to more active support (removing trash, pulling weeds, conducting and/or monitoring small experiments, writing and publishing a newsletter or website, and instituting art projects).

It is usually not difficult to find volunteers for most restorations. This is true because the act of restoring the land gives people a sense of accomplishment and of making a difference. Many restorations have the happy problem of having too many volunteers for the available work or the staff capacity to manage them (Figure 13.1).

People volunteer for restorations for a variety of reasons. Sometimes people volunteer as part of an educational experience; other times, volunteering becomes an avocation. Some people volunteer to learn new skills or to share their knowledge and expertise with new friends. Other people volunteer to contribute to the community, or for the social experience and the chance to meet new people. Some volunteers just want hard work in the outdoors and enjoy the food that is typical at work parties. Many people are attracted to restoration volunteer work by the realization that they are part of nature and part of the solution. Regardless of the reason for their participation, volunteers build support for restorations and for the environment. Whatever their motivations, the trick is to keep volunteers engaged, interested, and perhaps most importantly, returning.

13.3.2 Training and Keeping Volunteers

Professional volunteer coordinators emphasize that sustained civic engagement is all about building relationships among the land, the project, and the people involved in it. Professionals who work with volunteers and community groups offer this advice: Engaging people in general, and staff and volunteers in particular, boils down to relationship building, recruitment, recognition, and retention. This relationship building is done gradually and parallels the restoration planning process itself—goal setting, site analysis, stakeholder analysis, master plan/program outreach, implementation, monitoring, and management/volunteer program evaluation—except that people, instead of the land, are the key concern. (See the discussion of the logic model in Chapter 6.)

FIGURE 13.1. **Volunteers at an Urban Restoration.** These community members are planting a native plant garden along a public bike path right-of-way within the City of Madison, Wisconsin. Collaborators on the project included the Dudgeon-Monroe Neighborhood Association, the City of Madison Engineering, Friends of Lake Wingra, and Madison Gas & Electric. (Photo reprinted by permission from Steve Glass.)

Here are some things to consider in cultivating volunteers:

- Provide a variety of types of volunteer opportunities to attract different kinds of volunteers, such as website management, physical labor, and both individual and team projects.
- Match the skills and interests people bring to the project (from none to highly experienced) with specific restoration tasks.
- Treat volunteers with respect, and remember to provide food.
- Provide training, evaluation, and feedback to improve the restoration program and the volunteers' performance and satisfaction. Sponsor lectures on restoration skills, plant and animal identification, and ecological concepts; sponsor guided field trips to established restorations or natural areas.
- Give volunteers the opportunity to perform well and to achieve their goals. Be sure to schedule some volunteer work that is feasible to complete in a set amount of time. Document the "before" and "after" with photos.
- Reward them with recognition.

The cultivation of volunteers involves three main jobs: communication that speaks to peoples' heads, outreach that affects peoples' hearts, and education that changes peoples' behavior. Paying attention and engaging with others in this way takes a lot of time and energy; it requires putting the same kind of effort into building the human community as you expend in restoring the plant and animal communities and ecosystems. The following case study describes the TREAT program in Australia, an exemplary model of local, state, regional, and national cooperation and collaboration.

CASE STUDY | A Volunteer-Based Restoration in Queensland

To achieve restoration goals, local restoration ecologists in Queensland, Australia, use a variety of restoration strategies or methodologies to fit different situations. Regardless of the restoration plan, the work relies on partnerships, collaboration, and civic engagement. Here, restoration as practiced by the Trees for the Evelyn and Atherton Tablelands (TREAT) is a cooperative government/volunteer effort.

The Atherton Tablelands, a high plateau on the edge of the Outback, is former tropical rainforest that is now mostly given over to dairy and beef cattle, as well as sugarcane and high-income horticultural crops for the nursery industry. The Tablelands host a growing ecotourism industry and are attractive to both retirees and those looking for a summer home. These land uses have destroyed, degraded, and

fragmented the rainforest, leaving small patches of original forest isolated in an agricultural landscape. The remaining small patches of tropical rainforest are increasingly susceptible to invasion by weed species, thus creating poor habitat for threatened and endangered species. Thus, one restoration goal in the Atherton Tablelands is to provide habitat and a migration corridor for the endemic Lumholtz tree kangaroo (*Dendrolagus lumholtzi*), a small marsupial.

TREAT is an organization of volunteers powered by partnerships with local, state, and federal governments and other volunteer organizations. TREAT and its governmental partners have formed a full-service restoration practice. TREAT secures funding from state and federal governments and partners with local and regional groups to restore plant communities in Queensland's World Heritage Area tropical rainforests. It also assists private landowner applicants by cost sharing as a partner in the grant application process. TREAT coordinates, sponsors, and finds funding for a range of other programs, such as the Threatened Species Network and the Weed Spotters Network, which trains people to identify, locate, and map outbreaks of invasive species.

Land use changes in the Tablelands raise concerns about water quality because of increased nutrients from fertilizer and manure runoff, chemicals from pesticide and herbicide applications, the erosion and silting of streams, and trampling of stream corridors by cattle. Because of this, TREAT officials view their work as much a land management exercise as it is a tree-planting program. TREAT promotes improved water quality through a series of workshops that offer talks on monitoring and water quality sampling. TREAT also provides landowners with incentives to do simple things—such as installing silt traps, reha-

bilitating wetlands, and planting corridor buffers and barriers with appropriate tree species that exclude cattle from waterways—to improve the quality of water in streams on their own property.

In a typical year TREAT nurseries produce, and volunteers plant, thousands of tropical rainforest trees. Staff members of the Queensland Environmental Protection Agency use their technical knowledge to identify plants, collect seed, and design the seed and planting mixes. Still other TREAT volunteers clean the seeds, sow them, and care for the seedlings until they are ready for out-planting (Figure 13.2). In February and March of 2006, for example, TREAT volunteers planted just over 7,000 trees along Peterson Creek, a stream corridor restoration designed to serve as a habitat corridor between two tropical forest remnants. This work is typical of the shire-based Wet Tropics Tree Planting Scheme and has been widely practiced since the wet tropics received World Heritage status in 1988. ■

FIGURE 13.2. **Seed Collection and Preparation.** Fruit of the brown boxwood (*Homalium circumpinnatum*) of the *Flacourtiacea* family, collected on June 13, 2006, is ready to be cleaned and prepared for propagation by TREAT volunteers in Queensland, Australia. (Photo reprinted by permission from Steve Glass.)

13.3.3 Matching Skills and Interests with the Job

Regardless of their reasons for volunteering, people bring valuable life experiences to the job, and their skills and knowledge should be incorporated into the restoration project. The key is to match their skills and interests, their values and motivations to the work that has to be done.

The restoration plan will indicate the requisite skills for implementing, managing, and monitoring the project. You can make a list of the skills of your current staff, and then recruit volunteers to fill the gaps. If the project team does not include a staff photographer, writer, or budget person, for example, a member of the community may have the skills to help. If the project would benefit from scientific advice, perhaps a retired botanist, geologist, or wetland scientist would be willing to volunteer. Someone from a local nursery could help with plant propagation. People with other, more specialized skills, such as herbicide application or prescribed fire burn supervision, are also often out there ready to help. If volunteers with organizational or leadership skills are available, you would be smart to enlist them, too.

13.3.4 Treating Volunteers (and All People) with Respect

Volunteers are no different from other people. They like to know that their contributions are appreciated, their opinions are respected, and their advice is sought out. Ask their opinions, and check in frequently about how the restoration project is going—is it meeting their expectations, is it what they thought it was going to be? If not, how could things be changed for the better? Ask veteran volunteers to help lead newer volunteers by telling their stories and sharing their experiences with the group.

Two other aspects motivate volunteers: They like to eat and they like to learn. Always bring treats to work parties, and include in the work party a special hike or learning opportunity.

Provide opportunities for volunteers to perform well and to achieve their personal goals. Some people do not want to waste their time on mundane tasks, and nobody wants to have the sense that their efforts are futile or in vain. Volunteers are the same. The volunteer job must be meaningful and one that makes a contribution, over time, to achieving the restoration goals. Volunteers need to believe that the restoration is progressing. They want the opportunity to grow in the job, and to become trusted and respected members of the team (see Sidebar).

One common mistake to avoid is assigning make-work, tasks that are not essential to the restoration, or that are not efficient and effective—just to keep the volunteers busy and coming back. They will soon catch on that they are not making a contribution; they will lose interest and fade away.

Some restoration projects host an annual volunteer recognition banquet or spotlight a special volunteer in a newsletter story or photo essay on their websites. Volunteers are the best recruiters, and featuring a few hardworking and happy volunteers in volunteer recruitment brochures is a win-win approach. Since restoration is a team effort, what better way to recognize the group than by providing certificates, badges, team T-shirts, and/or hats emblazoned with the project's logo.

At the Presidio in San Francisco, workdays end with a short hike or nature tour (see Chapter 12). Volunteers are also occasionally offered special classes and field trips not available to the general public, a bonus for both volunteers and staff members. For example,

| Sidebar | TREAT's Keys to Success |

Trees for the Evelyn and Atherton Tablelands (TREAT) is a collaborative government and volunteer restoration project in Queensland, Australia. It has been extremely successful in its stream corridor restoration, one of its goals being the restoration of a wildlife habitat as a migration route for endangered marsupials. TREAT has also been successful at engaging people in the restoration program by focusing on recruitment, relationship building, recognition, and retention.

Specifically, TREAT tapped into the variety of motivations of the residents of the Tablelands to achieve restoration results on their own property by conducting a series of workshops that offered training on monitoring, sampling, and other things property owners could do to improve the quality of water in streams on their own property.

TREAT matched the skills and interests of volunteers with specific restoration tasks. It used the local expertise and traditional knowledge of indigenous peoples of the tropical forest trees. TREAT relied on other volunteers with specific training or experience in cleaning and sowing seeds and caring for the young plants to take on these tasks. Other volunteers were encouraged to use their professional skills or to gain new knowledge in advancing the TREAT restoration goals.

Much of what TREAT has accomplished has been facilitated by the variety of strong partnerships it has forged with shire, state, and federal agencies and the funding opportunities that these partnerships have opened up. Through newsletter articles and at weekly workdays, TREAT spotlights the knowledge and achievements of individual volunteers.

TREAT has showered its volunteers with respect and has encouraged them to advance through a variety of training workshops, continual evaluation, and positive feedback. All these efforts combine to improve the restoration and enhance the performance and satisfaction of the volunteers and other partners.

Visit the website of TREAT and see if you can identify other reasons (e.g., organizational structure, restoration team, scientific knowledge) for the success TREAT has had in its restoration programs.

volunteers were given the chance to attend a Park Academy class and work project that gave them a behind-the-scenes look at the site of a restoration project focusing on restoring habitat for the threatened California red-legged frog. An activity like this also helps build camaraderie, thereby integrating volunteers and staff.

KEY CONCEPTS

- To ensure the long-term survival of a restoration, it is important to include as many current and potential stakeholders as possible in all stages of the process.

- People often disagree about restoration for a variety of reasons. For example, they might differ about the definition of nature or whether it is right to hunt animals or cut down trees. People have varying levels of concern about the use of herbicides or fire, or the potential that species from the restoration will interfere with them.

- Public involvement in restoration projects can be encouraged through the use of participatory management tools in the context of public meetings and workshops, and by providing volunteer opportunities.

- Volunteers are recruited through efforts to build relationships and provide training and learning opportunities, and are retained by rewarding and recognizing their contributions.

FOOD FOR THOUGHT

1. Building partnerships with landowners of property adjoining the restoration project area is a key factor in maintaining sustainable restoration projects in urban and suburban areas. Think about how you would engage these different groups to discuss and develop shared understandings of the concept of your restoration project.

2. Land managers need to work at geographic and temporal scales that represent a broader context than the boundaries of their project areas; this requires management across ecological, political, generational, and ownership boundaries. Consider the geographic and temporal scales that would be applicable for a restoration project with which you are familiar.

3. Describe the methods you would use to maintain a high level of civic interest and participation in a restoration project. What would be the costs and benefits to your restoration project? What skills and talents would your restoration team need to have or acquire to be able to use these methods?

4. What opportunities would you provide for volunteers to help you in the planning and goal establishment phases of your restoration project?

5. Think of a restoration project that's familiar to you. Consider its role in the community, and define the community of interest. What perceptions does the local community have of the restoration project? How does the restoration draw upon the resources of the community to aid its work? What contributions does the restoration make to the community?

Resources for Further Study

1. This book provides practical advice and guidelines for organizing and conducting focus groups:

    Stewart, D. W., P. N. Shamdasani, and D. W. Rock. 2006. *Focus Groups: Theory and Practice*. Thousand Oaks, CA: Sage Publications.

2. Here is a good guide to collaboration and cooperation within and between natural resource management agencies:

    Wondolleck, J. M., and S. L. Yaffee. 2000. *Making Collaboration Work: Lessons from Innovation in Natural Resource Management*. Washington, DC: Island Press.

3. For information about how Chicago Wilderness is doing today, check out its website:

    http://www.chicagowilderness.org/

4. This is an excellent example of a community-based watershed restoration effort in Northern California that has been in existence since 1983:

    Mattole Restoration Council website: http://www.mattole.org/

5. This is the website for TREAT: http://www.treat.net.au/index.html)

14

Case Studies

The discipline of restoration ecology continues to grow and develop. Increasingly, restoration ecologists are being called on to respond to new challenges and adapt to new situations around the globe. In previous chapters, we have provided an introduction to the discipline and practice of restoration ecology, using a variety of examples and brief case studies to illustrate our points about the restoration process, and the opportunities and challenges restoration practitioners face. In this chapter, we present a series of three longer case studies, chosen to provide more insights about different aspects of restoration practice.

Our first example is the University of Wisconsin–Madison Arboretum, recognized as the birthplace of restoration ecology. Because it is one of the oldest restorations in the world, the Arboretum's history provides the opportunity for us to track changes in context and practice through time. Next come a series of projects from The Nature Conservancy's Great Rivers Partnership Program. This program works with river systems around the world that need conservation and restoration. In contrast to the UW–Madison Arboretum, this project is relatively new. It provides us with a good example of a coordinated global effort to restore biodiversity, while at the same time restoring ecosystem services for the survival of diverse human cultures. The third case study is the Kootenai River Restoration, a project initiated to restore wildlife habitat and sustain a food source and way of life for the Kootenai Tribe of Idaho.

▶ 14.1 THE UNIVERSITY OF WISCONSIN–MADISON ARBORETUM

The 509.9-ha (1,260-ac) University of Wisconsin–Madison Arboretum is located in the city of Madison, Wisconsin. The city of Madison has a population of 233,000 and a metropolitan population of approximately 560,000. The Arboretum includes a mix of small remnant and restored communities—woods, savannas, prairie, and marsh (Figure 14.1). It also has two display gardens, one devoted to native (mostly herbaceous) plants and one to woody plants. The Arboretum is located within the 1,831-ha (4,525-ac) Lake Wingra

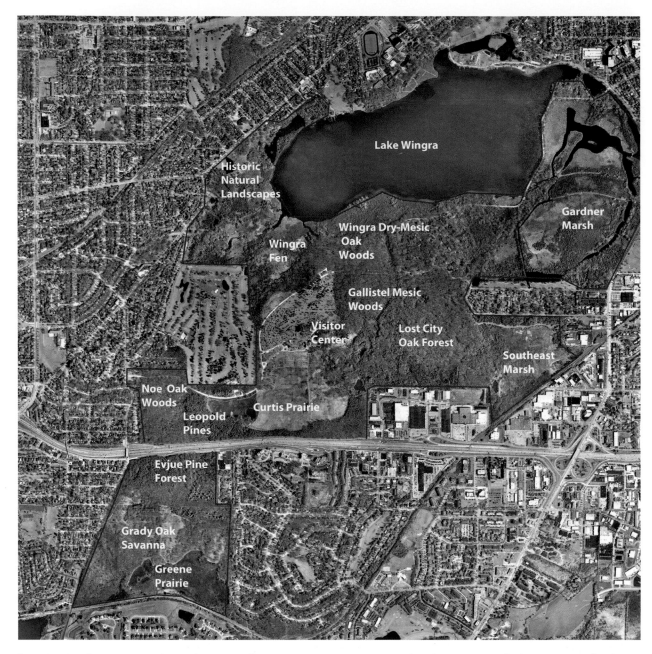

FIGURE 14.1. The University of Wisconsin–Madison Arboretum. Tallgrass prairie restorations intermingle with restorations of oak savanna, woodlands, and wetlands throughout the 1,200 acres of the Arboretum. The Arboretum supports restoration research and education. (Photo reprinted with permission of www. Wisconsinview.org through http://creativecommons.org/publicdomain/zero.1.0)

watershed, in a landscape context of residential, commercial, and industrial activity, about 75% of which has become urbanized since the Arboretum's founding in 1934.

Lake Wingra borders the Arboretum on the north, and residential properties abut its eastern and western borders and portions of its southern border. The southern border is shared with commercial and industrial areas that align a major highway that borders and bisects parts of the Arboretum. The case study begins with an overview of the restoration, and continues with a discussion of some of the major issues that the restoration team has dealt with over the years.

14.1.1 Project Overview

The UW–Madison Arboretum is part of the university's Graduate School, the branch of the campus administration that oversees research centers. It was established to preserve and restore samples of Wisconsin's plant communities for education and research.

The Arboretum director is a member of the faculty under the supervision of the Dean of the Graduate School. A committee comprised of university faculty and staff advises the director on policy issues. The Arboretum employs a number of staff, including those devoted to land care, research and ecological restoration studies, outreach programs, and fund-raising (development). In addition, an active "friends" group, among other things, raises funds in support of Arboretum programs and produces a newsletter. The Arboretum celebrated its seventy-fifth anniversary in 2009.

Land Use History

Historically, Native Americans made extensive use of the land, which included permanent settlements, seasonal camps, and effigy mound complexes. With settlement by Europeans, the land was further altered through agriculture, housing, road building, and wetland dredging. The legacy of these human impacts continues to pose restoration challenges.

Early in the twentieth century, Lake Wingra's large eastern marsh was divided in two and cut off from the lake by residential development. Subsequently, the marshes were ditched, dredged, and filled, and the disturbed soil became overgrown with invasive species.

As the cities of Madison and the surrounding metropolitan area grew, residential and commercial development and the associated urban infrastructure of highways, utility lines, and stormwater conveyances fragmented and perforated the watershed. In the past few decades, these forces have had significant impacts on the Arboretum's restorations and remnant plant and animal communities.

The Use-Policy, Past and Present

Originally, the intended use of the Arboretum was for research and teaching by university faculty, staff, and students (see Chapter 1). The university classifies the Arboretum as a research center, and since the beginning, it has functioned as a large outdoor teaching and learning laboratory. Arboretum research use-policy establishes guidelines for where manipulative or observational research is appropriate; management use-policy establishes guidelines for mechanical (mowing, cutting, and pulling), and where cultural (fire and mulching) and chemical tools are appropriate and inappropriate.

However, the original university-focused research/education use-policy has evolved and now accommodates passive recreation (such as walking, hiking, running, bicycling, and bird-watching) by the general public. The use-policy describes where each of these activities is acceptable and not acceptable—some sensitive areas like Native American effigy mounds, seasonally wet soils, and research plots are off-limits—and describes where levels of footpaths and wider service lanes are suitable. (See Chapter 6 for a discussion of use-policy and the Arboretum's visitor use policy.) This evolution has occurred for a variety of reasons, primarily the idea that, with the increasing urbanization of the landscape surrounding the property, it is better to control activities than to prohibit visitors. In other words, it is better that city residents view the Arboretum as an asset and an amenity, rather than as a liability. Indeed, the Arboretum now ranks high on the list of favorite Madison locations for a wide variety of activities.

The current use-policy has created many opportunities for the Arboretum. The ability to build community support and a donor network have led to the construction of a visitor center, with modern offices for staff and facilities for hosting educational events, and to funds for supporting student research. The use-policy has also presented challenges, such as the need to maintain hiking and cross-country skiing trails, staff the visitor center, and employ a ranger to roam the grounds to monitor visitors. These roles are often filled by volunteers or by student interns, thus turning challenges into opportunities.

Although they are necessary, use-policies, volunteer networks, and community support are not sufficient to deal with the challenges the Arboretum now faces—challenges that go beyond trying to achieve restoration goals. Today, in fact, the Arboretum assumes a much wider and more important role in the community than its founders could have imagined. Culturally, the Arboretum has developed multiple personalities within the community, and it is viewed as a site for recreation, birdwatching, nature study, and relaxation. The Arboretum has become a mecca for outdoor education and outreach, and is regarded as a model for leadership in sustainable environmental management. Ecologically, the Arboretum is a principal steward of the Lake Wingra watershed and an important link between lakeshore environments, wetlands, grasslands, and wooded communities of the uplands. Meeting public and staff expectations that come with these different public and professional roles has created a dense thicket of challenges for the Arboretum—a thicket through which the Arboretum is still trying to find its way.

Initial Goals

The initial vision for the entire property was, according to Leopold's vision, to re-create "a sample of what Dane County [Wisconsin] looked like when our ancestors arrived here" (Sachse 1974, 27). This would have been prairie, oak savanna, and oak opening. The idea of restoring a prairie in Madison was first proposed by UW faculty member Norman Fassett, and the goal was to re-create samples of prairie to be used for teaching and research, in which "the process of establishing the prairie would point up areas where research was required" (Cottam 1987, 261). This is a great example of the link between research and restoration (see Chapter 3).

One of the first restoration goals had strong support: reintroducing prairie to the landscape. Over the years, the Arboretum restoration vision expanded, and the goals and objectives for a time included attempts to restore all the major plant communities in Wisconsin.

14.1.2 The Master Plan

The Arboretum has been restored in stages, starting in 1934, with Curtis Prairie. In the beginning, the restorationists worked without a formal site inventory and analysis—at least in the sense that they did not leave a written record; and they did not create a formal written plan before beginning. It is important to note that the absence of a formal plan does not mean that the restoration lacked planning. In fact, the early restorationists were excellent ecologists and landscape architects who brought extensive field experience and then modern ecological theory to the task. As you will see below, they used what we would now call an adaptive approach, carefully documented their work, and built research experiments into the project.

A group of faculty members did the planning, assisted by students and later by members of the Civilian Conservation Corps (CCC). There was little involvement from the general university or the community of greater Madison. In more recent times, planning has been more formal, with input from a variety of stakeholders.

Early Years

Professor Fassett; John Thompson, who at the time was Fassett's graduate student and later UW–Madison Professor of Botany; and CCC supervisor Theodore M. Sperry were responsible for the initial planting of Curtis Prairie, the first restoration attempted at the Arboretum. When Fassett and Thompson began their work in 1934 and Sperry began his in 1936, they did so without benefit of a base map. By the time the CCC camp was closed in 1941, there were still no survey monuments or corner posts (Sperry 1990). Sperry used wooden corner posts to produce hand-drawn maps of the planting areas and to locate planting sites. As an additional handicap, the initial plantings were done without benefit of a soils analysis, as the first soils map of the Arboretum was not produced until 1939.

Curtis Prairie was planted on a 60-acre former prairie, turned horse pasture. It is not known what process was used to conduct the site inventory and analysis (as it would be described now), but the site was apparently dominated by bluegrass (*Poa* sp.) and ruderals, which persist today. The Curtis Prairie site was chosen because it was available, had good soils, and was close to headquarters and the workforce. The site also took advantage of proximity to existing roads and footpaths, so little was required to establish circulation infrastructure.

Professor Thompson, who supervised the early plantings in what was to become Curtis Prairie, used as his reference systems the remnant patches of native prairie in Dane County, southwestern Wisconsin, and the high bluffs along the Mississippi River. These native prairie remnants also served as source materials (seeds, sods, and transplants) for the early plantings. The team chose sites that were slated for destruction because of highway and building construction, or farmland expansion.

Recent History

In 1951, John T. Curtis wrote the first master plan for the Arboretum, which established the general layout of plant communities you can see today. In 1992, then Arboretum Ecologist Dr. Virginia Kline wrote an updated version of the Curtis plan, entitled "The Long-Range Management Plan for Arboretum Ecological Communities" (Kline 1992). The Kline Plan, as the Long-Range Plan has come to be known, was intended to review the status of the Arboretum's then several different restorations, and to recommend what implementation or management steps were still needed. In addition, it reviewed and updated the original Arboretum vision to reflect the evolution of the use-policy since 1951.

The 1992 plan is essentially a master plan in that it establishes the use-policies for research, management, and visitors, and, as was the case with the original Curtis plan, lays out the boundaries of the different communities. The Kline Plan does not include a detailed site plan, implementation plan, management plan, or monitoring plan—in the sense that those plans were discussed in the chapters of this book. As we describe below, it does include a general site inventory and analysis and ecosystem/community models, including a discussion of stressors of interest to management.

The Kline Plan has six sections. An introductory section provides an overview of the plan—its scope, organization, general considerations, critical management needs, and strategies for meeting management needs. The remaining five sections are organized around the major Wisconsin plant community groupings: southern forests, northern forests, prairies and savannas, and open wetlands, plus a section comprising buffer areas.

The prairie and savanna section begins with a description of the ecology and range of prairies and savannas in Wisconsin, and typical species composition—in other words, descriptive community/ecosystem models. The section discusses each of the Arboretum's nine prairie and savanna implementation units and includes subsections with site descriptions (size, slope, and aspect of soils; vegetation history and past management; a general description of present vegetation, plus special features and values); management (general management objectives, general management of prairies and savannas, site-specific management recommendations); research (research completed, research opportunities, and actions required for continued long-term research projects); teaching (value and accessibility); and public use (sensitivity and protective measures required). The plan also lists goals for each restoration. For example, the restoration goal (called an objective in the Kline Plan) for Curtis Prairie is: "To develop a good example of tallgrass prairie, with appropriate species throughout the moisture continuum" (Kline 1992, 107). No other specifications are given, as if the details of "a good example of tallgrass prairie" are self-evident. The Kline plan was reviewed and adopted by the Arboretum staff, director, and oversight committee, with invited input from several campus ecologists. It remains in effect as of 2011.

14.1.3 Implementation

At the time the Arboretum's first restoration project, Curtis Prairie, was initiated, virtually nothing was known about establishing prairie. Early restorationists tried experimenting with various establishment methods, including removing surface soil, plowing, seeding, transplanting sod, and mulching with prairie hay to transform the horse pasture into prairie. They judged that transplanting prairie sod dug from nearby remnants was the most effective, but that seeding was more affordable.

From 1935 to 1939, a CCC camp was based at the Arboretum, and its 300 male enrollees, under the supervision of Sperry, continued the plantings and experimentations. This labor was essentially free, as far as the university was concerned. Leopold had recruited Sperry—a newly minted doctoral student of Arthur Vestal's from the University of Illinois—to oversee the planting of the Arboretum prairie. Although plant introductions were discontinued during World War II, they were resumed around 1950 under the direction of Grant Cottam and David Archbald, who used seed mixtures and various planting methods, including burning before seeding, disking after seeding, and mulching with prairie hay (Cottam and Wilson 1966).

Besides his legacy of guiding the initial work in the first-ever prairie restoration, Sperry, who died in 1996, left behind a rich and detailed record of his work, including observations, planting records, and analysis of the results. In notebook after notebook, in tight script and precise prose, he tells where and when each species was collected—seed, sod, or plant—and where, when, and how it was planted. We know the weather on the planting day, whether or not the species was watered, and a general estimation of its survival rate by the end of the planting season. He also left supplemental accounts of the operation of the nursery, special experiments, and photography; the labor costs; and observations of species growth and survival field maps.

Because the Kline Plan included neither prioritization of actions nor an implementation plan or monitoring protocols, current Arboretum land care staff have adapted to this situation. For example, land care and research are now strongly linked through an Adaptive Restoration Task Force (ARTF). The ARTF has prioritized land care (restoration) needs, research opportunities, and instances where the two intersect. This process has resulted in a number of adaptive restoration projects—essentially, restorations designed as experiments (see Chapter 3). Thus, an iterative process develops, in which research findings result in implementation recommendations that the land managers try out. Subsequently, implementation and management outcomes are reported back in the form of recommendations for adapting the research approach, or an entirely new line of questions is proposed.

14.1.4 Research and Contributions to Restoration

As one of the research centers of UW–Madison, the Arboretum has a long legacy of research and scholarship. The Arboretum's work has resulted in, among other things, the development of prairie planting and management techniques—including research confirming that prescribed fire is the primary management procedure for tallgrass prairies—that are in use today throughout the Midwest. Generations of faculty, staff, and students have contributed to restoration studies at the Arboretum. The work has generated hundreds of research papers, resulting in the development of many restoration innovations and research methods.

Early Years

Historically, neither the Arboretum project in general, nor Curtis Prairie in particular, was set up as an experiment, because no one knew enough at the time about restoration or prairie ecology and all the variables involved in conducting a rigorous experiment (Cottam 1987). Nonetheless, experimentation was involved. Fassett encouraged his graduate student John Thompson to begin experimenting with methods of establishing prairie. Starting with the Fassett/Thompson studies on prairie establishments, the Arboretum has continued to be a major research site. Instead of adopting a pure trial-and-error approach, every step in the initial prairie restoration project was turned into a research project. Fassett and Thompson set up small experimental plots for the first major plantings of 42 species. Additional plantings and experimentation were carried out by Dave Archbald between 1950 and 1957.

Experiments carried out in the late 1940s by Max Partch and John Curtis showed that burning favored prairie species. This finding led to the first prescribed management burn in 1950.

Current Research

Current research takes advantage of grant opportunities, management problems, and faculty and graduate student research interests, and it relies upon an adaptive restoration framework. Ongoing studies focus on linking research and land care by developing adaptive management solutions to urban restoration challenges, such as stormwater runoff and invasive species.

14.1.5 Monitoring

The Arboretum has been involved in monitoring off and on for many years. Ted Sperry apparently conducted the first monitoring survey of the original plantings. His initial work at Curtis Prairie ended in 1941 when the CCC camp was closed, but he made three return trips—1946, 1982, and 1990—to evaluate the success of the project. In 1984, he published the following account of the prairie's progress:

> *The Curtis Prairie of the Arboretum, initiated between 1936 and 1941, originally consisted of 46 species separately planted in 237 plantings. Forty-six percent of the plantings were successful and 38 percent unsuccessful. Nine common successful species spread widely over the prairie while 9 others persisted well, but with little*

or no spread between 1941 and 1982. These latter are called documentary species since they precisely pinpointed their planting locations for the 1982 re-survey. From these key locations, all 237 planting areas could be determined within a meter or two. The plant persistence and succession in each of these plantings could thus be accurately documented after the 40-year interval. Some spreading species were also documentary. Starting with old farmland, this prairie has been spectacularly successful. It has been designated by the Arboretum as "the world's oldest restored prairie." (1984, 140)

Curtis as a faculty member became director of research at the Arboretum and supervised the first formal vegetation surveys of the prairie in 1951 and 1956. He established a 16.8- × 16.8-m (50- × 50-ft) grid and recorded species presence in randomly located 1- m^2 quadrats, one per grid cell. There are more than 1,000 quadrats in all. The grid is marked by 3-ft-tall metal posts that are now GPS located. These vegetation surveys were conducted every 5 years until 1976, and were resumed in 2002 and 2008. (For a recent summary of the results of the 2002 survey, see Looking to the Future: Curtis Prairie, below.)

14.1.6 Management

The Arboretum uses the word "management" to describe the land care activities that are applied to all of its restorations, regardless of whether the sites have achieved their restoration goals and objectives. Thus, the protocols described here are a mix of implementation and management activities as described in previous chapters.

There are more than 30 plant community types undergoing restoration in the Arboretum, divided into more than 50 management/burn units. (See the discussion of the difference between implementation units and management units in Chapters 8 and 10.) The landscape design of the Arboretum has created numerous sharp boundaries between plant communities of different types (e.g., northern pine forest next to prairie; see Figure 14.1). Sharp boundaries create much more edge habitat than smoother transitions would, such as between grassland and savanna or wooded grasslands.

Long-range restoration and daily management are guided by the Kline Plan. (The 1992 management plan is now undergoing review and revision to prepare the Arboretum for the next 25–30 years.) The Kline Plan sets forth the guiding principle that the Arboretum's restorations—except for Curtis and Greene Prairies, because of their legacy status—do not aim for replication of some former state, but rather use the knowledge of remnant ecosystems as a guide to create dynamic communities for the future. This approach recognizes the fact that natural communities are variable through space and time. Moreover, one may not be able to return historical assemblages of plant and animal communities to the way they were 10, 50, or 100 years ago, nor should we try—in every case—to do so. What we can do, however, is strive to create communities and ecosystems that can be sustained into the future. Wegener and Zedler suggest that instead: "It might be necessary to aim to 'keep all the parts' as recommended by Aldo Leopold somewhere in the Arboretum" (2009, 5).

The basic prairie and savanna management tool is prescribed burning, supplemented with brush cutting (Figure 14.2). The use of herbicides is limited to those species for which there is no other control option. The frequency of burning depends on the type of prairie, what species need to be encouraged or discouraged, and research needs. In general, the staff try to schedule the burns often enough to keep shrubs suppressed, yet irregular enough to

avoid a repeating pattern. Curtis Prairie consists of five management units, from 4 acres to 30 acres, each of which is burned on a rotation of 3 out of every 5 years.

Permanent field staff and volunteers pull herbaceous weeds, cut shrub pest species, and apply herbicide to pest trees. Supplemental seeding is done in areas of bare soil after shrub clones are removed.

14.1.7 Looking to the Future: The Master Plan

The Arboretum's current restoration master plan has provided a modern, flexible, and forward-looking restoration vision that has guided management for nearly 20 years. The plan is official policy and recognizes the historical objectives to develop good examples of restored and remnant native plant and animal communities. It identifies critical threats, such as pest species, and specific needs for reintroducing some native species.

However, the plan stopped short of developing detailed implementation and management plans, and monitoring protocols. Nor did the current master plan anticipate new threats, such as stormwater inflows, deer predation, and pine tree shading of Curtis Prairie. To respond to these previously unanticipated events, the Arboretum director has established ad hoc committees to draft policies that are then approved by the Arboretum Committee and director. Although this approach provides guidance, because these new policies have not been incorporated into the plan, it has also caused difficulties when administrative staff turnover results in the policies being forgotten over time. The master plan also does not provide guidance on dealing with the potential impacts of climate change, which only recently became a topic of discussion.

The lack of prioritization, as well as the absence of implementation and monitoring plans, has proven to be a significant handicap to progress. Without a consistent direction, ad hoc methods of determining restoration needs, and assigning management authority and responsibility, have held sway, resulting in a periodic shifting focus on the work, if not duplication of effort. For these reasons, the Arboretum has begun an extensive review and revision of the current master plan, which is scheduled to be completed by summer of 2012.

14.1.8 Looking to the Future: Curtis Prairie

The generic restoration goals and objectives for the prairies (and the Arboretum as a whole) have had their advantages and disadvantages. The vagueness of the desired outcomes has encouraged generations of stakeholders—researchers, managers, faculty, staff, students, volunteers, and the general public—to stamp the project with their own views of what is desirable. This freedom to interpret the prairie/Arboretum is, on the one hand, beneficial because it has encouraged, in the social realm, a good deal of public interest and involvement—a citizenry of diverse voices and opinions. On the other hand, it has created the challenging task of reconciling the competing viewpoints of what is socially desirable with the reality of what is ecologically attainable.

Lack of precision in goals and objectives has meant that it is difficult to know all the attributes to measure and monitor, let alone when the desired outcomes are reached. Nonetheless, an assessment of long-term restoration success in Curtis Prairie has recently been compiled (Wegener et al. 2008). The assessment asked and answered four basic questions: Is Curtis Prairie uniformly rich in species? Is Curtis Prairie restored? Does Curtis Prairie match natural prairies in species richness? What are the persistent restoration issues?

Is Curtis Prairie Uniformly Rich in Species?

The 2002 vegetation survey revealed that Curtis Prairie had 265 species, 230 of which are native (Snyder 2004). Diversity varied across the prairie, with the highest diversity in the unplowed remnant and lower diversity in wetland areas dominated by the invasive *Phalaris ardundinacea* (reed canary grass) or woody native pest species such as *Cornus racemosa* (grey dogwood) and *Salix interior* (willow).

Is Curtis Prairie Restored?

Any restoration is a process rather than a single event; it's possible it will never be completed, and Curtis Prairie is no exception. For example, the site is too small to provide habitat for some species of native grassland birds that would be expected in a midwestern prairie restoration. Likewise, there are some native mammal species missing, although one success is that a breeding pair of *Grus canadensis* (sandhill crane) has nested in the prairie for the past several years.

Does Curtis Prairie Match Natural Prairies in Species Richness?

This is a more difficult issue to evaluate. The main reason is that there are so few remaining native prairie remnants with similar soil and other site conditions, comparable management histories, and existing data sets. Yet, when looking at the entire prairie ecosystem and remnants within Wisconsin, Curtis Prairie is unlikely to have the representation of species that some of the better remnants contain (some of which have 400 or more recorded species).

What Are the Persistent Restoration Issues?

Native and introduced woody and herbaceous pest species continue to be widespread, despite 75 years of active management (McGaw 2002; Snyder 2004). The density and frequency of these pests reduce native species diversity, suppress flammable fuel loads, and make the conducting of prescribed fires difficult. In some cases, the early lack of understanding of the use of fire and cutting resulted in stimulating species into spreading via root systems. The management of aspen (*Populus tremeloides*), for example, was attempted early in the Arboretum history but resulted in its greater spread, and has yet to be brought under full control.

Stormwater runoff has a tremendous impact on the prairie. The estimated annual volume of storm water that enters Curtis Prairie is 64,141 cubic meters or 52 acre-feet (the prairie itself is just over 60 acres). The storm water carries with it sediment, pollutants, and seeds of pest species. It has been shown to facilitate the spread of reed canary grass and has eroded a ditch through the center of the prairie (see Figure 1.1).

Smoke management issues and public safety precautions to protect dense commercial and residential developments in the urban setting require that prescribed fires be conducted under relatively safe and cool conditions. These conditions might mean the fires are not as effective at controlling pest species as naturally set fires might have been historically.

Despite these issues, the UW Arboretum has been an active and engaging player in the promotion, practice, and study of restoration.

▶ 14.2 THE NATURE CONSERVANCY GREAT RIVERS PARTNERSHIP PROGRAM

The Nature Conservancy (TNC) is a nongovernmental conservation organization that is dedicated to protecting ecologically important lands and waters for nature and people all over the world. The organization works with scientists and local groups to fulfill its mission. The Nature Conservancy's approach to conservation is to set priorities and goals, develop strategies, take action, and measure results. Its effectiveness is measured using these questions: How is the biodiversity doing? Are our actions having the intended impact? Its strategy is to provide solutions (conservation, restoration) that meet the needs of the ecosystem, as well as the people who depend on that ecosystem. To do so, TNC has a 5-S framework for conservation project management that focuses on the following:

- Systems: The focal conservation targets and their key ecological attributes.
- Stresses: The most serious types of destruction or degradation.
- Sources of stress: Causes of destruction or degradation.
- Strategies: Actions to abate threats and enhance conservation target viability.
- Success measures: Monitoring progress in abating threats and improving biodiversity.

TNC depends on the large diversity of others for its success. This is particularly true for developing regions, where resources are often limited, and people's livelihoods depend on access to the same lands that TNC wishes to protect and restore to functioning ecosystems. For most conservation efforts, TNC programs consider stakeholders as well as local economic, political, and social conditions (see Chapters 12 and 13). The three projects below represent TNC's efforts with other partners to accomplish strategies that will eventually lead to the restoration and management of sensitive ecosystems, and at the same time engage the people who depend on these systems for sustenance and their livelihoods. Each of these Great Rivers Partnership projects illustrates the need for cooperative partnerships, science, and citizen involvement.

14.2.1 The Atlantic Forest and Central Savannas Program

Through the Great Rivers Partnership, The Nature Conservancy is working to protect the Paraguay-Parana River system, which encompasses parts of Brazil, Bolivia, Paraguay, Uruguay and Argentina. Planting trees and otherwise working to restore forest habitat in Brazil's Atlantic Forest, as well as in large grasslands like the Cerrado region of Brazil, not only benefit the forests, grasslands, and the wildlife that depend on them, but these efforts help maintain and/or improve the water quality in tributary streams, such as the Piracicaba, Capivari, and Jundiaí Rivers, which flow into the Paraguay and Parana.

TNC's Atlantic Forest and Central Savannas Program is restoring 865 acres of tropical forest located in Brazil's Atlantic Forest. Over the past 20 years, these lands were grazed by livestock and converted to agriculture, leaving approximately 7% of the primary forest intact. TNC is working with the local community, the state government, and the São Paulo State Basic Sanitation Company (SABESP) to restore these lands. Restoration efforts in 2010 were focused on 117 acres along the shoreline of the Cachoeira Reservoir, located in the Piracicaba-Capivari-Jundiaí watershed. Efforts to preserve and restore forest habitat along streams within this watershed are expected to improve water quality and quantity. These efforts are also tied to the lives of people in the watershed, as the reservoir is one of six in the Cantareira Water System, which provides water for the São Paulo metropolitan area.

To assist implementation, TNC is exploring several self-regeneration techniques. The first enlists natural seed dispersers like birds once cattle have been removed. Artificial perches are built in locations that encourage seed dispersal from remnants to the restoration sites. A second technique is to remove pest species, both plant and animal, that compete with the native species. The third method calls for transporting soils that may contain microorganisms, seeds, and pollen from remnant areas. And a final technique calls for transporting the "seed rain" collected in nearby remnants to the restoration site. TNC will also transplant tree seedlings from 70 species as needed.

The program benefits from and brings benefits to the inhabitants of the nearby community of Piracaia. More than 30% of the citizens have some type of involvement in this project, contributing their agricultural experience with the landscape, while benefiting from the employment it provides. Long-term funding for the restoration may come in several ways. In one proposed plan (not yet implemented), carbon credits would be purchased to offset corporate emissions and support continued restoration efforts. Additional funding for expanding restoration efforts may come from another proposal to have SABESP's customers contribute to restoration through their water bills.

14.2.2 The Yunnan Great Rivers Project

The Yunnan Great Rivers Project in China's northwest Yunnan Province worked with 40 public and private agencies to develop an area conservation and economic development plan. The project area comprises approximately 1,600 km² and is located within what is referred to as the Three Parallel Rivers World Heritage Site. This region is topographically and biologically diverse. It contains subtropical shrub-scrub, temperate coniferous forests, alpine meadows, old-growth alpine forests, and arid canyons. At least 30 endangered wild-life species exist here. The plan identifies the highest-quality habitats and their threats, and then provides ways to abate these threats. The Meili Snow Mountains in western China, along the border of Tibet, is one of these habitats. This is a region of 20,000-ft mountain peaks that contains a network of rivers and is home to a rich biodiversity that includes lesser pandas, golden monkeys, snow leopards, and rhododendron forests.

As in the previous TNC partnership project, restoring and maintaining the high quality of the Yunnan Great River landscape has implications for the 500 million people who live in its basins. Although efforts include restoration and protection, many of these activities focus on human management: the continued ability to cultivate medicinal plants by indigenous communities, incorporating traditional knowledge into the protection of conservation targets, managing tourist and national park development, abating fuel wood collection and timber harvests, and improving water quality through agricultural practices. The rivers—Yangtze, Irrawaddy, Salween, and Mekong—provide food, transportation, water, and livelihoods to the people of the region. Efforts to restore, protect, and manage the forests through which these rivers traverse are ongoing. The forests help contain and filter floodwaters and the sediments, nutrients, and pollutants that would otherwise go directly into these rivers. Collecting wood for fuel, by itself, is responsible for the loss of hundreds of thousands of acres of forest.

Several projects within Yunnan Province are occurring—mapping and studying the region for increased protection of its preserves and national parks, developing conservation plans among partners to reduce threats to its biodiversity and cultural diversity, protecting habitat for the Yunnan golden monkey (*Rhinopithecus bieti*) and other endangered flora and fauna, finding and teaching sustainable methods for local landowners to use the preserves' resources, finding alternatives to wood for energy and ways to provide loans to villagers to pay for these, and developing environmentally based education programs.

TNC is also working with Chinese partners on the restoration and management of the Yangtze River, the third longest river in the world (Figure 14.3). This program has resulted in partnerships between scientists in both China and the United States to apply advanced

FIGURE 14.3. The Yangtze River at Xiling Gorge. This is the exit of the Three Gorges, several miles upstream of Three Gorges Dam. The water looks aqua blue; its impounded and algae blooms give the water its color. In the area surrounding these terraced slopes are a water quality monitoring station, aquaculture pens in adjacent small tributaries, and very large truck ferries navigating the reservoir. (Photo provided courtesy of Brian S. Ickes, Fish Ecologist, Environmental Management Program, and Long-Term Resource Monitoring Program on the Upper Mississippi River System, La Crosse, Wisconsin.)

monitoring technology to better understand the fish populations of the Yangtze and the success of existing protection and restoration programs (see Chapter 9). Overfishing, dams, and agricultural runoff—each of these has impacted the Yangtze River, affecting its water quality and quantity, river dynamics, adjacent wetlands, and aquatic populations, especially the crash of fish populations that many citizens depend on for food. The pest species *Spartina alterniflora*, a wetland grass, has spread rapidly in the river's tidal flats, further damaging habitat.

Plans are for China to build 12 more hydropower dams on the Yangtze, in addition to the Three Gorges Dam. Several nonprofit organizations, government agencies, and hydropower companies are working to develop alternatives to dam design and operation, with these goals:

- Optimizing electricity production and flood control while reducing the impacts on the river's ecology.

- Minimizing the impact on river flows and fish populations.

- Restoring critical wetlands.

- Securing funding for conservation and floodplain restoration and programs that will reduce the spread of waterborne diseases.

Successfully achieving each of these goals requires cooperation and learning from the experience of other partners. The Chinese are also releasing Chinese sturgeon (*Acipenser sinensis*) and mitten-hand crabs to rebuild populations, and releasing benthos, particularly oyster (*Crassostrea* sp.) along concrete dam foundations to build artificial oyster reefs.

One of TNC's strategies is to facilitate collaboration between ecologists from around the world to exchange ideas on how to approach river restoration. For example, ecologists from the USGS Upper Midwest Environmental Sciences Center in LaCrosse, Wisconsin, met with Chinese scientists in the U.S. to demonstrate strategies that have been successfully used in the Mississippi River Basin for monitoring management and restoration efforts. They then traveled to the Yangtze to work with scientists there in their efforts to restore the Yangtze River ecosystem, including its fish populations, and to assist with the establishment of monitoring systems to assess the success of different strategies in use (Figure 14.4).

FIGURE 14.4. Monitoring the Ecological Health of the Yangtze River. This photograph was taken in October 2009 on the Jinsha River (Upper Yangtze River) near Yibin, Sichuan Province. On the right is Dr. Duan Xinbin of the Chinese Academy of Fisheries Science, flanked on the left by two of his students, and a commercial fisherman on the far left. They are deploying a hoop net, one of several fisheries sampling techniques used on the Mississippi River to monitor the ecological health of the Upper Mississippi for the last 20 years. This was the first deployment of a Mississippi River sampling method in the Yangtze Basin, representing a major advance in technical and scientific collaboration among great river scientists in Asia and North America. (Photo provided courtesy of Brian S. Ickes, Fish Ecologist, Environmental Management Program, and Long-Term Resource Monitoring Program on the Upper Mississippi River System, La Crosse, Wisconsin.)

14.2.3 The Mississippi River Basin

Similar to China's Yangtze River, the 3,734-km-long Mississippi River, as the world's fourth longest river and sixth largest drainage area, is of significant ecological and economic value to the United States. Nonprofits, farmers, government agencies, and citizens interested in the health of the Mississippi are working collaboratively throughout its watershed to implement conservation projects and restore tributaries, floodplain forests and swamps, and wildlife habitat, while finding ways to sustain the ability of people to utilize its resources. Many of these efforts include reducing nutrients, sediment, and runoff from disturbed uplands into the river and its tributaries, each of which degrades water quality and habitat for fish and other aquatic life (Figure 14.5). The elevated nutrient levels alone result in algae blooms that eventually lead to oxygen depletion and loss of aquatic plants and animals.

TNC's Upper Mississippi River Program partners include the U.S. Army Corps of Engineers and the U.S. Fish and Wildlife Service, as well as a large array of state, regional, and local government agencies, industry representatives, nonprofit organizations, and citizen groups. The focus is on developing a basinwide strategy to rebuild the river's ecological and economic value. Initial work began with drawing down water levels in pools and backwaters. The goal is to reestablish aquatic plants—emergents, such as river bulrush (*Schoenoplectus fluviatilis*) and wild rice (*Zizania palustris*), as well as submergents, such as wild celery (*Vallisneria americana*) and water star grass (*Heteranthera dubia*).

One effort concentrates on working with farmers in Minnesota, Iowa, Wisconsin, and Illinois to reduce the amount of nutrients and sediment that run off agricultural fields into nearby rivers, eventually making their way to the Gulf of Mexico. Results are being monitored, which will lead to the adaptation of strategies. Another project involves surveying the fish and other aquatic species (Figure 14.6). The survey of fish populations includes looking for pest species that disrupt populations of native aquatic species, both plant and animal, as well as degrade water quality. The health of native fish populations is also

FIGURE 14.5. Restoring Habitat in the Mississippi River. This small island is being constructed along the west bank of the Mississippi River near Brownsville, Minnesota. The Army Corps of Engineers, working with state and federal agencies, is in the process of building small islands to trap river sediment and re-create the wetlands and habitat that existed before the lock-and-dam system was established. (Photo provided courtesy of Mark Godfrey, The Nature Conservancy.)

FIGURE 14.6. Surveying Fish Populations in the Mississippi River. Researchers from the Upper Midwest Environmental Sciences Center (UMESC, part of USGS), located near La Crosse, Wisconsin, are using a specially equipped boat to briefly stun fish by electroshock in a process of surveying fish species, stocks, and locations on the Mississippi River near La Crosse. (Photo provided courtesy of Mark Godfrey, The Nature Conservancy.)

indicative of water quality and the recovery of the aquatic ecosystem. In addition, partners such as the National Audubon Society will help educate people about how their land use practices can affect the river system.

The lower Mississippi basin has lost nearly four-fifths of its bottomland forests (dropping from 24 million acres to fewer than 5 million acres). A goal of the Great Rivers Partnership is to restore 1–2 million of these acres. Doing so depends on ensuring that the restoration benefits the landowners. For example, TNC is working with local communities in the Big Woods bottomland forest of Arkansas to promote sustainable land uses such as ecotourism, and to provide incentives for participation in conservation programs.

The Atchafalaya River Basin is a second front in the efforts to restore the Mississippi River watershed. The Atchafalaya is a distributary; it takes water from the Mississippi and then delivers it to the Gulf of Mexico. The basin contains one of the country's largest freshwater cypress swamps and is home to a diverse group of plant and animal species. This natural resource supports families and people's livelihoods, from loggers to commercial fishermen. Past flood events on the Mississippi River that endangered human communities and livelihoods led to the creation of levees that altered the flow of waters in the Atchafalaya basin. This led to the infilling of swamps with sediments and unnaturally high water levels in other areas, along with reduced water quality, the depletion of oxygen, the decline of the cypress, and large fish kills. Efforts are under way to restore a more natural water flow by working with partners to establish a single set of goals for the Atchafalaya:

- Identify conservation focus areas in the basin.
- Refine procedures to monitor restoration projects and their impacts on the basin.
- Develop hydrodynamic and sediment models to explore what would happen if the percentage of water allowed into the basin from the Mississippi was modified.
- Assist basin landowners with enrollment in conservation easement programs.

The Upper Ouachita River in northern Louisiana is a third area of focus in the lower Mississippi region. The trees at the Mollicy Farms Unit, a 16,000-acre bottomland hardwood forest, were cut long ago, and levees were built to keep water out so that the land could be planted for agriculture. Over the past decade, the U.S. Fish and Wildlife Service has planted more than 3 million bald cypress (*Taxodium distichum*), oak, and ash trees on nearly 11,000 acres of the refuge. Working with partners, the intent is to reconfigure the levees and restore the connection between the Ouachita and the floodplain to allow for seasonal flooding. Monitoring of the project will be done with Louisiana State University, the University of Louisiana at Monroe, and the U.S. Geological Survey. The long-term goal is to restore 25 square miles of fish and wildlife habitat, while at the same time alleviating downstream flooding and improving water quality.

▶ 14.3 THE KOOTENAI RIVER HABITAT RESTORATION PROJECT MASTER PLAN

During the last century, the Kootenai River sub-basin was modified by agriculture, logging, mining, and flood control. The Kootenai sub-basin covers an area encompassing southern British Columbia, northern Idaho, and northwestern Montana. The Kootenai River was impounded by Libby Dam, in northwest Montana, which created Koocanusa Reservoir in the U.S., and by Corra Linn Dam, B.C., which created Kootenay Lake. Regulation of both dams cut annual peak flows in half and eliminated the annual spring season water flush that supported many ecosystem processes. The conversion of more than 50,000 acres of floodplain to agricultural fields also resulted in the loss of riparian and wetland plant and animal species, and related functions that normally support a healthy ecosystem. Native fish stocks and other wildlife populations declined, including many that are important for subsistence and cultural uses by the Kootenai Tribe. One of the fish species, the Kootenai River white sturgeon, is listed as endangered by the U.S. government due to the loss of spawning and rearing habitat.

The Kootenai River Habitat Restoration Project Master Plan provides detail on the project's purpose, goals, objectives, and site inventory and analysis. It is a broad overall restoration scheme that includes the "development of a conceptual adaptive management and monitoring program, identification of anticipated environmental compliance and consultation requirements, a funding analysis, and identification of specific design and implementation actions to be accomplished in subsequent project phases" (Kootenai Tribe of Idaho 2009, iii–iv).

The Kootenai Tribe of Idaho sets an overriding philosophy for the plan of "developing and implementing innovative, scientific approaches to guardianship of the land that consider the whole ecosystem at the watershed/sub-basin scale, are socially and economically responsible, are supported by the local community and other partners within the watershed, and that incorporate adaptive management principles" (Kootenai Tribe of Idaho 2009, iii).

Here are the stated purposes of the Kootenai River Habitat Restoration Project:

- The Kootenai River habitat addresses ecological limiting factors and constraints related to river morphology, riparian vegetation, aquatic habitat, and river management. The desired result is a more resilient ecosystem, capable of sustaining diverse native plant and animal populations, and that is tolerant of natural disturbances and altered regimes.

- The Kootenai River habitat conditions support all life stages (i.e., migration, occupancy, spawning, incubation, recruitment, and early rearing) of endangered Kootenai River white sturgeon (*Acipenser transmontanus*) and other aquatic focal species.
- The Kootenai River landscape sustains the tribal and local culture and economy, and contributes to the health of the Kootenai sub-basin as both an ecological and a socio-economic region.

14.3.1 Major Goals

The four goals of the project address restoring, or creating opportunities for restoring and enhancing, the following:

- River and floodplain morphology. Reduce the negative effects to river and floodplain ecological processes caused by river responses to the altered landscape.
- Riparian vegetation. Establish streambank and floodplain conditions that sustain plant community development processes.
- Aquatic habitat. Have conditions that support all life stages of native fish and promote sustainable populations.
- River stewardship. Ensure the existence of opportunities for river and floodplain stewardship in the community.

Objectives are assigned to each of these goals and are organized around habitat types (mainstream, tributaries, and reservoir); focal species (including bull trout [*Salvelinus confluentas*], sturgeon, burbot [*Lota lota*]); and biomes (regulated mainstem, wetland, riparian, grassland/shrub, xeric forest, and mesic forest).

14.3.2 Site Analysis

The Kootenai River Habitat Restoration Project focuses on a 55-mile stretch of the river. This stretch is divided into three major river reaches, based on unique geomorphic properties, that are identified as the Braided, Meander, and Straight Reaches. A suitability analysis (see Chapter 4) was used that factored in numerous variables, including relationships to other projects, floodplain elevations relative to the river stage at average peak flows (Figure 14.7), historic water-surface elevations, geology and soils, basin and channel morphology, individual reach locations, riffle and pool distribution, sediment depths, daily discharge and flow rates, water temperatures, land cover (Figure 14.8), dam operations, and river history. The limitations to achieving goals and objectives were determined. The suitability analysis was performed (Figure 14.9), and then strategies and implementation scenarios were proposed in the master plan.

As a result of the site analysis, the limiting factors for restoration were noted for the physical, biological, and ecological conditions with the project area that (1) limit the ability of the ecosystem to sustain diverse native plant and animal populations, and to accommodate natural disturbances; (2) limit the quality or availability of habitat that supports all life stages of endangered Kootenai sturgeon and other focal species; and (3) limit the ability of the ecosystem to sustain the local tribal culture, subsistence needs, and the economy. From Table 14.1 and the background information and reference models, a table of limitations that includes corresponding implementation scenarios was developed (see Table 8.1).

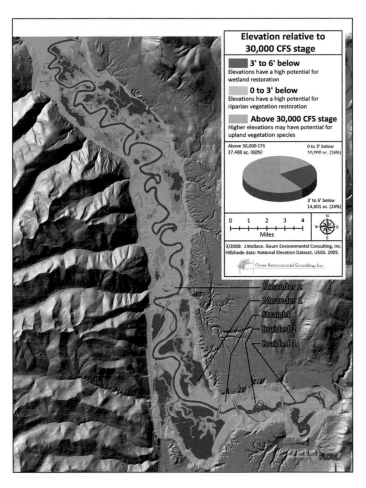

FIGURE 14.7. Floodplain Restoration Potential Relative to Flood Levels. The elevation categories are related to hydroperiod regimes. River stage is based on a water-surface elevation of 1,754 feet at Porthill, Idaho, near the Canadian border. Reprinted by permission from Kootenai Tribe of Idaho, "Kootenai River Habitat Restoration Project Master Plan: A Conceptual Feasibility Analysis and Design Framework." Bonners Ferry, ID, 2009 (with funding provided by Bonneville Power Administration).

Table 14.1 • Limiting Factors of the Kootenai River Habitat Restoration Project

Morphology Limiting Factors	Riparian Vegetation Limiting Factors	Aquatic Habitat Limiting Factors	Constraints Related to River and Floodplain Stewardship
Factors that have altered natural and historical morphological processes	Factors that have altered the native riparian vegetation community	Factors that have altered the aquatic habitat conditions	Several aspects of the ecosystem have been affected by river and floodplain management
• River response to altered hydraulics • River response to altered sediment transport • Loss of channel and floodplain connection • Reduced channel boundary roughness (smooth banks lacking vegetation or woody debris) • Bank erosion	• Lack of surfaces that support riparian vegetation recruitment • Lack of outer bank vegetation • Frequent scouring and deposition of floodplain surfaces • Altered hydroperiod • Invasive plant species • Lack of native plant diversity and seed sources • Lack of nutrient sources for primary productivity • Altered carbon balance	• Insufficient depth for Kootenai sturgeon migration • Insufficient velocity for successful Kootenai sturgeon spawning • Lack of coarse substrate for egg attachment • Lack of pool-riffle complexity • Simplified food web from lack of nutrients • Insufficient pool frequency • Lack of fish passage into tributaries • Lack of off-channel habitat for rearing • Altered water quality	• Dam-controlled flow regime • Dam-controlled sediment regime • Dam-controlled thermal regime • Dam-controlled nutrient regime • Bank armoring • Levees and diking districts • Transportation corridors • Floodplain land use

Adopted from Kootenai Tribe of Idaho, "Kootenai River Habitat Restoration Project Master Plan: A Conceptual Feasibility Analysis and Design Framework." Bonners Ferry, ID, 2009 (with funding provided by Bonneville Power Administration)

FIGURE 14.8. **Restoration Classes.** The restoration classes represent cover types suitable for restoration (Restoration Classes) and landcover types that may contain native vegetation, but may require some additional hands-on actions (Enhancement Classes). Reprinted by permission from Kootenai Tribe of Idaho, "Kootenai River Habitat Restoration Project Master Plan: A Conceptual Feasibility Analysis and Design Framework." Bonners Ferry, ID, 2009 (with funding provided by Bonneville Power Administration).

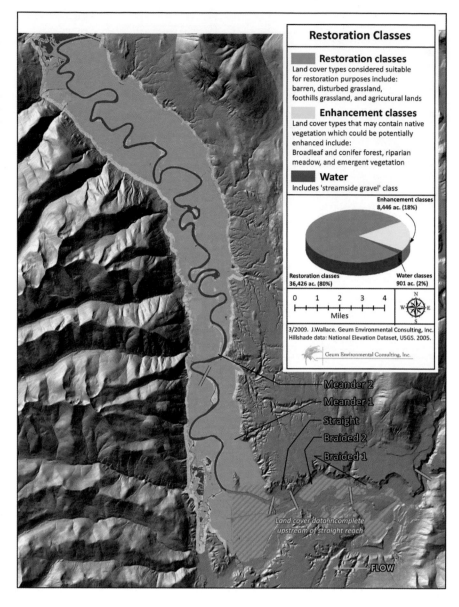

14.3.3 Environmental Compliance and Consultation

The master and site plans identified federal, state, and local laws and administrative requirements that apply to the project (see Chapter 7). Some of these are the U.S. National Environmental Policy Act, the Clean Water Act, Idaho water quality requirements, local floodplain permits, the Endangered Species Act, the National Historic Preservation Act, and the Federal Farmland Protection Policy Act.

14.3.4 Estimated Costs

A construction firm, with experience implementing similar types of projects, developed cost estimates based on the master plan scenarios. Costs included minimum, maximum, and moderate implementation scenarios for each of the project reaches, and were based on

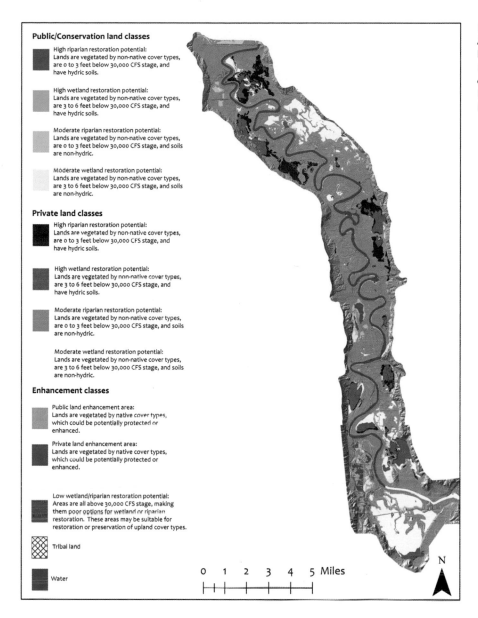

Public/Conservation land classes

High riparian restoration potential:
Lands are vegetated by non-native cover types,
are 0 to 3 feet below 30,000 CFS stage, and
have hydric soils.

High wetland restoration potential:
Lands are vegetated by non-native cover types,
are 3 to 6 feet below 30,000 CFS stage, and
have hydric soils.

Moderate riparian restoration potential:
Lands are vegetated by non-native cover types,
are 0 to 3 feet below 30,000 CFS stage, and soils
are non-hydric.

Moderate wetland restoration potential:
Lands are vegetated by non-native cover types,
are 3 to 6 feet below 30,000 CFS stage, and soils
are non-hydric.

Private land classes

High riparian restoration potential:
Lands are vegetated by non-native cover types,
are 0 to 3 feet below 30,000 CFS stage, and
have hydric soils.

High wetland restoration potential:
Lands are vegetated by non-native cover types,
are 3 to 6 feet below 30,000 CFS stage, and
have hydric soils.

Moderate riparian restoration potential:
Lands are vegetated by non-native cover types,
are 0 to 3 feet below 30,000 CFS stage, and soils
are non-hydric.

Moderate wetland restoration potential:
Lands are vegetated by non-native cover types,
are 3 to 6 feet below 30,000 CFS stage, and soils
are non-hydric.

Enhancement classes

Public land enhancement area:
Lands are vegetated by native cover types,
which could be potentially protected or
enhanced.

Private land enhancement area:
Lands are vegetated by native cover types,
which could be potentially protected or
enhanced.

Low wetland/riparian restoration potential:
Areas are all above 30,000 CFS stage, making
them poor options for wetland or riparian
restoration. These areas may be suitable for
restoration or preservation of upland cover types.

Tribal land

Water

0   1   2   3   4   5 Miles

N

FIGURE 14.9. Floodplain Restoration Suitability Analysis Results for Areas Outside Levees. Reprinted by permission from Kootenai Tribe of Idaho, "Kootenai River Habitat Restoration Project Master Plan: A Conceptual Feasibility Analysis and Design Framework." Bonners Ferry, ID, 2009 (with funding provided by Bonneville Power Administration).

the following categories: planning and coordination, data collection, preliminary and final design, construction, as-build documentation and establishment of monitoring baselines, postconstruction effectiveness monitoring, biological monitoring, postconstruction maintenance, permitting and environmental compliance, and management of land use practices.

14.3.5 Adaptive Management and Monitoring

In order for the Kootenai Tribe to determine whether the master plan goals were being accomplished, the framework for an adaptive management and monitoring program was provided to:

• Evaluate the effectiveness of the implemented habitat actions in terms of achieving the project goals.

- Identify project maintenance needs.
- Identify any potential unforeseen negative impacts on infrastructure.
- Support decisions to modify restoration treatments.
- Refine or modify restoration treatments that might be implemented in later phases of the project.

The actual success criteria will be developed once the design is created. They will be based on:

- Baseline monitoring (documents the prerestoration condition).
- Implementation monitoring (documents the restoration project as completed).
- Effectiveness monitoring (addresses whether project objectives are being met, determines maintenance needs, and provides inputs into decision pathways).

An interdisciplinary adaptive management team consisting of representatives from management agencies, various disciplines (biologists, hydrologists, etc.), and other experts in the field of restoration will review monitoring data in the context of developing ecosystem functions and processes. The master plan has been adopted and implementation of the restoration is anticipated to begin in the fall of 2011.

15

Conclusion

As a student of restoration ecology, you have chosen a forward-looking, multifaceted discipline that has both deep roots in the past and a future-oriented perspective. Although the term "restoration" implies that you will be "putting something back," modern restoration ecology looks to the near-distant past primarily because our understanding of natural communities comes in part from historic records and past experience. As a member of the restoration team, your goal will likely not be to duplicate the past—you cannot return an ecosystem to what it was a decade ago, let alone hundreds of years ago—but rather to create for the sustainable future. In other words, your job will be to provide the conditions under which the products of evolutionary and environmental history—plants, animals, communities, and ecosystems—can continue to thrive into the future.

This book has had two major purposes: (1) to explore the theories, principles, and assumptions you will use in restoration and (2) to explore how you can apply these underlying theories and principles to the practice of restoration. Our focus has been to provide you with a framework that you can use to guide your restoration decisions anywhere on the globe—both now and in the future. The framework is designed to work with the uniqueness, uncertainty, messiness, and constraints inherent in any restoration.

Although there are many kinds of restoration projects of various sizes, shapes, and contexts, and their purposes are diverse, you will find they all use an iterative problem-solving approach that we have described in this book. No matter what kind of restoration you do or where in the world it is, if you follow the restoration planning process we presented, your results will constitute a unique, feasible, and appropriate restoration solution.

The promise of restoration is that the process can be used with great success to enlarge existing preserves, create buffer zones, provide habitat links, revitalize damaged remnants, provide ecosystem services, serve as sites for research and education, and reconnect people to nature as well as to one another. But restoration is not easy. You will find there are many challenges to achieving your desired restoration outcomes. As we have seen, these challenges include a lack of documentation and experimentation, a scarcity of information about natural systems, the fact that natural systems are dynamic and always changing, the use of restoration as an excuse for destroying remnants, and the fact that there is no single formula for success, meaning that each situation requires a unique solution. You will also find that project stakeholders and restoration team members often have conflicting desires, and that finding the time and resources necessary to support long-term projects is often a challenge.

We have emphasized several themes running throughout the restoration process: (1) the importance of working with people during the establishment, planning, implementation, and monitoring of restoration projects; (2) the use of an adaptive management approach; (3) the importance of building into each project the opportunity to learn about restoration practice and/or ecological theory by documenting the reasoning behind each decision you make, as well as the day-to-day project activities; (4) the value of establishing field trials and more formal research projects; and (5) the responsibility of disseminating the knowledge you accumulate.

▶ 15.1 THE FUTURE OF RESTORATION

As you step out into the world of restoration ecology, you might be wondering what you can look forward to over the course of your career—say, the next 25–50 years. Even if you have been working in restoration for a while, you want to plan for the future and try to anticipate the changes can you expect; you are probably curious about what new and exciting challenges and opportunities await you. This anticipation and planning is to be expected and, as we have said throughout the book, a big part of the job of restoration ecology professionals.

As you look ahead and ponder what you can expect, it might be helpful to start with a few questions. We view these questions as a starting point for a discussion of restoration in the years to come.

15.1.1 What Will Be the Future Challenges?

As noted in Chapter 1, the United Nations Millennium Ecosystem Assessment Project (MA) published a report evaluating the impacts of human activities on ecosystem services.

The MA report issued four major findings, one of which is that "over the last 50 years, humans have changed ecosystems more rapidly and extensively than in any comparable period of time in human history, largely to meet growing demands for food, fresh water, timber, fiber, and fuel" (MA 2005, 1). The degradation of ecosystem services caused by habitat fragmentation, overexploitation of resources, invasive species, pollution, and climate change could become significantly worse in the next 50 years. Under some MA scenarios, this would be the case if existing natural resources continue to be squeezed by a growing human population. The heightened competition for natural resources may then lead society to view restoration as a luxury rather than a necessity. The result could be that people increasingly convert land for the production of agricultural commodities at the expense of resource protection and conservation. Such short-term thinking may push species toward at least local extinction, thus making restoration difficult to achieve.

15.1.2 Will Restorations Persist?

You may wonder, in the face of these challenges, whether individual restoration projects (or even restoration itself) will be able to survive and achieve their desired outcomes. Will restorationists be able, over the long haul, to carry out the management and monitoring required to sustain dynamic ecological communities? We believe the answer is yes.

To meet the challenges, your projects must do several things: maintain society's political support and institutional funding for restoration, do what is necessary to cultivate an engaged and well-informed community of stakeholders to support the restoration, and develop a dedicated restoration team of both volunteers and professionals to oversee the restoration into the future. Meeting these social-ecological challenges is just as important and difficult as carrying out an actual restoration. To do so, it is important that you continue to rely on a rigorous scientific basis for your work, utilize sound ecological principles, and adhere to high professional standards.

15.1.3 What Are the Future Opportunities for Restoration?

There is good news from the MA report. The authors of the report believe that future continued degradation of ecosystem services does not have to happen. Degradation could lessen under some scenarios, depending on how effectively society in general, and restoration in particular, responds.

Restoring ecosystem services is seen by the MA as a cross-cutting technological response to head off further ecosystem degradation in the future. The MA points to restoration ecology's track record of achievement of preserving and enhancing biodiversity and ecosystem services, and it extolls restoration ecology's adaptive management framework as being "particularly valuable given the high levels of uncertainty surrounding coupled socioecological systems" (MA 2005, 24).

Another cross-cutting response will be restoration ecology's continued role in generating and disseminating knowledge. The MA points to restoration ecology's use of the knowledge of traditional ecological practicioners as a model for using place-based knowledge to effectively manage ecosystems. It is exciting to think of the possibilities.

The Millennium Ecosystem Assessment sees a vital, essential role for restoration ecology over the next 50 years if society is to avoid further ecosystem degradation and produce benefits for human well-being. More importantly, the Society for Ecological Restoration (SER) has the same view. In a recent policy document, SER states, "Ecological restoration and conservation should not be considered as last resort activities, but rather as vitally important investments in the future sustainability of the planet" (SER 2009, 3). This statement emphasizes the role of restoration in restoring ecosystem services that are necessary for sustainable human populations.

▶ 15.2 QUESTIONS YOU WILL YOU BE ASKING IN THE FUTURE

As a restorationist, the ultimate questions on your mind are always: (1) Toward what standards should I restore? (2) What is the best way to achieve the desired outcomes? Our best advice for approaching the answers to these questions is what we have emphasized throughout the book—setting clear, measurable goals and objectives and identifying a range of alternative approaches and a range of acceptable outcomes. Also, consider enlarging the restoration target and extending the time frame that the restoration will take. Be flexible and ready to change course as new conditions arise and new information becomes available.

Restoration plans should be feasible and realistic—attainable and sustainable—both in the present and into the future.

The restoration planning process we have presented is designed to help you plan for the future; if you follow it, you will be ready to anticipate and respond to whatever lies ahead. Just like any preparedness planning (such as prescribed fire planning), there are a few underlying principles that provide the planning foundation:

- *Know what you are doing and why.*
- *Be intimately familiar with your surroundings.* If you are, you will be sensitive to important changes.
- *Anticipate the range of possible eventualities.* In other words, plan for everything you can foresee happening—flood, fire, species loss, or whatever it might be in your part of the world—and be ready to respond to it. The human ability to make predictions is imperfect, so you will not be able to anticipate everything; but you can't ignore what you do expect to happen.
- *Plan for the worst but hope for the best.* Develop a list of if/then scenarios to guide your actions for a range of possible situations. If you anticipate that a particular species will become a serious management problem in the near future, develop a response (management) plan. Because you have planned ahead, you will be ready to respond. If you expect that your site will loose species or see species shifts, develop a policy (if/then scenarios) for which species to replace and which (if any) to let go.
- *Perhaps most importantly, build the capacity within your team and organization to respond to changes.* Recruit a team with a variety of skills and talents that works well together. Invest in education and training to increase current skill levels and acquire new skills. Celebrate the achievement of desired outcomes.

Two challenges that will become particularly important in the next several decades concern the restoration of severely disturbed land through the creation of novel communities/ecosystems, and anticipating global climate change.

15.2.1 What Novel Communities Will Be Needed?

Novel ecosystems may require novel approaches. Novel ecosystems are those that contain combinations of species that have not previously been found to be coexisting in a given setting. Here are some examples of disturbed landscapes that may require restoration with novel ecosystems and communities:

- Industrial brownfields.
- Mine tailing deposits.
- Abandoned properties in urban areas with contaminated soil.
- Native systems with significant components of pest species.
- Systems with altered or missing natural disturbance regimes.
- Ecosystems in which the key predators or pollinator species are extinct or impossible to restore.

Although restorationists already have some experience dealing with novel ecosystems, you may have to consider additional unusual approaches in the future. For example, including pest or non-native species in a restoration was once unthinkable, but it could become a necessary approach in situations where removing pest species may have a negative impact or where there are other desired restoration outcomes. When dealing with pest species assemblages, you will have to ask yourself under what circumstances does one simply modify the existing situation or replace the introduced species. In other words, should some restorations accommodate pest species instead of insisting on a pure assemblage of natives? We looked at the example of Kings Bay, Florida, in which four non-native aquatic weeds were found to perform functions that were consistent with project restoration goals (see Chapter 11).

You may also have to ask and answer similar questions about managing for rare native species that have established themselves (or been planted) outside of their historical ranges. Managing for rare species that are shifting their ranges in response to climate change may prove to be an important conservation tool.

15.2.2 How Should We Take Climate Change Predictions into Account?

Indicators of a changing global climate include increases in the frequency and severity of extreme events such as precipitation, flooding, and drought, and a more variable climate. But these factors signal global changes; they cannot tell you what to expect where you work. You will have to develop a monitoring checklist of events that are relevant to your geographic area.

Pay attention to local weather patterns and especially to extreme events, such as drought, storms, and flooding, for potential impacts on species distributions. Changed drought patterns are likely to result in some species mortality or increases in fire frequency or severity. Changed flooding patterns may result in shifts in species distributions because of mortality, and a consequent influx of new species. Look downstream to low-lying areas and wetlands that are likely to be impacted first by extreme precipitation events because they are the ultimate recipient of floodwaters.

Coastal wetlands will also be vulnerable to potentially rising sea levels. If you work in a coastal system, SER (2009) warns that rising sea levels could result in the loss or alteration of intertidal zones and the intrusion of salt water into freshwater systems.

Rely on the monitoring protocols of your pest species management plan to alert you to new intruders. Continually monitor the effectiveness of your management activities for pest species as well as other impacts, for changes in the reliability of your tools and approaches. Unexpected management results (such as decreased effectiveness of a traditionally effective tool) could indicate the need for a management "tune-up" to check the suitability of management procedures (given the current biotic and abiotic situation) and the assumptions behind your procedures.

SER sees an important role for restoration in managing to maintain and enhance rare species in response to climate change. The possible responses by species to one of the potential effects of climate change—global warming—range from migrating to suitable habitat at higher elevations, to expanding their geographic ranges to higher latitudes, to moving inland, away from coastal areas likely to be inundated by increases in sea levels set in motion by the melting of polar ice. SER recommends that restoration ecology can respond by "increasing habitat area and reconnecting fragmented landscapes" (SER 2009, 2). Other SER recommendations for rare species management include restoring coastal wetlands to increase habitat diversity, managing rare species outside of their historic ranges, and reducing fire and fuel loads in fire-prone ecosystems to avoid possibly devastating fires.

▶ 15.3 FINAL THOUGHTS

We believe the future of restoration ecology is bright. In a rapidly changing world, restoration has proven to be a model of how to manage uncertainty, find the connections between social and ecological issues, and turn challenges into opportunities. Restoration is already managing across watersheds and landscapes that are shifting from rural to urban, and taking what SER terms a holistic approach to restoration. A continuation of these approaches will be needed in the future.

Restoration can build upon this expertise by developing management solutions for maintaining and managing resilient social-ecological systems in fragmented and densely populated landscapes, and for resource-demanding societies. Restoration efforts anywhere and at any time—but especially in the future—require not only the ability to manage ecological systems but also the skills to cultivate engaged, well-informed, and ecologically literate civic communities. This is the partnership needed to compose the landscape of the future.

References

Chapter 1

Bradshaw, A. D. 1987. "Restoration: An Acid Test for Ecology." In *Restoration Ecology: A Synthetic Approach to Ecological Research*, edited by W. R. Jordan, 24–29. New York: Cambridge University Press.

Falk, D. 1990. Discovering the future, creating the past: Some reflections on restoration. *Restoration and Management Notes* 8:71–72.

Hull, R. B., and D. P. Robertson. 2000. "Conclusion: Which Nature." In *Restoring Nature: Perspectives from the Social Sciences and Humanities*, edited by P. H. Gobster and R. B. Hull, 299–307. Washington, DC: Island Press.

Leopold, A. 1949. *A Sand County Almanac and Sketches Here and There.* New York: Oxford University Press.

Longenecker, W. G. 1941. University of Wisconsin Arboretum. *Parks & Recreation* Sept.:1–8.

Maloney, C. J. 2008. *Chicago Gardens: The Early History.* Chicago: University of Chicago Press and Center for American Places.

Marsh, G. P. 1965. *Man and Nature.* Edited by D. Lowenthal. Cambridge, MA: Harvard University Press. First published 1864.

McDonald, T. 2008. Evolving restoration principles in a changing world. *Ecological Management & Restoration* 9:165–167.

Mendelson, J., S. P. Aultz, and J. D. Mendelson. 1992. Carving up the woods: Savannah restoration in Northeastern Illinois. *Restoration and Management Notes* 10:127–131.

Millennium Ecosystem Assessment (MA). 2005. *Ecosystems and Human Well-Being: General Synthesis.* Washington, DC: Island Press.

Muir, J. 1896. Proceedings of the minutes of the Sierra Club, held Nov. 23, 1895. *Sierra Club Bulletin* 1 (7):275–276.

Restoring our wetlands. 2005. *Irish Times* (October 1):9. http://www.rncalliance.org/WebRoot /rncalliance/Shops/rncalliance/45C3/ECC1/6854/A8D2/A33C/CA94/8D31/601B /PWSERconference1Oct05.pdf

Society for Ecological Restoration International Science & Policy Working Group. 2004. *SER International Primer on Ecological Restoration.* Version 2. http://www.ser.org/pdf/primer3.pdf

Chapter 2

Anderson, D. C., and J. MacMahon. 1985. Plant succession following the Mount St. Helens volcanic eruption: Facilitation by a burrowing rodent, *Thomomys talpoides. American Midland Naturalist* 114:62–69.

Both, C., S. Bouwhuis, C. M. Lessells, and M. E. Visser. 2006. Climate change and population declines in a long-distance migratory bird. *Nature* 441:81–83.

Boyer, K. E., J. C. Callaway, and J. B. Zedler. 2000. Evaluating the progress of restored cordgrass (*Spartina foliosa*) marshes: Belowground biomass and tissue nitrogen. *Estuaries* 23 (5):711–721.

Clements, F. E. 1936. Nature and structure of the climax. *Journal of Ecology* 24:252–284.

Connell, J. H., and R. O. Slatyer. 1977. Mechanisms of succession in natural communities and their role in community stability and organization. *American Naturalist* 111 (982):1119–1144.

Cowles, H. C. 1901. The physiographic ecology of Chicago and vicinity. *Botanical Gazette* 31 (3):145–182.

Darwin, C. R. 1979. *The Origin of Species.* New York: Avenol Books. First published 1859.

Egler, F. E. 1954. Vegetation science concepts. 1. Initial floristic composition, a factor in old-field vegetation development. *Vegetation* 4:412–417.

Fischer, J., and D. B. Lindenmayer. 2000. An assessment of the results of animal relocations. *Biological Conservation* 96(1):1–11.

Intergovernmental Panel on Climate Change (IPCC). 2007. "Climate Change 2007 Synthesis Report." http://www.ipcc.ch/publications_and_data/ar4/syr/en/main.html

Tilman, D., P. B. Reich, and J. M. H. Knops. 2006. Biodiversity and ecosystem stability in a decade-long grassland experiment. *Nature* 441:629–632.

Chapter 3

Muir, J. 1965. *The Story of My Boyhood and Youth.* Madison: University of Wisconsin Press. First published 1913.

Chapter 4

Bailey, R. G. 1995. *Descriptions of the Ecoregions of the United States,* 2nd ed. Misc. Pub. No. 1391, Map scale 1:7,500,000. Washington, DC: USDA Forest Service.

Harrington, J. A. 2008. "Vegetation Patterns and Land Cover Change for the Cross Plains Ice Age National Scientific Reserve, 1937–2007." Report to the Ice Age National Scenic Trail Office, U.S. National Park Service. University of Wisconsin–Madison.

Kabbes, K. C. 1996. "Regulatory Issues in Stream Restoration." In *Proceedings of the 2006 World Environmental and Water Resource Congress: Examining the Confluence of Environment and Water Concerns.* doi:10.1061/40856(200)340.

National Park Service DO-12 Handbook and Director's Order. 1982. U.S. Department of the Interior.

Read, C. 2008. "Spatial patterns and the underlying heterogeneity of remnant praries in Southern Wisconsin." Master Thesis. University of Wisconsin-Madison.

The Nature Conservancy. 2004. *Terrestrial and Marine Ecoregions of the United States.* Sources: Bailey, R. G. 1995; Wiken, E. B. 1986; ESRI 2002. Map produced by L. Sotomayor.

University of Kentucky Cooperative Extension Service. 1961. "Liming Acid Soils." Publication AGR-19. University of Kentucky.

Vick, J., S. McBain, and B. Di Gennaro. 1997. "Merced River Corridor Restoration Plan." http://www.calwater.ca.gov/Admin_Record/I-005286.pdf

Wiken, E. B. 1986. *Terrestrial EcoZones of Canada.* Ecological Land Classification Series No. 19. Lands Directorate, Environment Canada, Hull, Quebec.

Chapter 5

Beilfuss, K. G. 2001. "Habitat Use and Movement of the Adult Regal Fritillary Butterfly in Southwestern Wisconsin." Master's thesis. University of Wisconsin–Madison.

Hardell, J. A. 1980. "Response of Prairie Species Planted on Iron Ore Tailings Under Different Fertilization Levels." Master's thesis. University of Wisconsin–Madison.

Hengst, G. E. 1982. "The Response of Native Plant Species to Alternative Fertilizer Applications During Second and Third Growing Seasons on Iron Ore Tailings." Master's thesis. University of Wisconsin–Madison.

Ludwig, J., T. Hunt, and J. Broughton. 2000. "Badger State Pioneers Mine Reclamation Techniques." http://www.landandwater.com/features/vol43no4/vol43no4_1.html

Chapter 6

Biohabitats, Inc. 2007. "Hog Island and Newton Creek Ecological Restoration Master Plan." http://www.biohabitats.com/hogisland/HogIslandMasterPlanFinal.pdf

Carlson, B., A. Anders, A. Baker, J. Craig, and J. Solomon. 2004. "Kishwauketoe Nature Conservancy Restoration and Management Plan." Student report. Department of Landscape Architecture, University of Wisconsin–Madison.

Kagle, R. 2007. "An Evaluation of the Restoration of the East Preserve, UW–Madison Lakeshore Nature Preserve." Master's thesis. University of Wisconsin–Madison.

Kootenai Tribe of Idaho. 2009. "Kootenai River Habitat Restoration Project Master Plan: A Conceptual Feasibility Analysis and Design Framework." Bonners Ferry, ID. http://www.kootenai.org/documents/KRHRP-MP-0709-SV_001.pdf

Taylor-Powell, E., and E. Henert. 2008. "Developing a Logic Model-Teaching and Training Guide." University of Wisconsin–Extension Cooperative Extension Program Development and Evaluation. http://www.uwex.edu/ces/pdande/evaluation/pdf/lmguidecomplete.pdf

University of Wisconsin–Extension. n.d. "Logic Model." http://www.uwex.edu/ces/pdande/evaluation/evallogicmodel.html

Chapter 7

Acker, E., D. Gudex-Cross, R. Kay, and S. Shivy. 2007. "Ice Age National Scientific Reserve Cross Plains Ice Age Trail Unit 2 Site Plan." Student report. Landscape Architecture 666: Restoration Ecology, University of Wisconsin–Madison.

Bouressa, E., H. Kummel, J. Liu, and K. Vinyeta. 2007. "Restoration Plans for the Ice Age National Scientific Reserve: Cross Plains State Park Unit." Student report. Landscape Architecture 666: Restoration Ecology, University of Wisconsin–Madison.

Boyer, W. D., 1990. Growing season burns for control of hardwoods in longleaf pine stands. Re. Pap. SO-256. U.S. Department of Agriculture, Forest Service, Southern Forest Experiment Station, New Orleans, LA.

Curtis, J. T. 1959. *The Vegetation of Wisconsin.* Madison: University of Wisconsin Press.

Galatowitsch, S. M., and A. G. van der Valk. 1994. *Restoring Prairie Wetlands.* Ames: Iowa State University Press.

Harrington, J. A. 2008. "Vegetation Patterns and Land Cover Change for the Cross Plains Ice Age National Scientific Reserve, 1937–2007." Report to the Ice Age National Scenic Trail Office, U.S. National Park Service. University of Wisconsin–Madison.

Larson, M. A., W. D. Dijak, F. R. Thompson III, and J. H. Millspaugh. 2003. "Landscape-Level Habitat Suitability Models for Twelve Wildlife Species in Southern Missouri." General Technical Report NC-233. St. Paul, MN: U.S. Department of Agriculture, Forest Service, North Central Research Station.

Olson, K., A. Brown, and R. de Regnier. 2009. "Ice Age Scientific Reserve-Cross Plains State Park." Student report. Landscape Architecture 666: Restoration Ecology, University of Wisconsin–Madison.

Read, C., H. Hillhouse, S. Kraszewski, D. Cassidy, J. Jacobsen, and A. Bennett. 2004. "Cross Plains Ice Age Reserve and Ice Age National Scenic Trail." Student report. Landscape Architecture 666: Restoration Ecology, University of Wisconsin–Madison.

Riley, A. L. 1998. *Restoring Streams in Cities.* Washington, DC: Island Press.

Van Lear, D. H., W. D. Carroll, P. R. Kapeluck, and R. Johnson. 2005. History and restoration of the longleaf pine-grassland ecosystem: Implications of species at risk. *Forest Ecology and Management* 211(1–2):150–165.

Chapter 8

Bainbridge, D. 2007. *A Guide for Desert and Dryland Restoration: New Hope for Arid Lands.* Washington, DC: Island Press.

Biohabitats, Inc. 2007. "Hog Island and Newton Creek Ecological Restoration Master Plan." http://www.biohabitats.com/hogisland/HogIslandMasterPlanFinal.pdf

Blumenthal, D., N. Jordan, and M. Russell. 2003. Soil carbon addition controls weeds and facilitates prairie restoration. *Ecological Applications* 13:605–615.

Chapman, C. A., and L. J. Chapman. 1999. Forest restoration in abandoned agricultural land: A case study from East Africa. *Conservation Biology* (6)13:1301–1311.

Duncan, R. S., and C. A. Chapman. 1999. Seed dispersal and potential forest succession in abandoned agriculture in tropical Africa. *Ecological Applications* 9:998–1008.

Fink, R. D., C. A. Lindell, E. B. Morrison, R. A. Zahawi, and K. D. Holl. 2009. Patch size and canopy cover of planted tree patches influence birds' duration and rate of visit. *Restoration Ecology* 17:479–486.

Fondow, L., B. Mann, L. Sorenson, and A. Wells. 2006. Student project in Restoration Ecology. University of Wisconsin–Madison.

Harrington, J. A., and E. Kathol. 2009. Responses of shrub midstory and herbaceous layers to managed grazing and fire in a North American savanna (oak woodland) and prairie landscape. *Restoration Ecology* 17:234–244.

Harris, J. A., R. J. Hobbs, E. Higgs, and J. Aronson. 2006. Ecological restoration and global climate change. *Restoration Ecology* 2:170–176.

Kinyua, D., L. E. McGeoch, N. Georgiadis, and R. P. Young. 2010. Short-term and long-term effects of soil ripping, seeding, and fertilization on the restoration of a tropical rangeland. *Restoration Ecology* 18:226–233.

Kootenai Tribe of Idaho. 2009. "Kootenai River Habitat Restoration Project Master Plan: A Conceptual Analysis and Design Framework." Bonners Ferry, ID.

Matías, L., R. Zamora, I. Mendoza, and J. A. Hódar. 2010. Seed dispersal patterns by large frugivorous mammals in a degraded mosaic landscape. *Restoration Ecology* 18:619–782.

Nellemann, C., I. Vistnes, P. Jordhøy, O. Støen, B. P. Kaltenborn, F. Hanssen, and R. Helgesen. 2010. Effects of recreational cabins, trails and their removal for restoration of reindeer winter ranges. *Restoration Ecology* 18:873–881.

Papanastasis, V. P. 2009. Restoration of degraded grazing lands through grazing management: Can it work? *Restoration Ecology* 17:441–445.

Parrotta, J. A., J. W. Turnbull, and N. Jones. 1997. Catalyzing native forest regeneration on degraded tropical lands. *Forest Ecology and Management* 99:1–7.

Poschlod, P., C. Meindl, J. Sliva, U. Herkommer, M. Jäger, U. Schuckert, A. Seeman, A. Ullmann, and T. Wallner. 2007. Natural revegetation and restoration of drained and cut-over raised bogs in southern Germany: A comparative analysis of four long-term monitoring studies. *Global Environmental Research* 11:205–216.

Rowe, H., C. Brown, and M. Paschke. 2009. The influence of soil inoculums and nitrogen availability on restoration of high-elevation steppe communities invaded by *Bromus tectorum*. *Restoration Ecology* 17:686–694.

Sliva, J. and J. Pfadenhauer 1999. Restoration of cut-over raised bogs in southern Germany: A comparison of methods. *Applied Vegetation Science* 2:137–148.

Williams, Mary I., Gerald E. Schuman, Ann L. Hild, and Laurel E. Vicklund. 2002. Wyoming big sagebrush density: effects of seeding rates and grass competition. *Restoration Ecology* 10:385–391.

Chapter 9

Curtis, J. T. 1959. *Vegetation of Wisconsin*. Madison: University of Wisconsin Press.

Chapter 10

Pyne, S. J. 1982. *Fire in America: A Cultural History of Wildland and Rural Fire*. Princeton, NJ: Princeton University Press.

Chapter 11

Evans, J. M., A. C. Wilkie, J. Burkhardt, and R. P. Haynes. 2007. Rethinking exotic plants: Using citizen observations in a restoration proposal for Kings Bay, Florida. *Ecological Restoration* 25:199–210.

Hiebert, R., and J. Stubbendieck, 1993. *Handbook for Ranking of Exotic Plants for Management and Control*. Natural Resources Report NPS/NRMWRO/NRR93/08. Denver, CO: U.S. National Park Service Natural Resources Publication Office.

Jelinski, N., and N. Anderson. 2007. "Diversity and Productivity of Faville Prairie." *Arboretum Leaflet* 14. http://www.botany.wisc.edu/zedler/images/Leaflet_14.pdf

Kercher, S. M., and J. B. Zedler. 2004. Multiple disturbances accelerate invasion of reed canary grass *(Phalaris arundinacea* L.) in a mesocosm study. *Oecologia* 138:455–464.

Kline, V. M. 1984. "Response of Sweet Clover (*Melilotus alba* Desr.) and Associated Prairie Vegetation to Seven Experimental Burning and Mowing Treatments." *Proceedings of the Ninth North American Prairie Conference*, 149–152.

Kline, V. M., and E. A. Howell. 1987. Prairies. In *Restoration Ecology: A Synthetic Approach to Ecological Research*, edited by W. R. Jordan, 75–84. New York: Cambridge University Press.

Nordhaus, H. 2009. *Bark Beetle Outbreaks in Western North America: Causes and Consequences*. Bark Beetle Symposium, Snowbird, Utah, November 15–18, 2005. Salt Lake City: University of Utah Press.

Sax, D. F., and Gaines, S. D. 2008. "Species Invasions and Extinction: The Future of Native Biodiversity on Islands." *Proceedings of the National Academy of Sciences* 105:11490–1197.

Wittenberg, R., and M. J. W. Cook, eds. 2001. *Invasive Alien Species: A Toolkit of Best Prevention and Management Practices*. Wallingford, England: CABI Publishing.

Chapter 12

Bader, B. J., and D. Egan. 1999. Community-based ecological restoration: The Wingra Oak Savanna Project. *Orion Afield*, Spring:30–33.

Golden Gate National Park Stewardship Program. http://www.parksconservancy.org/about/

Janzen, D. H. 1986. "The Eternal External Threat." In *Conservation Biology*, edited by Michael Soule, 286–303. Sunderland, MA: Sinauer Associates.

Water Resources Management (WRM). 2008. "Report from the 2007 Water Resources Management Practicum." Madison: University of Wisconsin–Madison Arboretum.

Chapter 13

Biohabitats, Inc. 2007. "Hog Island and Newton Creek Ecological Restoration Master Plan." http://www.biohabitats.com/hogisland/HogIslandMasterPlanFinal.pdf

Friederici, P. 2006. *Nature's Restoration: People and Places on the Front Lines of Conservation*. Washington, DC: Island Press.

Gobster, P. H. 2000. "Restoring Nature: Human Actions, Interactions, and Reactions." In *Restoring Nature: Perspectives from the Social Sciences and Humanities*, edited by P. H. Gobster and R. B. Hull, 1–19. Washington, DC: Island Press.

Miller, J. R., and R. J. Hobbs. 2007. Habitat restoration: do we know what we're doing? *Ecological Restoration* 15:382–390.

Stevens, W. K. 1995. *Miracle Under the Oaks: The Revival of Nature in America*. New York: Pocket Books.

Chapter 14

Blewett, T. J., and G. Cottam. 1984. "History of the University of Wisconsin Arboretum Prairies." *Transactions of the Wisconsin Academy of Sciences, Arts and Letters* 72:130–144.

Cottam, G. 1987. "Community Dynamics on an Artificial Prairie." In *Restoration Ecology: A Synthetic Approach to Ecological Research*, edited by W. R. Jordan, 257–270. New York: Cambridge University Press.

Cottam, G., and H. C. Wilson. 1966. "Community Dynamics on an Artificial Prairie." *Ecology* 47:88–96.

Kline, V. M. 1992. "The Long-Range Management Plan for Arboretum Ecological Communities." University of Wisconsin–Madison Arboretum.

Kootenai Tribe of Idaho. 2009. "Kootenai River Habitat Restoration Project Master Plan: A Conceptual Feasibility Analysis and Design Framework." Bonners Ferry, ID with Funding provided by the Bonneville Power Administration.

Longenecker, W. G. 1941. University of Wisconsin Arboretum. *Parks & Recreation* Sept.:1–8.

McGaw, M. 2002. "The Response of Gray Dogwood (*Cornus racemosa*) to Prescribed Fire and the Effects of Invasion on Fuel Loading and Plant Community Composition at Curtis Prairie." Master's thesis. University of Wisconsin–Madison.

Sachse, N. D. 1974. "A Thousand Ages: The University of Wisconsin Arboretum." Madison: Regents of the University of Wisconsin. Revised edition, originally published in 1965.

Snyder, T. A., III. 2004. "A Spatial Analysis of Grassland Species Richness in Curtis Prairie." Master's thesis. University of Wisconsin–Madison.

Sperry, T. M. 1984. "Analysis of the University of Wisconsin–Madison Prairie Restoration Project." In *Proceedings of the Eighth North American Prairie Conference*, 140–147.

Sperry, T. M. 1990. "Report on the 1990 Curtis Prairie Survey for the University of Wisconsin–Madison Arboretum." University of Wisconsin–Madison Arboretum.

The Nature Conservancy. Great Rivers News. http://www.nature.org/greatrivers.

Wegener, M., and J. Zedler. 2008. "Taking Stock: Status Report on Our 75th Anniversary." *Arboretum Leaflets*, no. 18. http://www.botany.wisc.edu/zedler/images/Leaflet_18.pdf

Wegener. M., P. Zedler, B. Herrick, and J. Zedler. 2008. "Curtis Prairie: 75-Year-Old Restoration Research Site." *Arboretum Leaflets*, no. 16. http://www.botany.wisc.edu/zedler/images /Leaflet_16.pdf

Chapter 15

Millennium Ecosystem Assessment (MA). 2005. *Ecosystems and Human Well-Being: General Synthesis*. Washington, DC: Island Press.

Society for Ecological Restoration (SER). 2009. "Policy Position Statement: Ecological Restoration and Rare Species Management in Response to Climate Change." https://www.ser .org/pdf/PPS_on_Rare_Species_Final2.pdf

Glossary

adaptive management An approach to management that encourages periodic changes in management objectives and protocols as needed, in response to monitoring data and other new information. *See also* adaptive restoration.

adaptive restoration The adaptive approach to restoration includes an expectation that projects be flexible and able to switch strategies or desired outcomes in the face of changing conditions or new information. It incorporates three steps: (1) Restoration research informs and leads to improvements in restoration practice, and contributes to theory and the understanding of community/ecosystem ecology. (2) Project documentation involves recording notes and observations, compiling work journals, and team reports. (3) The evaluation and communication of information generated by research and documentation are essential for advancing the restoration profession.

alien species A species native to other areas that was introduced into a locality as a consequence of the activities of Neolithic or post-Neolithic humans and their domestic animals. Also known as non-native species, nonindigenous species, foreign species, exotic species.

biodiversity The diversity of life at all levels of hierarchical organization (genetic, individual organism, population, community, ecosystem), and according to all taxonomic categories (species, genus, family, order, class, phylum, kingdom), at a specified location or in the biosphere.

carbon sequestration The capture and storage of carbon in sinks such as forests, soils, ocean sediments, and wood products.

climate change Changes in global temperature and precipitation patterns that are largely attributable to increasing atmospheric concentrations of carbon dioxide and other greenhouse gases (e.g., methane, nitrous oxides) since the mid-nineteenth century.

climax The ecosystem or plant community that develops as the final successional stage. The term is based on largely superceded ecological theory of linear, predictable successional trajectory, which assumes environmental stability and discounts random results from ecological processes. Although "climax" in the sense of an inevitable and eternally appropriate community for each ecosystem is misleading, there are identifiable mature communities that represent the relatively stable outcomes of succession. These can be called climax communities for convenience, although not with their previously held, deterministic connotations. In reference to forests, the terms "old-growth" and "primary" forest connote a climax condition, prior to human management.

collaboration The partnership of two or more stakeholders who combine resources to solve a problem, or problems, that neither can solve by themselves.

complete restoration A restoration that aims to restore the complexity of all the attributes (composition, structure, functions, processes, services, and aesthetics) of a plant community, ecosystem, or landscape to the best extent possible. The most idealized type of restoration, it may be the most difficult, if not impossible, goal to achieve.

cooperation The working together of individual or group stakeholders for a common purpose, in which no party has the power to command the behavior of the others.

cultural landscape A landscape that has developed under the joint influence of natural processes and human-imposed organization and resource use.

community A group of interacting species living together in the same place at the same time.

degradation The simplification and loss of biodiversity caused by disturbances that are too frequent or severe to allow natural ecosystem recovery. Degradation generally reduces the flow of ecosystem goods and services.

disturbance A natural event or human-mediated activity that changes the structure, content, and/or function of an ecosystem.

ecological services restoration A restoration that aims to establish the structural and functional attributes of community/ecosystems (native species patterns and distributions, flows and cycles, natural capital, social capital, animal habitat, and other services) that best serve the goals of humans (rehabilitated working landscapes, such as ranches, mined areas, and urban sites) or those that serve non-human entities that are valued by people, such as endangered species. Also known as process-based, structural, or functional restoration.

ecology The study of the interactions of organisms and their environment. The science of ecology covers a wide scale of phenomena, moving from an individual molecule to the entire global system.

ecosystem All the living and dead organisms in an area, together with the atmosphere, the hydrosphere (water systems), the lithosphere (rocks and minerals), and their interconnections. The focus here is not on the organisms per se, but rather on the roles they perform; in other words, the spotlight is on functions and processes.

ecosystem goods and services Natural products and processes generated by ecosystems that sustain and fulfill human life. The Millennium Ecosystem Assessment recognizes four categories of benefits to people: provisioning, regulating, supporting, and cultural functions. Examples are the provisioning of clean water; the regulation of flood waters; soil protection and erosion control; climate maintenance (carbon sequestration); and crop pollination; and in cultural terms, fulfilling recreational, intellectual, and spiritual needs. Also known as environmental goods and services.

ecosystem management The determination of strategies and the application of techniques to guide (not prevent) ecological change within a range of predetermined boundaries.

emerging ecosystem An ecosystem without analogs in the natural environment that develops in response to radically altered environmental changes caused by social, economic, and cultural activities. Also known as novel ecosystem, no-analog ecosystem.

endangered species A plant or animal species considered to be in danger of extinction.

enhancement Actions that improve ecosystem function in a manner that may contribute partially toward ecological restoration.

enrichment planting A strategy for expanding the diversity of species by introducing species that enhance commercial or conservation values or that help recover biodiversity relative to a reference model.

environmental services *See* ecosystem goods and services.

eradication The removal of pest species, with the intent of extirpating their populations completely from a certain locality.

erosion The removal of soil by the action of wind, water, and gravity. *See also* runoff.

eutrophication Nutrient enrichment of soil or a water body, commonly by human activity.

experiential restoration A restoration that aims to revitalize landscape attributes that provide pleasure to humans. Such restorative activities typically simplify the number of species and exaggerate the more charismatic qualities of an ecosystem or plant community.

fragmentation The separation of a formerly continuous natural area into smaller natural units that are isolated from one another by lands that were converted for economic production or the development of infrastructure such as road building. *See also* habitat fragmentation.

functional group A group of species that have a common set of attributes and play a particular role in ecosystem processes. Examples are nitrogen-fixing plant species and microbe-feeding nematodes. When the species rely on a similar set of resources, functional groups are also called guilds.

habitat fragmentation The fragmentation of a formerly continuous living space or geographic range of a species population into smaller units, resulting in a metapopulation or in genetically separated subpopulations.

habitat restoration An ecological restoration with respect to the living conditions for a particular species.

habitat suitability model (index) A description of the resources thought to be needed by a particular species (generally animal) for survival; used to evaluate and rank sites for their ability to sustain a (usually endangered) species. Helps specify restoration goals and objectives for projects designed to support particular species, or to predict the kinds of species that might be able to inhabit a site.

human well-being A condition in which members of communities are able to determine and meet their needs and reach their potentials.

implementation A five-step phase in the restoration process that involves (1) defining implementation units and boundaries; (2) developing a strategy to accomplish the site plan given the physical, economic, and social resources available; (3) preparing the site in accordance with the strategy; (4) implementing the strategy; and (5) resolving logistics such as resource needs and permits.

intervention A specific action or intentional strategy that will bring about an action, such as site preparation, invasive species removal, desirable species introductions, biomanipulation, altering canopy structure, or reintroducing fire.

invasive alien A non-native species that has become naturalized in an area, and whose distribution and/or abundance is in the process of increasing regardless of habitat. If this process threatens ecosystems, habitats, or species with economic or environmental harm, it can be called a noxious invasive alien. Such species are addressed under Article 8(h) of the Convention on Biological Diversity. However, a native species whose distribution and/or abundance in the wild is increasing should be described simply as increasing, colonizing, or spreading, not as invading. *See also* pest species.

keystone species A species that has a substantially greater positive influence on other species in an ecosystem than would be predicted by its abundance or size.

landscape An assemblage of ecosystems that are arranged in recognizable patterns and that exchange organisms and materials such as water. Currently interpreted as a land area mosaic of interacting natural ecosystems, production systems, and spaces dedicated for social and economic use. The size of the landscape is determined mostly by the scale of the restoration initiative and the likely or desired geographic extent of its impacts.

management The management phase of a restoration project begins once the initial restoration goals and objectives have been achieved. Management includes planning, implementation, monitoring, and revising, and directs, but does not prevent, change.

metapopulation One of a set of local populations of a plant or animal species that fluctuate independently but interact within a larger area of space. The long-term survival of the species depends on a shifting balance between local extinctions and recolonizations in the patchwork of fragmented landscape or habitat.

mitigation Measures required by government agencies, or international agreements, for permission to undertake development projects that cause unavoidable environmental harm. Such measures include compensation in terms of enhancement, rehabilitation, ecological restoration; steps taken during a project to reduce impacts; or modifications in the scope of the project.

monitoring The monitoring phase of the restoration process provides information necessary to determine how well a restoration is meeting project goals and objectives at any point in time.

native species A plant or animal species in a specified area that has arrived independent of human activities, or that was already present before the Neolithic period. Also known as indigenous species. *See also* alien species, invasive alien species.

nature conservation/management An intended complex of human activities aimed at the maintenance or regeneration of ecosystems, and particularly their species composition.

novel ecosystem *See* emerging ecosystem.

opportunities and constraints analysis Part of the site inventory and analysis phase of a restoration process; a list of opportunities and constraints is created and linked to specific site locations using a map. Opportunities are site features (or resources) that contribute to or aid the restoration process; constraints are features (or resources) that limit the success of the project.

outcomes In a restoration plan, outcomes describe in measurable terms what a site will be like when the restoration has been achieved. Site plan objectives are expressed as outcomes. In the Logic Model, outcomes are the anticipated program results.

partnership An association of two or more stakeholders working toward a common goal.

pest species A native or introduced plant, animal, insect, bacterium, or virus that interferes with restoration and management goals. Under different circumstances, different times, or in different places, a species that behaves as a pest on one site may be a beneficial component of the ecosystem being restored. *See also* invasive alien.

performance standard For restoration implementation or management, a clear description of what is to be accomplished and the deadline that must be met.

provenance The geographic place of origin or source of seeds or other propagules.

reclamation The conversion of land, wetland, or shallow seas, which are perceived as being unusable to a productive condition, commonly for agriculture, aquaculture, or silviculture. Sometimes used as a synonym for rehabilitation, particularly with respect to mines.

redundancy A situation in which several species play a particular role in ecosystem processes when only one or a few species could in appearance fulfill that role. The other species thus might appear to be dispensable or redundant, in terms of function and process, even though they contribute to biodiversity.

reference ecosystem One or more natural ecosystems, ecological descriptions thereof, or, if these are unavailable, characteristics of presumed natural ecosystems that serve as models or targets for planning ecological restoration and rehabilitation projects. Also known as reference model.

reforestation The reestablishment of forest on forested or wooded land that was recently cleared with native or non-native species by means of artificial, natural, or assisted natural regeneration techniques.

reintroduction The intentional introduction of a plant or animal species in an ecosystem from which it earlier had been removed.

resilience The capacity of an ecosystem to tolerate disturbance and recover autonomously by natural regeneration, without collapsing or shifting into a qualitatively different state that is controlled by a different set of processes. A resilient ecosystem can withstand major disturbance and rebuild itself, or persist on a given developmental trajectory or within a given configuration of states (a regime) in systems where multiple regimes are possible.

restoration ecology The process of assisting the recovery of an ecosystem that has been degraded, damaged, or destroyed.

restoration goals General descriptions of what the site will be like in physical, biological, and cultural terms after it has been restored.

restoration objectives Measurable quantitative or qualitative statements that further define and clarify the restoration goals.

restoration purpose A simple statement of the reason for a restoration project; a summary of the problems to be solved by the restoration.

restoration use-policy A description of who (or who will not) be able to visit or otherwise directly interact with a restoration site, the kinds of activities that will or will not be allowed, and the nature of the desired experiences through engaging with the site.

revegetation The establishment of plant cover on open land, often with one or few species, irrespective of their origin or provenance.

riparian Pertaining to rivers, such as a forest that occupies a river floodplain.

runoff Rainfall or other water that moves toward lower elevations by spreading across the land surface, rather than flowing within a defined channel.

seed bank The stock of viable seeds, spores, and other plant propagules in the soil.

site inventory and analysis This phase of the restoration process informs the restoration team of the physical, biological, and cultural conditions within and around the site. The site analysis uses the inventory to determine the suitability of a site or areas within a site for a specific purpose or objective, such as plant community types, ecosystem functions, social experiences, or research.

social-ecological system An ecological system, usually perceived at the landscape scale, that consists of a mixture of natural ecosystems, production systems, and sometimes developed land that is dedicated to residences, farm buildings, and even light industry, all units of which are functionally interrelated in both ecological and socioeconomic terms.

stakeholder Any individual or group directly or indirectly affected by, or interested in, actions pertaining to a given resource.

suitability analysis Part of site inventory and analysis phase of the restoration process. An approach to evaluating the degree to which the resources of a site support, or pose potential problems for, the implementation of a restoration target (such as plant community, animal species, or ecosystem function). Each resource is inventoried and rated as being positive ("suitable" or "few limitations") or negative ("unsuitable" or "severe restrictions") for the proposed restoration, after which a composite evaluation is created.

sustainability In an economic context, the capacity of a system to remain productive indefinitely for the benefit of future generations. Principles of sustainability: stocks of renewable resources must not be used faster than they are renewed, waste emissions must not exceed waste absorption capacity, and essential nonrenewable resources cannot be depleted any faster than technology develops renewable substitutes. It is increasingly recognized that sustainability includes economic, social, and environmental components.

sustainable development Economic development that meets the needs of the present generation without compromising the ability of future generations to meet their own needs.

use-policy *See* restoration use-policy.

weeds Plants, indigenous and alien, that grow on sites where they are not wanted and that may have negative economic, aesthetic, or environmental effects. Also known as plant pests, harmful species, problem plants.

wick apply Often a woven nylon or cotton rope that is moistened with herbicide provided by capillary action via a reservoir system. The wick is wiped across the vegetation leaving minimal amounts of the herbicide. In theory, only vegetation touched by the wick will be exposed to the chemical. In reality, herbicide drift by volatilization will come in contact with other species.

Index